Stability Control of Underground Openings under High Stress and Deep Mining Environment

Stability Control of Underground Openings under High Stress and Deep Mining Environment

Guest Editors

Jianhang Chen
Zizheng Zhang
Qingliang Chang

MDPI

Basel • Beijing • Wuhan • Barcelona • Belgrade • Novi Sad • Cluj • Manchester

Guest Editors
Jianhang Chen
School of Energy and Mining Engineering
China University of Mining and Technology (Beijing)
Beijing
China

Zizheng Zhang
School of Resource & Environment and Safety Engineering
Hunan University of Mining and Technology
Xiangtan
China

Qingliang Chang
School of Mines
China University of Mining and Technology
Xuzhou
China

Editorial Office
MDPI AG
Grosspeteranlage 5
4052 Basel, Switzerland

This is a reprint of the Special Issue, published open access by the journal *Applied Sciences* (ISSN 2076-3417), freely accessible at: https://www.mdpi.com/journal/applsci/special_issues/X2324A1N7Q.

For citation purposes, cite each article independently as indicated on the article page online and as indicated below:

Lastname, A.A.; Lastname, B.B. Article Title. *Journal Name* **Year**, *Volume Number*, Page Range.

ISBN 978-3-7258-4333-6 (Hbk)
ISBN 978-3-7258-4334-3 (PDF)
https://doi.org/10.3390/books978-3-7258-4334-3

Contents

applied sciences

MDPI

Editorial

Stability Control of Underground Openings Under High-Stress and Deep Mining Environments

Jianhang Chen [1,2,*], Zizheng Zhang [3] and Qingliang Chang [4]

1 School of Energy and Mining Engineering, China University of Mining and Technology (Beijing), Beijing 100083, China
2 Engineering Research Center of Green and Intelligent Mining for Thick Coal Seam, Ministry of Education, Beijing 100083, China
3 School of Resource & Environment and Safety Engineering, Hunan University of Mining and Technology, Xiangtan 411201, China
4 School of Mines, China University of Mining and Technology, Xuzhou 221116, China
* Correspondence: jianhang.chen@cumtb.edu.cn

1. Introduction

Mining engineering plays a significant role in current society [1–3]. In fact, coal mining engineering significantly contributes to the development of the social economy [4].

Since raw coal production is abundant, mining depths are becoming larger. For in situ stress in rock masses before mining, when the buried depth becomes larger, in situ stress increases significantly, including vertical stress and horizontal stress. Under high in situ stress conditions, strata deformation and failure behavior become more apparent. For example, the deformation of roadways and chambers becomes larger. In certain cases, roof collapsing may occur. This poses a high risk to underground operators and equipment. Moreover, since the mining depth becomes larger, elastic energy accumulated in rock strata becomes higher. When accumulated high elastic energy releases suddenly, rock burst disasters may occur [5,6]. This creates more risks to the safety of underground staff [7].

To overcome the above issues, rock reinforcement and rock support methods should be used [8–12]. Although much research has been conducted to study rock reinforcement mechanisms and rock support mechanisms [13,14], the load transfer process between rock reinforcement bolts, rock support instruments and rock masses has not been fully understood.

Therefore, it is quite necessary to further study the rock reinforcement mechanism and rock support mechanism of bolts, cables, tendons, mesh and lining [15–17]. As such, we established this Special Issue named "Stability Control of Underground Openings Under High-Stress and Deep Mining Environments". For this Special Issue, we aim at attracting researchers and engineers to submit their latest research regarding stability control of underground openings, especially in a deep mining environment. Through reading these published academic papers, readers can further and better understand the latest stability control methods and corresponding mechanisms in ground control areas in deep mining engineering.

2. An Overview of Published Articles

This Special Issue received 27 academic papers. Among them, 12 were successfully published. These academic papers are from China and Poland, and their topics are related to roadway reinforcement, strata control of roadways that are near gobs, the adaptability of anchor support, rock burst issues and rock failure around roadways. These academic papers used different methods to reveal rock reinforcement mechanisms and rock failure mechanisms, including in situ observations, numerical simulation, laboratory experimental tests and in situ engineering practice. For example, Zhou, et al. [18] indicated that under thick and hard roofs, the stress environment of roadways that are near gobs became more complicated. This even induced coal bursts in underground mining engineering. To solve this issue, they used numerical simulation methods to analyze the stress–energy evolution law of working faces, coal pillars and rock masses around roadways. Their results are valuable for readers to understand coal burst mechanisms in similar geological conditions. Wang et al. [19] indicated that layered composite roofs were composed of developed bedding fissures. This may lead to severe deformation and the failure of rock bolts in roadway roofs. Based on this background, they conducted in situ observation and found that minor shear deformation existed in rock bolts at the center of roadway roofs. Significant shear deformation existed in rock bolts in roof corners and lateral deformation was more than 176 mm. Then, numerical simulation was conducted with the commercial software FLAC^{3D}. They found that the main factors affecting rock bolt deformation included the strength of geological joints and the strength of the rock mass around openings. Their research is beneficial for understanding instability mechanisms and supporting the treatment of layered composite roofs. Pytel et al. [20] indicated that maintaining the stability of underground openings buried at a great depth was challenging. They developed a simplified analytical method for ductile ground support design in underground mining. Their proposed algorithm made it possible to conduct tracking of the rock burst phenomenon. This information is valuable when designing support parameters for underground openings. To sum up, these published academic papers in this Special Issue will help readers to better understand the interaction between rock masses and rock reinforcement instruments.

3. Conclusions

The stability of underground openings is significant in maintaining the normal operation of underground mines. As mining depth increases, the difficulty of maintaining underground opening stability becomes larger. Researchers have proposed various methods, instruments and facilities to guarantee the stability of underground openings in a deep mining environment. However, there are still a number of challenging issues. Therefore, it is quite necessary to continue studying in this research area. With the development of fundamental research, it is expected that the stability issues of underground openings can be minimized to a great extent.

Author Contributions: J.C. original draft preparation; Z.Z. and Q.C. review and editing. All authors have read and agreed to the published version of the manuscript.

Funding: It is funded by National Natural Science Foundation of China (52174093, 52034009), Fundamental Research Funds for the Central Universities (2023ZKPYNY03, 2024ZKPYNY01, 2023YQTD02).

Conflicts of Interest: The authors declare no conflicts of interest.

References

1. Knox, G.; Hadjigeorgiou, J. Behaviour of yielding mechanical hybrid rockbolts under complex loading conditions. *Int. J. Rock Mech. Min. Sci.* **2024**, *182*, 105877. [CrossRef]
2. Fu, Y.; Song, X.; Xing, P. Study of the mining height of caving zone in large mining height and super-long face of shallow seam. *J. Min. Saf. Eng.* **2010**, *27*, 160–174.
3. Zuo, J.; Liu, H.; Liu, D.; Zhu, F.; Zhao, W.; Zhang, Q.; Li, J. Theoretical analysis and numerical simulation on the coupled support technology of concrete-filled steel tube and bolt-cable in deep roadway. *J. Cent. South Univ.* **2023**, *30*, 257–275. [CrossRef]
4. Rusman, M.R. Mine wastes and water, ecological engineering and metals extraction: Sustainability and circular economy. *Int. Geol. Rev.* **2024**, *67*, 1–2. [CrossRef]

5. Chen, J.; Zeng, B.; Hu, S.; Xu, W.; Wang, S.; Liu, Z. Mechanical behaviour and instability mechanism of sandstones with impact tendency under different loading paths. *Int. J. Min. Reclam. Environ.* **2024**, *38*, 1–20. [CrossRef]
6. Wei, C.; Zhang, C.; Canbulat, I.; Song, Z.; Dai, L. A review of investigations on ground support requirements in coal burst-prone mines. *Int. J. Coal Sci. Technol.* **2022**, *9*, 13. [CrossRef]
7. Zhang, C.; Wang, P.; Wang, E.; Chen, D.; Li, C. Characteristics of coal resources in China and statistical analysis and preventive measures for coal mine accidents. *Int. J. Coal Sci. Technol.* **2023**, *10*, 22. [CrossRef] [PubMed]
8. Chen, J.; Zeng, B.; Xu, W.; Hu, S.; Wang, S.; Wang, K.; Zhang, W.; Wu, S.; Song, Z. Mechanical behaviour and damage constitutive model of semi-circular arch tunnels with straight walls under soaking conditions. *Eng. Fail. Anal.* **2025**, *169*, 109137. [CrossRef]
9. Zhang, W.; Song, Y.; Huang, L.; Gong, S. Analytical relations for the mechanical properties of full-length bonded rock bolts when subjected to freeze–thaw processes. *Comput. Geotech.* **2022**, *145*, 104667. [CrossRef]
10. Fu, M.; Huang, S.; Fan, K.; Liu, S.; He, D.; Jia, H. Study on the relationship between the maximum anchoring force and anchoring length of resin-anchored bolts of hard surrounding rocks based on the main slip interface. *Constr. Build. Mater.* **2023**, *409*, 134000. [CrossRef]
11. He, M.; Ren, S.; Xu, H.; Luo, S.; Tao, Z.; Zhu, C. Experimental study on the shear performance of quasi-NPR steel bolted rock joints. *J. Rock Mech. Geotech. Eng.* **2023**, *15*, 350–362. [CrossRef]
12. Zhou, Y.; Lai, H.; Liu, Y.; Hong, Q.; Li, Z. Analysis and treatments of primary support failure in large-span tunnel under a soft-hard interbedded surrounding rock mass. *Eng. Fail. Anal.* **2024**, *156*, 107755. [CrossRef]
13. Staniek, A. Technical problems and non destructive testing of rock bolt support systems in mines. *Int. J. Coal Sci. Technol.* **2023**, *10*, 6. [CrossRef]
14. Yu, S.; Zhu, W.; Niu, L. Experimental and numerical evaluation on debonding of fully grouted rockbolt under pull-out loading. *Int. J. Coal Sci. Technol.* **2022**, *9*, 8. [CrossRef]
15. Wang, G.; Fang, W.; Han, W.; Jiang, Y.; Zhang, X.; Xu, F.; Zhang, S. Coupled rheological behavior of tunnel rock masses reinforced by rock bolts based on the non-hydrostatic stress field. *Tunn. Undergr. Space Technol.* **2023**, *137*, 105123. [CrossRef]
16. Tong, X.; Shan, R.; Liu, N.; Liu, D.; Wei, Y. Experimental study on the shear performance of an anchor cable with a C-shaped tube anchored flat structural plane. *Sci. Rep.* **2023**, *13*, 797. [CrossRef] [PubMed]
17. Thirukumaran, S.; Oliveira, D. Innovative design of slender rock pillar formed within large span road tunnels and cavern Y-junction in Hawkesbury Sandstone. *Tunn. Undergr. Space Technol.* **2023**, *141*, 105376. [CrossRef]
18. Zhou, J.; Pan, J.; Xia, Y.; Liu, W.; Du, T.; Wu, J. Investigation of load characteristics and stress-energy evolution laws of gob-side roadways under thick and hard roofs. *Appl. Sci.* **2024**, *14*, 9513. [CrossRef]
19. Wang, Z.; Fang, S.; Zhang, C. Deformation characteristics and response factors of rock bolt body in roadway with layered composite roof. *Appl. Sci.* **2024**, *14*, 6694. [CrossRef]
20. Pytel, W.; Fuławka, K.; Pałac-Walko, B.; Mertuszka, P. Numerical and analytical determination of rockburst characteristics: Case study from Polish deep copper mine. *Appl. Sci.* **2023**, *13*, 11881. [CrossRef]

applied sciences

MDPI

Article

Investigation of Load Characteristics and Stress-Energy Evolution Laws of Gob-Side Roadways Under Thick and Hard Roofs

Jinlong Zhou [1,2,3,*], Junfeng Pan [2], Yongxue Xia [2], Wengang Liu [4], Taotao Du [2] and Jianhong Wu [2]

1 Coal Mining Branch, China Coal Research Institute, Beijing 100013, China
2 Coal Mining Research Institute, China Coal Technology and Engineering Group, Beijing 100013, China; panjunfeng@yeah.net (J.P.)
3 School of Energy and Mining Engineering, China University of Mining and Technology (Beijing), Beijing 100083, China
4 Research Institute of Emergency Science, China Coal Research Institute, Beijing 100013, China; liuwgccteg@126.com
* Correspondence: zhoujlcumtb@163.com

Abstract: The stress environments of gob-side roadways (GSRs) are becoming increasingly complex during deep coal mining under thick and hard roofs. This leads to strong strata behaviors, including roadway floor heave, roof subsidence, and even coal bursts. Among them, coal bursts pose the greatest threat to production safety in coal mines. Coal bursts in a GSR strongly correlate with the load characteristics and stress-energy evolution laws of the roadway. This study analyzes the roof structures of double working faces (DWFs) during the initial weighting stage (IWS) and full mining stage (FMS) of gob-side working faces (GSWFs). This study also explores how varying roof structures affect the stability of GSRs. Three-dimensional roof structure models of DWFs and mechanical models of dynamic and static loads superposition on a GSR throughout the IWS and FMS of a GSWF were developed. An analysis identified the primary stress sources affecting the GSR throughout various mining stages of the GSWF. Subsequently, the principle of "three-load" superposition was developed. A novel method was proposed to quantify the stress state in the GSR surrounding rock across different mining stages of the GSWF. The method quantitatively characterizes the load of the GSR surrounding rock. Based on this, the criterion for judging the burst failure of the roadway was established. Numerical simulations are used to analyze the stress-energy evolution laws of the working face, coal pillar, and GSR surrounding rock during the mining process of the GSWF. These findings offer valuable references for studying and preventing coal bursts in GSRs under equivalent geological situations.

Keywords: deep coal mining; thick and hard roofs; gob-side roadways; load characteristics; stress-energy evolution laws

1. Introduction

Recently, some coal mines in Ordos region of China have reached depths of 600–750 m [1]. These deep coal mines feature high and wide working faces overlaid by multiple thick and hard roofs. Combining deep and high-intensity mining subjects the stope to complex stress conditions. Coal bursts frequently occur in the gob-side roadway (GSR) preceding the working face throughout the extraction of the gob-side working face (GSWF). Such occurrences present substantial hazards to the workforce engaged in underground operations. Coal bursts in GSRs are strongly influenced by the roof structure of double working faces (DWFs) under thick and hard roofs and their loading effect on the GSR. Mining-induced stress and energy evolution play a crucial role in coal burst occurrences at GSRs. To elucidate the mechanism of coal bursts in GSRs and develop effective prevention strategies, it is crucial to investigate the roof structure of DWFs and the stress-energy evolution laws and load characteristics of GSRs during the mining process of GSWFs.

Coal extraction induces fracturing and movement in the overlying strata, forming different overburden spatial structures. These overburden structures and their mechanical properties significantly influence the strata behavior in stopes [2–5]. Extensive research has focused on characterizing the overburden structure of working faces. Early research primarily originated from German, Soviet, and Belgian scholars. Their work introduced several hypotheses, including the "cantilever beam" [6], "pressure arch" [7], "hinging rock strata", and "pre-formed fissure" hypotheses. On this basis, Chinese scholars have conducted relevant research: Qian [8,9] put forward the "voussoir beam" theory and advanced the field by introducing the key strata theory, expanding on the "voussoir beam" theory [10]. Song [11] proposed the "transfer rock beam" theory, explaining stress distribution in mining-affected areas. Over the past 15 years, researchers have developed numerous roof structure models for different mining conditions based on these foundational theories. He [12–14] proposed the "cutting cantilever beam" theory, which paved the way for the cutting roof and stress relief technology. This innovation enables the creation of roadways without coal pillars through automated means. This advancement holds particular importance for roadway surrounding rock control and coal burst prevention. Dou et al. [15] proposed an overburden spatial structure model for single, double, and island working face mining boundary conditions, denoted as "O-X, F, and T" structures, respectively. They validated this model using microseismic monitoring techniques, providing empirical support for their theoretical framework. Huang et al. [16] proposed a structure model for main roofs due to insufficient gob filling in large mining height working faces. Yu et al. [17,18] modelled overburden in the large-space stopes of extremely thick coal seam mining, identifying distinct near-field and far-field key strata responses. Zuo et al. [19–21] proposed an "analogous hyperbola (hyperboloid)" theoretical model for overburden movement. This model integrates the mechanical analysis of breaking patterns in different strata and similar simulation experiments. These studies have significantly enriched ground control theory in coal mining.

Coal bursts in the roadway strongly correlate with the stress induced by mining operations and the energy evolution within the roadway surrounding rock [22–26]. Scholars have conducted extensive research grounded in theoretical modelling, physical experiments, numerical computations, and in situ measurements, laying the groundwork for analyzing the stability of roadway surrounding rock. Zhang et al. [27] measured the lateral abutment pressure distribution within coal pillars and, based on this, proposed a mechanical model validated through numerical simulation. To establish roof fracture criteria, Bai et al. [28] developed a dynamic roof structure model through monitoring, analysis, and simulation. Zhang et al. [29] examined the stress evolution properties of GSRs through numerical simulation. They elucidated the load transfer mechanism of the surrounding rock and developed a model for it. Lou et al. [30] conducted large-scale physical similar simulation experiments to study mining-induced stress evolution. Their results revealed an arch-shaped distribution for principal stress with high stress concentrations at the arch shoulders and varying vertical stress patterns in different areas. Wang et al. [31,32] conducted computational modelling analysis regarding the rotation characteristics of stress in deep mining. They proposed and field-tested a method to improve roof stability by controlling stress rotation. Ma et al. [33] highlighted that the abrupt discharge of accumulated energy is the leading cause of coal bursts. The intensity of coal bursts correlates closely with the degree of energy accumulation.

Previous research has focused on the roof structures of the working face and their mechanical analysis, examining the patterns of stress redistribution and energy accumulation in the roadway surrounding rock. However, few studies have quantitatively examined the relationship between roof structures and loads on GSRs at the initial weighting stage (IWS) and full mining stage (FMS) of GSWFs. This problem is particularly evident in thick and hard roofs in the Ordos region. The mechanism by which different roof structures load the

GSR surrounding rock remains to be determined. Coal bursts in GSRs are closely linked to roof structure morphology and its loading effect on roadway surrounding rock. Addressing this problem, we develop roof structure models of DWFs for various stages of GSWFs and analyze the load sources of the GSR surrounding rock. A stress calculation method for the roadway surrounding rock is proposed which can quantitatively evaluate the load patterns in the roadway surrounding rock. Concurrently, we analyze the distribution of stress and energy in roadway surrounding rock through numerical simulation. This study provides a valuable framework to understand the underlying mechanisms and mitigate the risk of coal bursts in GSRs.

2. Engineering Background

A multilayered thick and hard sandstone overburden overlays the working face in some deep coal mines within the Ordos region. The high static and dynamic load combination resulting from thick and hard roofs fracturing has led to frequent coal bursts in GSRs during coal mining. These events present substantial hazards to coal mine operations. This study takes the coal bursts in the GSR throughout the mining process of the 3^{-1}103 working face at Hongqinghe Coal Mine in Xinjie Town, located in Ordos, which serves as the engineering background. Then, this study analyzes the roof structure, load behavior of the GSR, and the distribution of stress and energy induced by the mining of the GSWF. This study aims to provide insights into the mechanisms of coal bursts in GSRs and contribute to developing effective prevention strategies.

The 3^{-1}103 working face, adjacent to the 3^{-1}101 working face gob, measures 210 m in working face length, 2480 m in mining length, and 707.15 m in average depth. The coal seam averages 6.36 m in thickness. A fully mechanized mining method with a large mining height is used for coal mining. The working face comprises three roadways: the 3^{-1}101 material roadway, the 3^{-1}103 belt roadway, and the 3^{-1}103 material roadway. The width of the coal pillars between adjacent roadways is 30 m. The roof and two sides of the roadway are supported by bolts and cables. Notably, coal bursts in the 3^{-1}103 working face primarily occur in the 3^{-1}101 material roadway, also known as the GSR, which is located near the advancing working face. These coal bursts manifest as dynamic phenomena, including instantaneous floor heave, roof subsidence, rib spalling, and broken bolts and cables. Figure 1 illustrates the arrangement of the 3^{-1}101 and 3^{-1}103 working faces and typical bursting failure characteristics of the roadway.

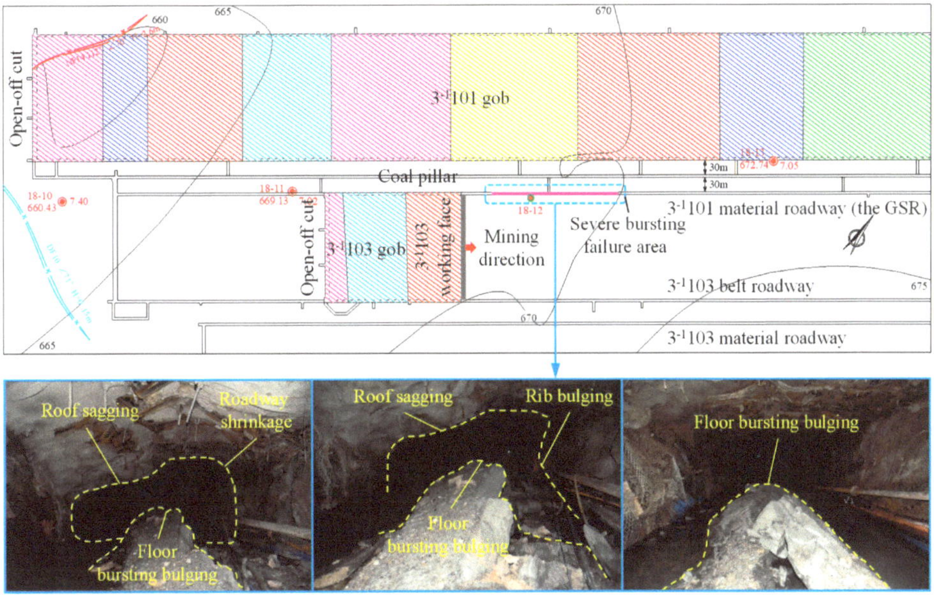

Figure 1. The arrangement of the 3^{-1}101 and 3^{-1}103 working faces and the typical bursting failure characteristics of the GSR [34].

Relevant studies have shown that the occurrence of coal bursts in GSRs under thick and hard roofs conditions is significantly influenced by the roof structure of the working face and the stress environment of surrounding rock [1,29,33,35,36]. Building on these findings, this study develops a roof structure model of DWFs for three critical stages during the mining process of GSWFs: the IWS, periodic weighting stage (PWS), and FMS. These models aid in examining the load characteristics of the GSR surrounding rock. A new stress determination method for roadway surrounding rock is developed. Mining-induced stress and energy evolution in the roadway are investigated by numerical simulation. Our objective is to gain insights into coal burst mechanisms in GSRs and inform prevention strategies.

3. Roof Structure of DWFs and GSR Load Analysis

The GSR experiences dynamic and static loads stemming from roof collapse in the gob adjacent to the first working face (FWF) throughout extraction. Subsequently, when mining the GSWF, the GSR is subjected to a more complex stress environment. This environment combines static forces adjacent to the gob with loads from the GSWF, comprising dynamic and static elements. The cumulative effect of two mining stages significantly increases the risk of severe damage to the GSR. Consequently, the likelihood of coal bursts rises substantially. To analyze these complex load characteristics, the roof structure model of DWFs is established for three mining stages: the IWS, PWS, and FMS of the GSWF. By applying this model, we can effectively assess the load characteristics of the GSR during the secondary mining stage.

3.1. Roof Structure and Stress Calculation of Roadway Surrounding Rock During the IWS

According to key strata theory [10], an analysis was conducted on the overlying strata exposed in the drilling histogram of No. 18–12 (Figure 2) near an area prone to coal bursts. Medium-grained sandstone (No. 16) and fine-grained sandstone (No. 15) with high strength and large thickness were classified as inferior key strata I (IKS I) and inferior key strata II (IKS II). The key strata fracture structure significantly influences the stress evolution of the surrounding rock. Accurate assessment of the stress state and the loading patterns on the roadway must be evaluated in relation to the fracture structure of the key strata, therefore enabling the analysis of coal burst mechanisms.

No.	Stratigraphic Column	Thickness /m	Depth /m	Lithology
1		2.70	311.81	Mudstone
2		21.55	333.36	Fine-grained sandstone
3		34.12	367.48	Medium-grained sandstone
4		47.30	414.78	Fine-grained sandstone
5		27.26	442.04	Pebblestone
6		36.65	478.69	Coarse-grained sandstone
7		39.29	517.98	Medium-grained sandstone
8		16.44	534.42	Fine-grained sandstone
9		28.83	563.25	Mudstone
10		15.49	578.74	Sandy mudstone
11		14.47	593.21	Siltstone
12		8.13	601.34	Sandy mudstone
13		19.15	620.49	Siltstone
14		10.67	631.16	Sandy mudstone
15		44.90	676.06	Fine-grained sandstone
16		20.94	697.00	Medium-grained sandstone
17		5.73	702.73	Pebblestone
18		14.68	717.41	Fine-grained sandstone
19		9.11	726.52	Siltstone
20		7.23	733.75	No. 3^{-1} seam
21		15.25	749.00	Sandy mudstone

Figure 2. The histogram of No.18–12 drillhole (Partial) [1].

When the FWF is fully mined and the GSWF reaches its IWS, the roof of the DWFs' gob forms an asymmetrical T-shaped structure (Figure 3). In the FWF area, the IKS I and II fractures give rise to a "double-inclined voussoir beam" along the working face. In contrast, the GSWF exhibits a "single voussoir beam" structure. Only IKS I fractures, while IKS II remains intact, resulting in an overhanging roof. At this time, the intensity of strata behavior of GSWF is relatively low. The roof load shifts to the strata at the working face ends, inducing localized stress concentration and energy accumulation, which provides conditions for inducing the coal bursts in the GSR.

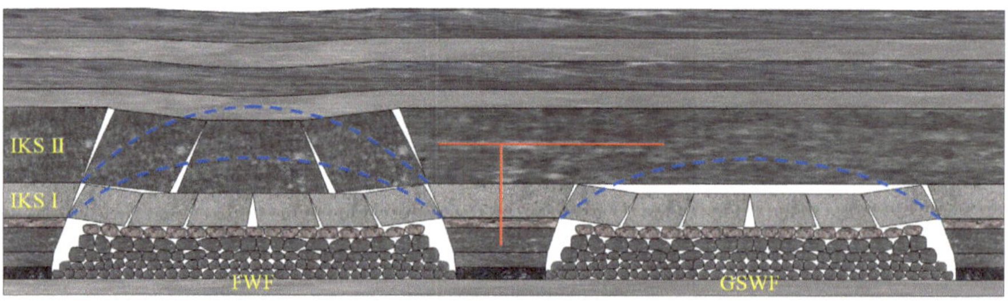

Figure 3. Asymmetrical T-shaped roof structure model of DWFs at the IWS of GSWF.

The GSR preceding the GSWF experiences lateral abutment pressure from the FWF gob that is basically the same across various mining stages. Figure 4 illustrates the distribution pattern of this load.

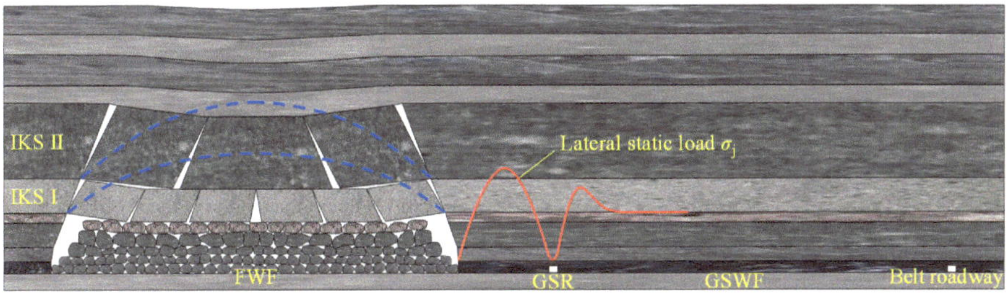

Figure 4. Schematic diagram of the gob side static load distribution in front of GSWF.

According to the literature [1], the lateral static load from the FWF gob on the GSR includes the foundation static load and the mining-induced additional static load:

$$\sigma_j = \sigma_y + \Delta\sigma_I + \Delta\sigma_{II} \quad (1)$$

where σ_j is the lateral static load from the gob on the GSR throughout the FMS of the FWF; σ_y is the foundation static load due to the self-weight of overlying strata; $\Delta\sigma_I$ is the mining-induced additional static load formed by the IKS I; and $\Delta\sigma_{II}$ is the mining-induced additional static load formed by the IKS II.

In Equation (1), σ_y, $\Delta\sigma_I$, and $\Delta\sigma_{II}$ are calculated by Equation (2):

$$\sigma_y = \begin{cases} \frac{\gamma H \tan\alpha}{2(h_0+h_1)+h_2}x & \left(0 \leqslant x \leqslant \frac{2(h_0+h_1)+h_2}{\tan\alpha}\right) \\ \gamma H & \left(x \geqslant \frac{2(h_0+h_1)+h_2}{\tan\alpha}\right) \end{cases} \quad (2)$$

where γ is the strata bulk density; H is the mining depth; α is the strata fracture angle, which can be taken as 70° [37]; and h_0, h_1, and h_2 are the respective thicknesses of the immediate roof, IKS I, and IKS II.

$$\Delta\sigma_I = \begin{cases} \frac{2\sigma_{Imax}\tan\alpha}{2h_0+h_1}x & \left(0 \leqslant x \leqslant \frac{2h_0+h_1}{2\tan\alpha}\right) \\ 2\sigma_{Imax}\left(1 - \frac{\tan\alpha}{2h_0+h_1}x\right) & \left(\frac{2h_0+h_1}{2\tan\alpha} \leqslant x \leqslant \frac{2h_0+h_1}{\tan\alpha}\right) \\ 0 & \left(x \geqslant \frac{2h_0+h_1}{\tan\alpha}\right) \end{cases} \quad (3)$$

where σ_{Imax} is the maximum mining-induced additional static load caused by IKS I.

$$\Delta\sigma_{II} = \begin{cases} \frac{2\sigma_{IImax}\tan\alpha}{2(h_0+h_1)+h_2}x & \left(0 \leqslant x \leqslant \frac{2(h_0+h_1)+h_2}{2\tan\alpha}\right) \\ 2\sigma_{IImax}\left(1 - \frac{\tan\alpha}{2(h_0+h_1)+h_2}x\right) & \left(\frac{2(h_0+h_1)+h_2}{2\tan\alpha} \leqslant x \leqslant \frac{2(h_0+h_1)+h_2}{\tan\alpha}\right) \\ 0 & \left(x \geqslant \frac{2(h_0+h_1)+h_2}{\tan\alpha}\right) \end{cases} \quad (4)$$

where σ_{IImax} is the maximum mining-induced additional static load caused by IKS II.

The formula for calculating σ_{Imax} in Equation (3) is as follows:

$$\sigma_{Imax} = \frac{\frac{1}{2}Q_{zI}}{\frac{2h_0+h_1}{2\tan\alpha}} = \frac{\gamma_1 h_1 l_1 \tan\alpha}{2h_0+h_1} \quad (5)$$

where Q_{zI} is the weight of the fragmented rock mass of IKS I; γ_1 is the bulk density of IKS I; and l_1 is the IKS I breaking interval.

The formula for calculating σ_{IImax} in Equation (4) is as follows:

$$\sigma_{IImax} = \frac{\frac{1}{2}Q_{zII}}{\frac{2(h_0+h_1)+h_2}{2\tan\alpha}} = \frac{(\gamma_2 h_2 + \gamma_3 h_3 K_z) l_2 \tan\alpha}{2(h_0+h_1)+h_2} \quad (6)$$

where Q_{zII} is the weight of the fragmented rock mass of IKS II and the load transferred from the overburden; γ_2 and γ_3 are the bulk density of IKS II and the overlying load layer; h_3 is the thickness of overlying load layer; K_z is the transfer coefficient of overlying load; and l_2 is the IKS II breaking interval.

The lateral static load from the FWF gob on the GSR σ_j can be calculated by Equations (1)–(6).

The GSR surrounding rock in front of the GSWF experiences combined load conditions. These include FWF gob-induced lateral abutment pressure, advanced abutment pressure of the working face, and disturbance dynamic loading caused by breaking thick and hard roofs. When the combined load exceeds the critical threshold for coal and rock bursting failure, coal bursts may occur. The loads at the front of the working face vary during the different stages, necessitating stage-specific analysis of the roof structure to accurately describe the loaded state of the GSR.

Figure 5 illustrates a three-dimensional model of the DWFs' roof during the IWS of the GSWF. It also shows the patterns of different load distributions: the lateral static load σ_j of FWF gob, the advanced static load σ_{jc} of GSWF, and the dynamic load σ_{dc} caused by the fracturing of thick and hard roofs.

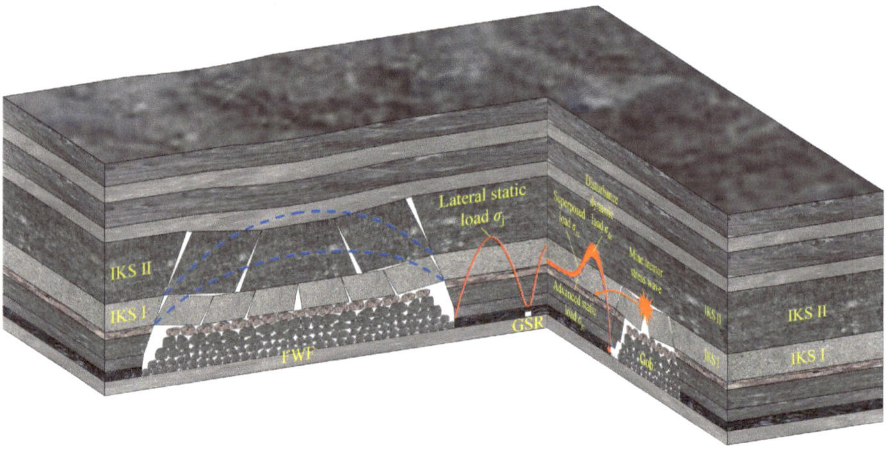

Figure 5. 3D model of the DWFs' roof and "three-load" superposition combined on the GSR at the IWS of the GSWF.

Figure 6 illustrates the roof structure and the combined loading profile in front of the GSWF, extending from the setup entry along the mining direction.

Two main sources of static load on the GSR surrounding rock are present at the IWS of the GSWF. The first source is the lateral static load σ_j, which is induced by the FWF gob. The second source is the advanced static load σ_{jc}, originating from the advanced abutment pressure of the GSWF.

The magnitude and distribution of the abutment pressure of the working face are influenced by various factors such as mining depth, mining height, roof lithology, and its structural characteristics. The distribution of abutment pressure, calculated using the limit equilibrium theory [38–40] in limit equilibrium and elastic zones, is as follows:

$$\sigma_a = \begin{cases} \delta(p + c\cot\varphi)e^{\frac{2\delta f}{m}x} - c\cot\varphi & (0 \leqslant x \leqslant x_0) \\ K'\gamma He^{\frac{2f}{m\lambda}(x_0 - x)} & (x \geqslant x_0) \end{cases} \quad (7)$$

where $\delta = \frac{1+\sin\varphi}{1-\sin\varphi}$, $x_0 = \frac{m}{2\delta f}\ln\frac{K'\gamma H + c\cot\varphi}{\delta(p+c\cot\varphi)}$; σ_a is the vertical stress; p is the support resistance; δ is the triaxial stress coefficient; and c and φ are the cohesion and friction angle.

Additionally, f is the friction coefficient; m is the mining height; x is the distance to coal rib; x_0 is the limit equilibrium zone width; K' is the stress concentration coefficient; γ is the strata bulk density; and λ is the lateral pressure coefficient.

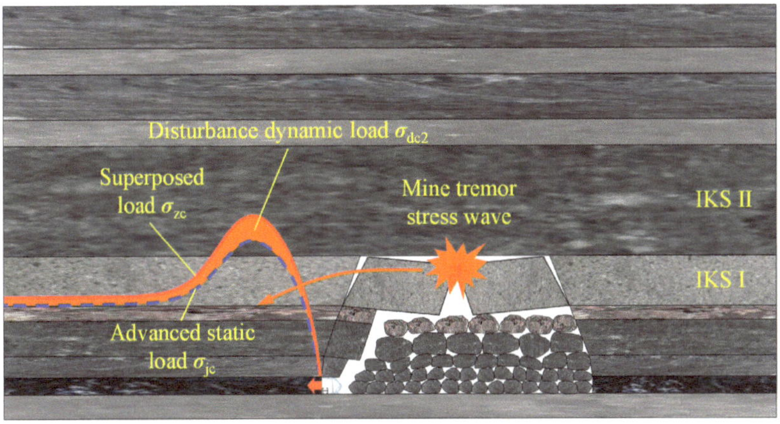

Figure 6. Schematic diagram of dynamic and static loads combined in front of working face at the IWS.

Equation (7) can be used to calculate the advanced static load σ_{jc}, σ_{jz} during the IWS and the PWS (FMS) of the GSWF.

The main source of dynamic load on the GSR surrounding rock at the IWS of the GSWF is the disturbance dynamic load that stems from the IKS I initial failure. Using a calculation approach for the disturbance dynamic load generated by the IKS I failure during the IWS, as per reference [41], the resulting disturbance dynamic load of the GSWF is

$$\sigma_{dc} = 0.0645 \rho V_s \left(\eta \frac{{q_c}^2 {l_c}^5}{576EJ} \right)^{0.3566} L^{-\beta} \tag{8}$$

where ρ is the medium density; σ_{dc} is the disturbance dynamic load caused by the IKS I fractures; V_s is the wave velocity; η is the energy conversion coefficient, which can be taken as 0.1% to 1% [42]; q_c is the overlying strata load of IKS I; l_c is the IKS I initial breaking interval; J is the moment of inertia; E is the roof strata elastic modulus; L is the propagation distance; and β is the attenuation index, which can be taken as 1.526 [43].

Based on the above analysis, the total stress on the GSR surrounding rock during the IWS of the GSWF is

$$\sigma_{zc} = \sigma_j + \sigma_{jc} + \sigma_{dc} \tag{9}$$

3.2. *Roof Structure and Stress Calculation of Roadway Surrounding Rock During the FMS*

The GSWF undergoes several stages as the working face progresses: the IWS, PWS, and FMS. During the PWS, the IKS II fractures and forms a "double-inclined voussoir beam" structure along the working face. This fracture creates a symmetrical T-shaped structure in the roof of the DWFs' gob (Figure 7). The IKS II failure generates dynamic loads on IKS I and the working face. These dynamic loads elevate the potential of coal bursts in the GSR, particularly in areas where the coal and rock mass already experience high-stress concentrations or significant elastic energy accumulation.

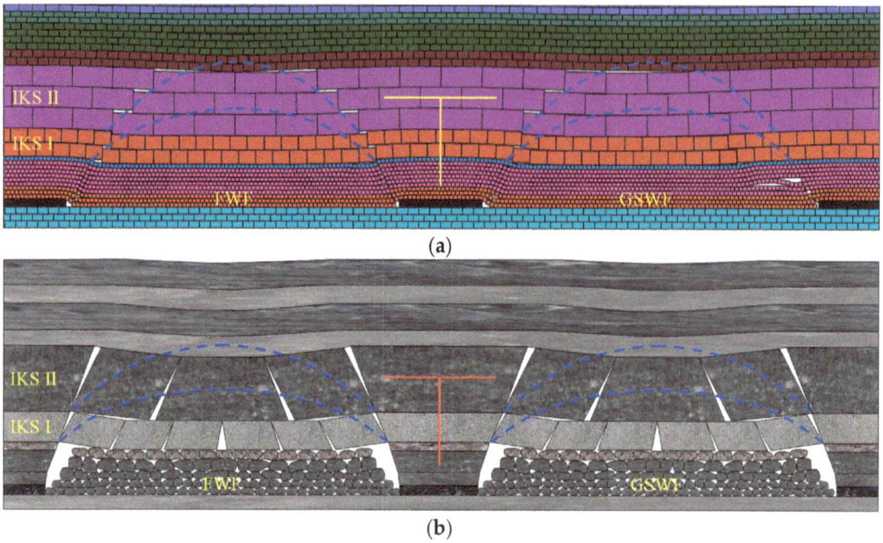

Figure 7. Symmetrical T-shaped roof structure model of the DWFs during the PWS of the GSWF. (**a**) UDEC numerical simulation results of the roof structure form of the DWFs; (**b**) symmetrical T-shaped roof structure model of the DWFs.

Throughout the FMS of the GSWF, the roof of the DWFs' gob forms a symmetrical T-shaped structure, as depicted in Figure 8. Meanwhile, this triggers large-scale movement in the overlying high-position thick and hard strata, leading to energy release within the roof, thereby intensifying the dynamic disturbance exerted by the IKS II on the underlying strata. This intensified disturbance could induce coal bursts in the GSR when the total stress or energy exceeds the critical threshold for burst failure of the coal and rock mass. Therefore, the FMS presents a higher propensity for coal bursts in the GSR compared to the PWS. This increased risk stems from the more intense dynamic loading effect due to the large-scale movement of the overlying strata at the DWFs.

Figure 9 illustrates a three-dimensional model of the DWFs' roof during the IWS of the GSWF. It also shows the patterns of different load distributions: the lateral static load σ_j of the FWF gob, the advanced static load σ_{jz} of the GSWF, and the dynamic load σ_{dz} caused by the fracturing of thick and hard roofs.

Analysis reveals a higher likelihood of coal bursts in the GSR throughout the FMS compared to the PWS. Consequently, we focus on the load status of the GSR in front of the working face at the FMS of the GSWF. Figure 10 exhibits the roof structure and the dynamic and static loads combined along the mining direction as the GSWF reaches the FMS.

At this stage, the static load on the GSR surrounding rock remains comparable to that in the IWS. However, the primary dynamic load source stems from the fracture-induced dynamic load of IKS I and II (Figure 10). Research indicates that the maximum disturbance dynamic load occurs when these two sets of inferior key strata break simultaneously [16]. The stress estimation should take into account the most adverse stress conditions for the surrounding rock, which is determined by the strong disturbance dynamic load generated during the simultaneous fracture of two sets of inferior key strata.

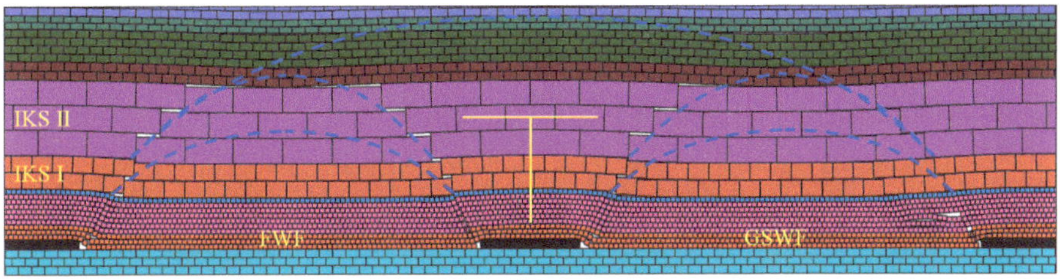

(**a**)

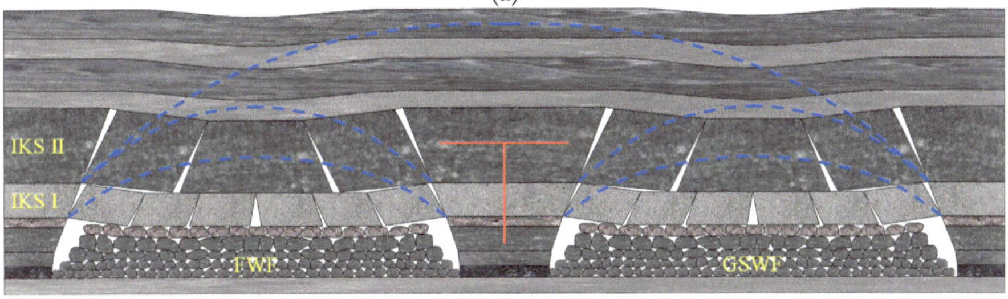

(**b**)

Figure 8. Symmetrical T-shaped roof structure model of the DWFs during the FMS of the GSWF. (**a**) UDEC numerical simulation results of roof structure form of the DWFs; (**b**) symmetrical T-shaped roof structure model of the DWFs.

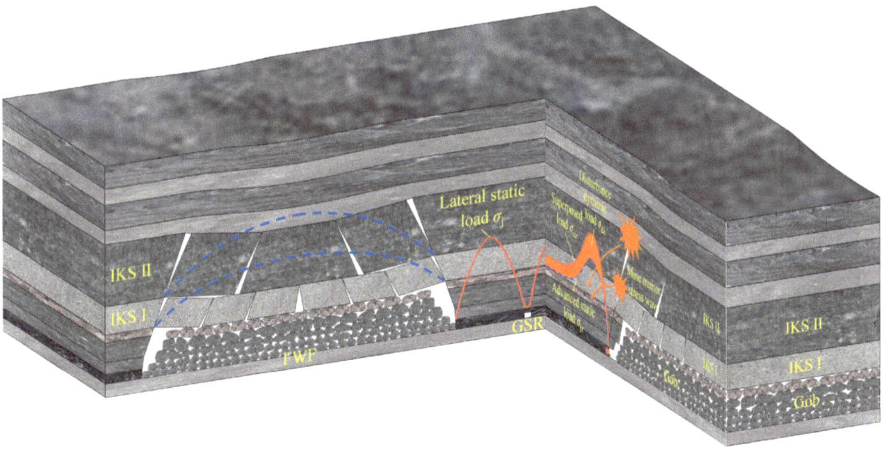

Figure 9. 3D model of the DWFs' roof and "three-load" superposition combined on the GSR at the FMS of the GSWF.

According to the literature [41], the approach of determining the disturbance dynamic load generated by the synchronous fracture of IKS I and II is presented as follows:

$$\begin{aligned}\sigma_{\mathrm{dz}} &= 0.0645\rho V_{\mathrm{s}}\left[\eta\left(U_{\mathrm{wzI}}+U_{\mathrm{wzII}}\right)\right]^{0.3566}L^{-\beta} \\ &= 0.0645\rho V_{\mathrm{s}}\left[\eta\frac{\left(q_{\mathrm{zI}}{}^{2}l_{\mathrm{zI}}{}^{5}+q_{\mathrm{zII}}{}^{2}l_{\mathrm{zII}}{}^{5}\right)}{8EJ}\right]^{0.3566}L^{-\beta}\end{aligned} \quad (10)$$

where q_{zI} and q_{zII} are the overlying strata load of IKS I and II; and l_{zI} and l_{zII} are the periodic breaking intervals of IKS I and II.

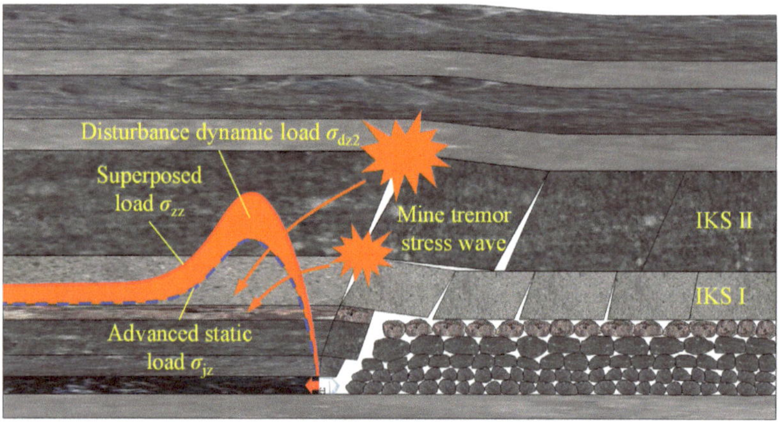

Figure 10. Schematic diagram of dynamic and static loads combined in front of the working face during the FMS.

Based on the above analysis, the total stress on the GSR surrounding rock during the FMS of the GSWF is

$$\sigma_{zz} = \sigma_j + \sigma_{jz} + \sigma_{dz} \quad (11)$$

3.3. Stability Analysis of Roadway Surrounding Rock

The critical stress required for burst failure in a coal and rock mass depends on the coal seam storage conditions and mining methods. Key factors influencing this stress include mining depth, coal and rock lithology, bursting liability, mining district layout, and mining techniques. Assuming that the critical stress for coal and rock burst failure under certain mining conditions as σ_{bmin} [44], the total stress on the GSR surrounding rock during the IWS and FMS of the GSWF, represented as σ_{zc} and σ_{zz}, can be calculated by Equations (9) and (11). Meanwhile, we define an index of burst failure of the GSR surrounding rock K_b, which is the ratio of the difference between the total stress on the GSR surrounding rock σ_{zc} (σ_{zz}) and the support stress of the roadway ψ to the critical stress for coal and rock burst failure σ_{bmin}. The stability of the GSR surrounding rock can be categorized into three cases, as expressed in Equation (12):

(1) Failure: When K_b is greater than 1;
(2) Critical state: When K_b equals 1;
(3) Safe: When K_b is less than 1.

$$K_b = \begin{cases} \frac{\sigma_{zc}(\sigma_{zz})-\psi}{\sigma_{bmin}} > 1 & \text{Failure} \\ \frac{\sigma_{zc}(\sigma_{zz})-\psi}{\sigma_{bmin}} = 1 & \text{Critial State} \\ \frac{\sigma_{zc}(\sigma_{zz})-\psi}{\sigma_{bmin}} < 1 & \text{Safe} \end{cases} \quad (12)$$

4. The Dynamic Changes of Stress and Energy Distributions Induced by Mining Activities

Section 3 analyzes the roof structure throughout the IWS and FMS of the GSWF. It identifies the load sources and characteristics affecting the GSR surrounding rock across different mining stages. A method for calculating the stress on the GSR was proposed, and load characteristics within the GSR surrounding rock during the mining process of the GSWF were quantitatively characterized. To further investigate the stress and energy evolution of the working face and GSR surrounding rock, FLAC^{3D} 5.01 software is employed to investigate the dynamic changes of stress and energy fields in the specified working face and GSR surrounding rock throughout the FMS of the GSWF, considering varying coal pillar widths.

4.1. Numerical Computational Approach

(1) Numerical modeling

A 3D numerical FLAC^{3D} model was constructed using the geological data from the coal mine. This model spans 750 m lengthwise, 800 m across, and 164 m in height, incorporating 1,464,960 grid cells and 1,545,495 nodes, as illustrated in Figure 11. To minimize boundary effects, a 100 m boundary coal pillar was implemented around the model. Due to the significant burial depth of the coal seam, the model may not simulate all overlying strata. To compensate, a vertical stress of 14.8 MPa was applied to the top of the model, calculated based on the depth of the coal seam and the model height. The model employs a fixed bottom boundary and utilizes roll-support boundaries for the remaining surfaces. Based on the in situ stress test results and the literature [45], the coefficient of lateral stress was set to 1.2. The model primarily simulates the stress and energy distribution characteristics of the surrounding rock after working face mining. According to relevant studies [46–48], the Mohr–Coulomb yield criterion was used to determine the failure behavior of the coal and rock mass in the model. Furthermore, the model includes the FWF and the GSWF. Material properties of geological strata employed for the numerical model were sourced from the literature [1] and are listed in Table 1.

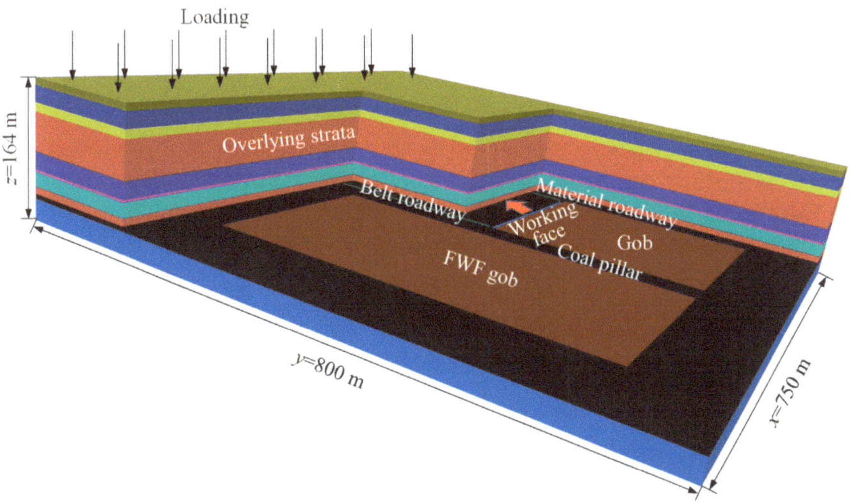

Figure 11. Illustrative representation of the 3D model and DWFs.

(2) Simulation scheme

The simulation process began with excavating the setup entry and roadway within the model. Subsequently, the FWF was excavated to the FMS. Next, the GSWF was excavated step by step at a distance of 10 m, concurrent with backfilling operations. Backfilling of the gob occurred at 10 m following the working face, with each backfill section measuring 210 m long, 10 m wide, and 6.4 m high. Material properties for the backfilling material can be found in Table 1.

Monitoring lines were established at 5, 10, 15, 20, 30, 50, 150, 200, and 250 m in front of the GSWF and the central region within the coal pillar. Concurrently, the monitoring surface was installed on the immediate roof above the working face. These monitoring systems were designed to measure the stress redistribution and energy accumulation in the surrounding rock during the mining process of the working face. By extracting stress values from the monitoring lines and surfaces, theoretical calculations enabled the quantitative characterization of the stress concentration and energy accumulation in the monitored areas. This study investigated the evolution laws of stress and energy at different advance distances of the GSWF with coal pillar widths of 5, 8, 10, 20, 30, and 40 m. A visual representation of the stress monitoring lines around the GSWF is provided in Figure 12.

Table 1. Material properties of the geological strata employed for the numerical model.

Lithology	Density/(kg/m^3)	Shear Modulus/GPa	Bulk Modulus/GPa	Cohesion/MPa	Friction/(°)	Tensile Strength/MPa
Sandy mudstone	2510	4.2	5.6	4.7	34	4.3
Siltstone	2750	5.0	6.9	3.0	32	3.6
Sandy mudstone	2510	3.5	5.2	5.0	33	4.5
Fine-grained sandstone	2650	2.9	3.9	2.8	36	3.5
Medium-grained sandstone	2850	2.4	4.2	3.0	38	3.7
Pebble stone	2600	2.5	4.2	3.2	32	3.0
Fine-grained sandstone	2640	3.3	4.4	3.1	35	2.9
Siltstone	2750	4.5	6.7	2.9	34	3.3
No. 3-1 seam	1380	1.2	2.3	2.6	25	1.3
Sandy mudstone	2470	3.8	4.8	5.0	34	4.6
Backfill material	1700	0.2	1.0	0.4	15	0

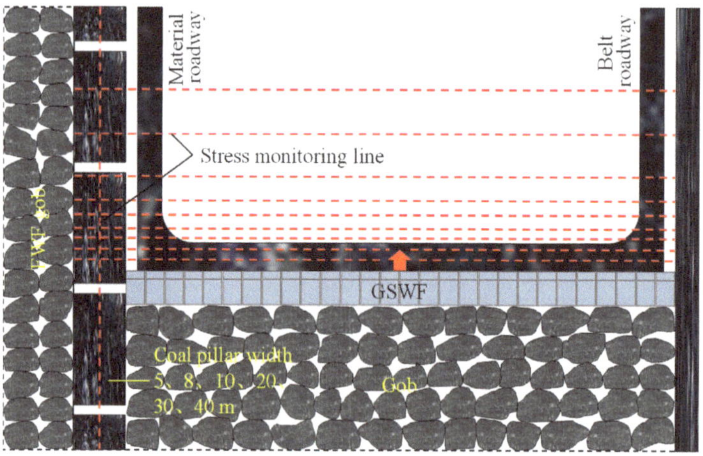

Figure 12. Illustrative representation of stress monitoring lines at GSWF.

4.2. Vertical Stress Variation of Coal Pillars and the GSWF Roof

(1) Influence of secondary mining on the vertical stress distribution within the coal pillar

We analyzed the vertical stress within the coal pillar at various advance distances, considering four typical coal pillar widths of 5, 10, 20, and 40 m. The stress distribution within the coal pillar during the mining process of the GSWF for various coal pillar widths is shown in Figure 13.

Figure 13 illustrates that the excavation of the GSWF has a profound influence on the vertical stress patterns within coal pillars. This impact varies depending on the coal pillar width. Notably, the stress experienced by the coal pillar during the secondary mining stage exceeds that of the first mining stage. After the excavation of the first working face, vertical stress concentrates in the coal pillar area, potentially causing plastic deformation. Subsequent excavation of the adjacent working face results in superposition of secondary mining-induced stress in the coal pillar area. Consequently, the vertical stress within the coal pillars increases. Furthermore, the adjacent DWFs form a symmetrical T-shaped structure in the gob area, leading to the complete fracturing of the overlaying strata. Consequently, these fractured layers impose a substantial load on the coal pillar. This leads to a marked increase in the vertical stress within the coal pillar. Meanwhile, the vertical stress in the coal pillar also rises within about 150 m in front of the working face. Consequently, high stress concentrates on the coal pillar side of the GSR. This stress concentration often leads to dynamic pressure behavior in the GSR and, in extreme cases, may trigger coal bursts. With a coal pillar width of 5 m, vertical stresses ranged from 7.42 to 9.45 MPa, exhibiting a maximum growth rate of 30.89% compared to the first mining stage. The coal pillar width being increased to 10 m led to higher vertical stresses, reaching 18.97 to 23.23 MPa, though the maximum growth rate decreased to 22.54%. Further widening to 20 m coal pillars caused a significant increase in vertical stress, ranging from 36.96 to 52.15 MPa, with the maximum growth rate rising to 27.92%. Notably, the 40 m coal pillar width showed lower vertical stresses of 31.40 to 42.28 MPa, with the lowest maximum growth rate of 22.33% compared to the first mining stage. Within the simulated range, vertical stress in the coal pillar side on the GSR under secondary mining initially increases with coal pillar width, then decreases. This pattern generates stress relief areas with lower vertical stress. Consequently, placing roadways in stress relief areas can create a low static load condition. Under the same conditions of dynamic load disturbance, the potential risk of coal bursts in the roadway can be reduced.

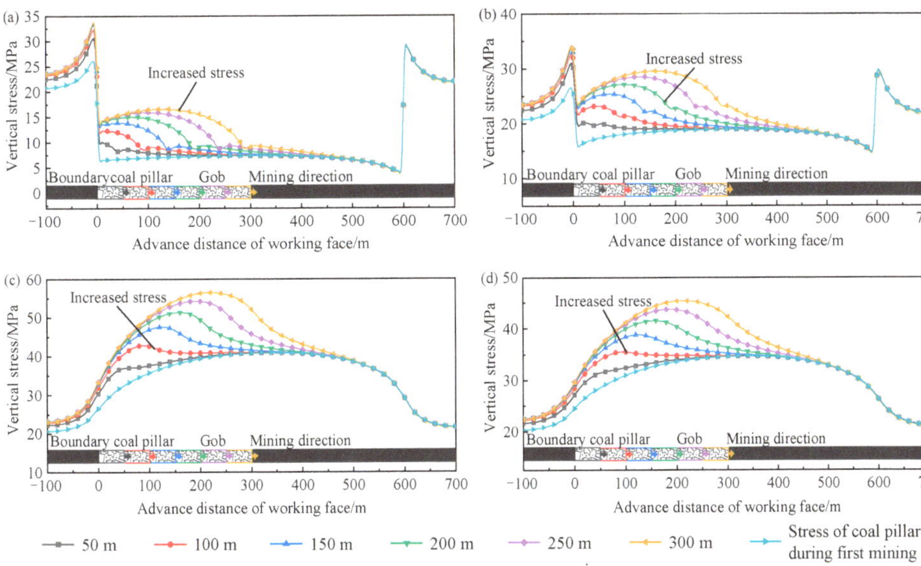

Figure 13. Vertical stress distribution within the coal pillar, influenced by varying coal pillar widths and advance distances under the secondary mining: (**a**) 5 m; (**b**) 10 m; (**c**) 20 m; (**d**) 40 m.

(2) Vertical stress variation of the working face roof

By arranging stress monitoring lines at different positions in front of the working face, we can observe the changes in the vertical stress of the roofs in front of the working face under varying coal pillar widths. Illustrative examples of this are presented in Figure 14, showcasing four typical simulation results with the coal pillar widths of 5, 10, 20, and 40 m.

The visual representation in Figure 14 shows an asymmetric pattern of vertical stress distribution in front of the GSWF with varying coal pillar widths. This asymmetry, characterized by higher stress on the left side, results from the influence of the adjacent FWF gob. This concentrated vertical stress in the roof at the gob-side end of the GSWF induces high static loads on the GSR surrounding rock within approximately 300 m in front of the working face. This concentrated stress significantly contributes to an increased risk of coal bursts in the GSR. With increasing distance from the working face, the vertical stress in the coal pillar and the immediate roof above the working face initially increases before subsequently decreasing. As observed, maximum vertical stress above the working face is located approximately 15 m in front of the working face. Beyond this point, this vertical stress decreases at a rate of about 0.06 MPa/m, as depicted in Figure 15a. With the coal pillar width increasing within this simulated range, the corresponding vertical stress initially rises, followed by a subsequent decline with a corresponding increase in the coal pillar widths. This non-linear relationship suggests that an optimal coal pillar width exists to minimize stress concentrations. The maximum vertical stress in roofs at the gob-side end in the GSWF spans from 40.18 to 49.80 MPa and exhibits a negative linear correlation with coal pillar width. The relationship follows the equation C), where B is the coal pillar width. This suggests a stress diminution rate of roughly 0.30 MPa/m expansion in coal pillar width. In a similar vein, the concentration coefficient of vertical stress on the roof fluctuates between 2.20 and 2.73, demonstrating a linear decline as the coal pillar width expands. This relationship is described by $K_S = -0.02B + 2.84$ ($R^2 = 0.98$), signifying a decrease of 0.02 units in K_S for every meter of expansion in the coal pillar width (Figure 15b).

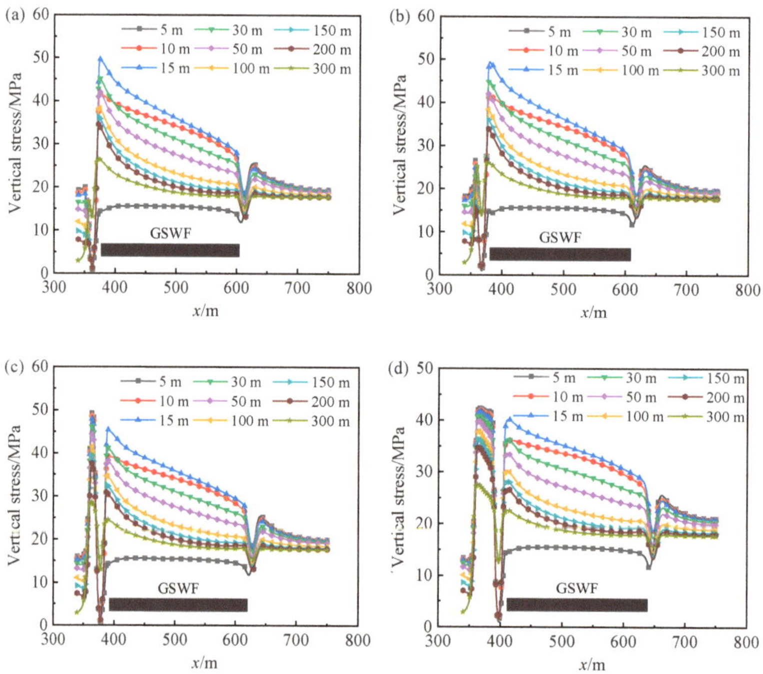

Figure 14. Changes in vertical stress in front of the working face influenced by varying coal pillar widths: (**a**) 5 m; (**b**) 10 m; (**c**) 20 m; (**d**) 40 m.

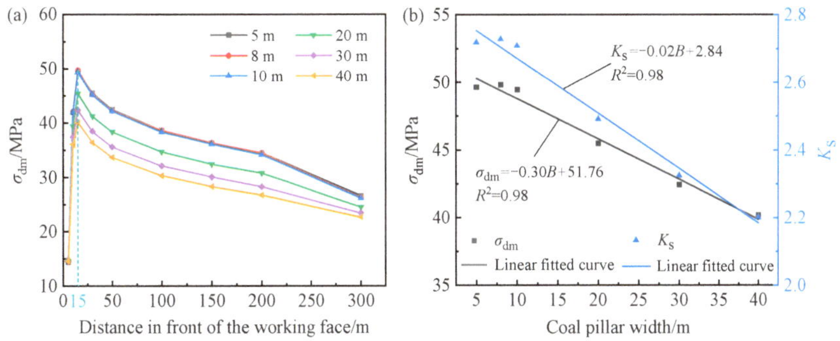

Figure 15. Changes in peak values and concentration coefficient of vertical stress on the roof, influenced by varying coal pillar widths: (**a**) Changes in peak values of vertical stress observed at various points in front of working face; (**b**) Changes in peak values and concentration coefficient of vertical stress.

4.3. Energy Density Distribution of GSWF Roof

A primary factor triggering coal bursts is the abrupt discharge of the substantial energy stored within the surrounding rock of the stope [13,22,33,49–51]. Multiple factors influence the energy distribution in the surrounding rock of the stope. While the characteristics and magnitude of ground stress are important factors [52,53] affecting the energy levels in the surrounding rock of the stope, horizontal stress does not significantly impact the overall energy distribution pattern under specific ground stress conditions. By strategically positioning stress monitoring surfaces within the roof to acquire the three principal stress data of the roof layers in the stope with varying coal pillar widths, the energy density value of the roof strata can be computed using Equation (13) [49,54–57].

$$u = \frac{1}{2E}\left[\sigma_1^2 + \sigma_2^2 + \sigma_3^2 - 2\mu(\sigma_1\sigma_2 + \sigma_2\sigma_3 + \sigma_3\sigma_1)\right] \quad (13)$$

where u is the energy density; E is the elastic modulus; and μ is Poisson's ratio.

The energy density of different areas in the roof under four typical coal pillar widths of 5, 10, 20, and 40 m is analyzed, and the distribution pattern is illustrated in Figure 16.

Figure 16 illustrates the varying degrees of energy accumulation in the roof of the gob-side end of the GSWF (the blue "triangle" area) for varying coal pillar widths. If the coal pillar width exceeds 20 m, energy accumulation also occurs within the coal pillar behind the working face. This energy accumulation pattern exhibits a notable trend: with the coal pillar width increasing, the amount of energy accumulated within the coal pillar behind the working face gradually decreases. The relationship between the peak values of energy density at the gob-side end of the working face and coal pillar widths demonstrates a linear negative correlation. Within the simulated coal pillar widths range, the peak values of energy density vary from 52.54 to 74.81 MJ/m^3. This relationship is described by $E_m = -0.67B + 78.66$ ($R^2 = 0.97$), where E_m is the maximum energy density and B is the coal pillar width. The peak energy density decreases at a rate of 0.67 (MJ/m^3)/m corresponding to an increasing coal pillar width (Figure 17). The coefficient of energy accumulation, defined as the ratio of the peak values of energy density in the roof after coal mining to the initial energy density in the roof before coal mining, also exhibits an inverse relationship with coal pillar width. This coefficient ranges from 3.20 to 4.56 and follows $K_e = -0.04B + 4.80$ ($R^2 = 0.96$), where K_e is the coefficient of energy accumulation. The coefficient of energy accumulation decreases at a rate of 0.04/m increase with the coal pillar width (Figure 17).

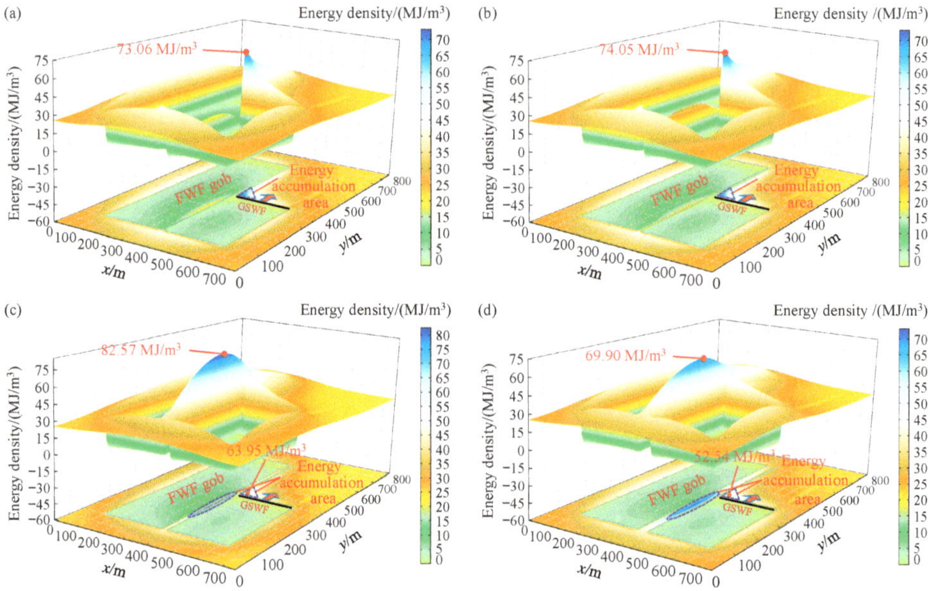

Figure 16. The energy density distribution at the working face roof under various coal pillar widths: (**a**) 5 m; (**b**) 10 m; (**c**) 20 m; (**d**) 40 m.

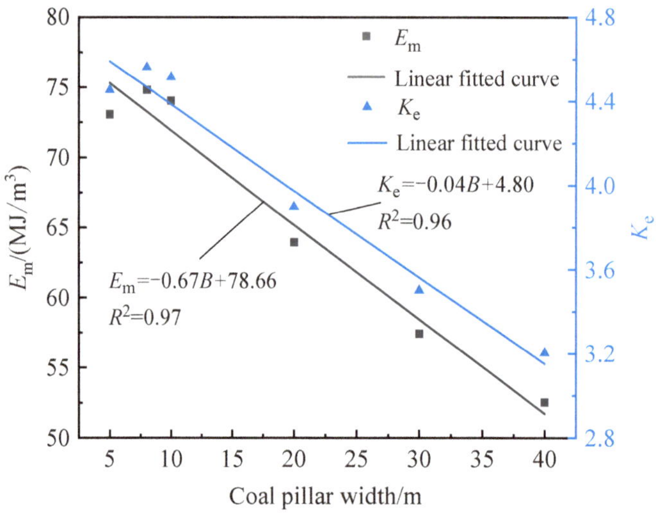

Figure 17. Changes in peak values of energy density and coefficient of energy accumulation across various coal pillar widths.

Changes in the peak values of both vertical stress and energy density within the roof, as well as the vertical stress within the coal pillar at the gob-side end of the GSWF, considering the coal pillar widths of 5, 8, 10, 20, 30, and 40 m, are illustrated in Figure 18.

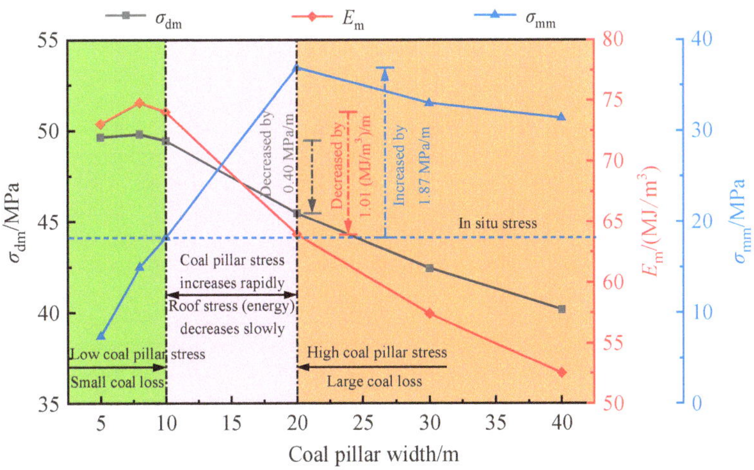

Figure 18. Changes in maximum vertical stress and energy density within the immediate roof, along with vertical stress in coal pillar, across varying coal pillar widths.

As illustrated in Figure 18, when the coal pillar width is less than 20 m, reducing the coal pillar width can reduce the overall pressure in the coal pillar and improve stress conditions within the GSR. However, both the vertical stress as well as the energy density in the immediate roof at the gob-side end above the working face will increase, affecting the overall loading status of the GSR. Therefore, the above factors should be considered when selecting the optimal coal pillar width to reduce the static load on the GSR surrounding rock and reduce potential burst failure within the roadway. For coal pillar widths of 5–10 m, the vertical stress in coal pillar remains below in situ stress levels, creating a reduced stress area. Such a condition facilitates effective management of the GSR surrounding rock while minimizing resource loss in the designated coal pillar. Within the range of coal pillar widths from 10–20 m, an increase in coal pillar width corresponds to a rise in the vertical stress in the coal pillar of approximately 1.87 MPa/m. Concurrently, the maximum vertical stress and energy density within the roof at the gob-side extremity of the working face decrease at rates of 0.40 MPa/m and 1.01 (MJ/m^3)/m, respectively. The rise in vertical stress attributed to the coal pillar width is more significant, and the overall stress conditions within the GSR become poor. For coal pillar widths ranging from 20–40 m, maximum levels of both vertical stress and energy density within the immediate roof, as well as peak values of vertical stress within the coal pillar, all decrease at varying rates with the coal pillar width increases. However, the peak vertical stress experienced by the coal pillar remains above the in situ stress levels, approximately 1.72 to 2.03 times these in situ stress levels. This is accompanied by significant resource loss within the reserved coal pillar. Considering the vertical stress within the coal pillar, vertical stress and energy density within the roof at the gob-side extremity of the working face, and resource loss caused by the reserved coal pillar, we can conclude that a reasonable coal pillar width is 5 to 10 m. At the same time, the roof support at the gob-side end of the working face should be strengthened in order to improve the overall stability of the GSR, thereby reducing the chance of coal bursts in the GSR.

5. Discussions

Section 3 analyzed the morphology of roof structures during the IWS and FMS of the GSWF. It explored how these varying roof structures apply loads to the GSR surrounding rock. Despite the different mining stages, the static loads of the advanced working face area in the GSR remain relatively consistent. These loads primarily consist of the lateral static load from the gob and advanced static load of the working face. The lateral static load from the gob increases initially and then decreases with distance from the gob. Consequently, advanced static load on the solid coal side of the GSR shows an asymmetric stress distribution [58]. Stresses are highest near the gob and decrease further away; this evolution law is consistent with the numerical simulation results of vertical stress distribution in front of the working face shown in Figure 14. Furthermore, the theoretical analysis of stress distribution based on the roof structure aligns with the numerical simulation results.

The load transmitted from the roof fracture of the adjacent gob to the advanced area of the solid coal side of the GSR is related to the coal pillar width and the distance to the GSWF. When the coal pillar width is small, the lateral roof of the adjacent gob transmits a large load to the solid coal side of the GSR. Additionally, when the distance to the GSWF is small, the load on the solid coal side of the GSR in front of the working face transmitted from the GSWF is relatively large [29,47,59]. These loads accumulate, leading to energy accumulation in this region (Figures 5 and 9). The accumulated energy decreases as the coal pillar width and the distance to the GSWF increase. This trend aligns with the numerical simulation results, which show the energy density distribution of the roof in front of the working face (Figure 16). The theoretical analyses are consistent with numerical simulation results, confirming the reliability of both approaches.

The analysis of the roof structures of the GSWF and its loading characteristics on the surrounding rock of the GSWF and GSR, as discussed in Section 3, is grounded in classic mine pressure theory and key strata theory [10]. While variations in stratigraphic configurations, lithology, and mining conditions lead to differences in the roof structures, they also make a difference in the load characteristics and stress-energy levels in the GSR. Despite these differences, the results presented in this paper are broadly applicable and reflect typical patterns. They offer valuable insights for studying and preventing coal bursts in GSRs under similar geological and mining conditions.

6. Conclusions

This study establishes a model of the roof structure during various mining stages within the GSWF, analyzes various load sources of the GSR, proposes a method for calculating stress within the GSR surrounding rock, and further discusses how the roof configuration form influences the stability of the GSR. Numerical simulations analyze dynamic changes in the mining-induced stress and energy fields within the working face, coal pillar, and GSR surrounding rock during the mining process of the GSWF. Key findings from this study include the following:

(1) Throughout the IWS in the GSWF, the roof of the DWFs forms an asymmetrical T-shaped structure along the working face orientation. During this stage, the intensity of the strata behavior of the GSWF is relatively low and the roof load is transferred to strata at the working face on both ends. Stress concentration, accompanied by energy accumulation, occurs in local areas, posing a risk of coal bursts in the GSR. Throughout the FMS of the GSWF, the roof of the DWFs forms a symmetrical T-shaped structure along the working face orientation. The breaking of IKS II causes dynamic loads on IKS I, the working face itself, and the working face on both ends, which can easily induce coal bursts in the GSR.

(2) A combined dynamic and static loading mechanical model for the GSR in front of the working face throughout both the IWS and FMS was developed. Additionally, a "three-load" dynamic and static load superposition principle applicable to the GSR was proposed, based on the lateral static load of the adjacent gob, the advanced static load originating from the working face, and the associated dynamic load. A calculation method for stress distribution within the GSR surrounding rock during various mining stages in the GSWF was obtained, and load characteristics within the GSR surrounding rock throughout the mining process of the GSWF were quantitatively characterized. The index of burst failure of GSR surrounding rock was proposed, and the criterion for judging the burst failure of roadways was established.

(3) Numerical calculation reveals that the vertical stress within the coal pillar in front of the GSWF increases compared to the first mining stage, causing a high stress concentration on the coal pillar side in the specified GSR in front of the working face. The elevated stress levels contribute to energy accumulation in the roof at the gob-side end of the GSWF, potentially leading to dynamic pressure in the roadway and, in extreme cases, coal bursts. As the distance to the working face increases, both the vertical stress within the coal pillar and the roof first rise, then decline. Furthermore, the vertical stress within the coal pillar initially rises and subsequently declines as coal pillar widths increase. Reducing the coal pillar width appropriately can improve its stress state.

(4) Considering the vertical stress within the coal pillar, the vertical stress and energy density within the roof at the gob-side end of the working face, and resource loss due to the presence of the coal pillar, the ideal coal pillar width should range between 5 and 10 m. Meanwhile, the roof support at the gob-side end of the working face should be strengthened to enhance the overall stability of the GSR surrounding rock, thereby reducing the potential for coal bursts in the GSR.

Author Contributions: J.Z. drafted the manuscript. J.P. and Y.X. provided the technical guidance. W.L., T.D., and J.W. provided writing guidance. All authors have read and agreed to the published version of the manuscript.

Funding: The research is financially supported by the National Key Research and Development Program of China (No. 2022YFC3004604) and the National Natural Science Foundation of China (No. 52174118).

Institutional Review Board Statement: Not Applicable.

Informed Consent Statement: Not Applicable.

Data Availability Statement: The data presented in this study are available on request from the corresponding author. The data are not publicly available due to privacy.

Acknowledgments: The first author would like to thank the Yixin Zhao at China University of Mining and Technology (Beijing), China.

Conflicts of Interest: The authors declare no conflict of interest.

References

1. Zhao, Y.; Zhou, J.; Zhang, C.; Liu, B.; Ling, C.; Liu, W.; Han, C. Failure Mechanism of Gob-Side Roadway in Deep Coal Mining in the Xinjie Mining Area: Theoretical Analysis and Numerical Simulation. *J. Cent. South Univ.* **2023**, *30*, 1631–1648. [CrossRef]
2. Qian, M.; Shi, P.; Xu, J. *Mine Pressure and Ground Control*; China University of Mining and Technology Press: Xuzhou, China, 2010.
3. Ju, J.; Xu, J.; Zhu, W. Longwall Chock Sudden Closure Incident below Coal Pillar of Adjacent Upper Mined Coal Seam under Shallow Cover in the Shendong Coalfield. *Int. J. Rock Mech. Min. Sci.* **2015**, *77*, 192–201. [CrossRef]
4. Ju, J.; Xu, J. Structural Characteristics of Key Strata and Strata Behaviour of a Fully Mechanized Longwall Face with 7.0m Height Chocks. *Int. J. Rock Mech. Min. Sci.* **2013**, *58*, 46–54. [CrossRef]
5. Wang, F.; Xu, J.; Xie, J. Effects of Arch Structure in Unconsolidated Layers on Fracture and Failure of Overlying Strata. *Int. J. Rock Mech. Min. Sci.* **2019**, *114*, 141–152. [CrossRef]
6. Unrug, K.S.B. Methods of Roof Cavability Prediction. *State Art Gr. Control Longwall Min. Min. Subsid.* **1982**, *3*, 13–29.
7. Kratzsch, H. *Mining Subsidence Engineering*; Springer Science & Business Media: Berlin, Germany, 2012.
8. Chien, M. A Study of the Behaviour of Overlying Strata in Longwall Mining and Its Application to Strata Control. *Dev. Geotech. Eng.* **1981**, *32*, 13–17.
9. Qian, M. Conditions Required for Equilibrium of Overlying Strata at Working Areas. *J. China Inst. Min. Technol.* **1981**, *2*, 31–40.
10. Qian, M.; Miao, X.; Xu, J. Theoretical Study of Key Stratum in Ground Control. *J. China Coal Soc.* **1996**, *21*, 225–230.
11. Song, Z. Rules for the Stope Bearing Pressure and Its Application. *J. Shandong Inst. Min. Technol.* **1982**, *1*, 1–25.
12. He, M.; Zhu, G.; Guo, Z. Longwall Mining "Cutting Cantilever Beam Theory" and 110 Mining Method in China-The Third Mining Science Innovation. *J. Rock Mech. Geotech. Eng.* **2015**, *7*, 483–492. [CrossRef]
13. He, M.; Wang, Q. Rock Dynamics in Deep Mining. *Int. J. Min. Sci. Technol.* **2023**, *33*, 1065–1082. [CrossRef]
14. He, M.; Wang, Q.; Wu, Q. Innovation and Future of Mining Rock Mechanics. *J. Rock Mech. Geotech. Eng.* **2021**, *13*, 1–21. [CrossRef]
15. Dou, L.M.; He, X.Q.; He, H.; He, J.; Fan, J. Spatial Structure Evolution of Overlying Strata and Inducing Mechanism of Rockburst in Coal Mine. *Trans. Nonferrous Met. Soc. China* **2014**, *24*, 1255–1261. [CrossRef]

16. Huang, Q.; Zhou, J.; Cao, J. Key Stratum Structure and Support Working Resistance of Longwall Face with Large Mining Height in the Shallow Coal Seams, China. *Adv. Civ. Eng.* **2020**, *2020*, 8834403. [CrossRef]
17. Yu, B.; Zhu, W.; Gao, R.; Liu, J. Strata Structure and Its Effect Mechanism of Large Space Stope for Fullymechanized Sublevel Caving Mining of Extremely Thick Coal Seam. *J. China Coal Soc.* **2016**, *41*, 571–580.
18. Yu, B.; Gao, R.; Kuang, T.; Huo, B.; Meng, X. Engineering Study on Fracturing High-Level Hard Rock Strata by Ground Hydraulic Action. *Tunn. Undergr. Sp. Technol.* **2019**, *86*, 156–164. [CrossRef]
19. Sun, Y.; Zuo, J.; Karakus, M.; Wang, J. Investigation of Movement and Damage of Integral Overburden during Shallow Coal Seam Mining. *Int. J. Rock Mech. Min. Sci.* **2019**, *117*, 63–75. [CrossRef]
20. Zuo, J.; Wang, J.; Jiang, Y. Macro/Meso Failure Behavior of Surrounding Rock in Deep Roadway and Its Control Technology. *Int. J. Coal Sci. Technol.* **2019**, *6*, 301–319. [CrossRef]
21. Zuo, J.; Xu, C.; Sun, Y.; Li, Y.; Wu, G.; Yu, M. Theoretical Model and Verification of "Analogous Hyperbola (Hyperboloid)" for the Overall Movement of Mining Rock Strata: From Two-Dimensional "Analogous Hyperbola" to Three-Dimensional "Analogous Hyperboloid" Models. *J. China Coal Soc.* **2024**, *49*, 1731–1751.
22. Pan, Y.; Song, Y.; Luo, H.; Xiao, Y. Coalbursts in China: Theory, Practice and Management. *J. Rock Mech. Geotech. Eng.* **2024**, *16*, 1–25. [CrossRef]
23. He, M.; Cheng, T.; Qiao, Y.; Li, H. A Review of Rockburst: Experiments, Theories, and Simulations. *J. Rock Mech. Geotech. Eng.* **2023**, *15*, 1312–1353. [CrossRef]
24. Dai, L.; Pan, Y.; Xiao, Y.; Wang, A.; Wang, W.; Wei, C. Parameter Design Method for Destressing Boreholes to Mitigate Roadway Coal Bursts: Theory and Verification. *Rock Mech. Rock Eng.* **2024**, 1–18. [CrossRef]
25. Zhang, J.; Zhang, Y.; Han, Y.; Fan, W.; Song, Z.; Yu, H.; Zhang, J.; Zhang, H.; Liu, J. Study on Stress Distribution Law of Surrounding Rock of Roadway under the Goaf and Mechanism of Pressure Relief and Impact Reduction. *Eng. Fail. Anal.* **2024**, *160*, 108210. [CrossRef]
26. Zhou, K.; Dou, L.; Li, X.; Song, S.; Cao, J. Coal Burst and Mining-Induced Stress Evolution in a Deep Isolated Main Entry Area–A Case Study. *Eng. Fail. Anal.* **2022**, *137*, 106289. [CrossRef]
27. Zhang, Z.; Li, Z.; Xu, G.; Gao, X.; Liu, Q.; Li, Z.; Liu, J. Lateral Abutment Pressure Distribution and Evolution in Wide Pillars under the First Mining Effect. *Int. J. Min. Sci. Technol.* **2023**, *33*, 309–322. [CrossRef]
28. Bai, J.; Shen, W.; Guo, G.; Wang, X.; Yu, Y. Roof Deformation, Failure Characteristics, and Preventive Techniques of Gob-Side Entry Driving Heading Adjacent to the Advancing Working Face. *Rock Mech. Rock Eng.* **2015**, *48*, 2447–2458. [CrossRef]
29. Zhang, G.; Wen, Z.; Liang, S.; Tan, Y.; Tian, L.; Zhao, Y.; Zhao, D. Ground Response of a Gob-Side Entry in a Longwall Panel Extracting 17 m-Thick Coal Seam: A Case Study. *Rock Mech. Rock Eng.* **2020**, *53*, 497–516. [CrossRef]
30. Lou, J.; Gao, F.; Yang, J.; Ren, Y.; Li, J.; Wang, X.; Yang, L. Characteristics of Evolution of Mining-Induced Stress Field in the Longwall Panel: Insights from Physical Modeling. *Int. J. Coal Sci. Technol.* **2021**, *8*, 938–955. [CrossRef]
31. Wang, Z.; Wang, J.; Yang, S. An Ultrasonic-Based Method for Longwall Top-Coal Cavability Assessment. *Int. J. Rock Mech. Min. Sci.* **2018**, *112*, 209–225. [CrossRef]
32. Wang, Z.; Yang, S.; Tang, Y.; Sun, W.; Shui, Y. A Stress Rotation-Based Method for Improving Roof Stability of a Deep Longwall Panel. *Int. J. Geomech.* **2023**, *23*, 04023085. [CrossRef]
33. Ma, Z.; Li, S.; Zhao, X. Energy Accumulation Characteristics and Induced Rockburst Mechanism of Roadway Surrounding Rock under Multiple Mining Disturbances: A Case Study. *Sustainability* **2023**, *15*, 9595. [CrossRef]
34. Zhou, J.; Pan, J.; Xia, Y.; Du, T.; Liu, W.; Zhang, C. Mechanism and Prevention of Coal Bursts in Gob-side Roadway Floor under Thick and Hard Roof in the Deep Mining Area of Ordos. *Int. J. Coal Sci. Technol.* **2024**, *11*, 60.
35. Han, Z.; Dou, L.; Gong, S.; Kan, J.; Chen, S.; He, X. Mechanism of Rock Burst in Deep Gob-Side Entry Based on Dynamic and Static Stress: A Case Study. *Geomat. Nat. Hazards Risk* **2023**, *14*, 2271636. [CrossRef]
36. Bai, J.; Dou, L.; Li, X.; Ma, X.; Lu, F.; Han, Z. Evolution Laws of Stress–Energy and Progressive Damage Mechanisms of Surrounding Rock Induced by Mining Disturbance. *Appl. Sci.* **2023**, *13*, 7759. [CrossRef]
37. Sun, Y.; Zuo, J.; Karakus, M.; Wen, J. A Novel Method for Predicting Movement and Damage of Overburden Caused by Shallow Coal Mining. *Rock Mech. Rock Eng.* **2020**, *53*, 1545–1563. [CrossRef]
38. Zhu, S.; Feng, Y.; Jiang, F. Determination of Abutment Pressure in Coal Mines with Extremely Thick Alluvium Stratum: A Typical Kind of Rockburst Mines in China. *Rock Mech. Rock Eng.* **2016**, *49*, 1943–1952. [CrossRef]
39. Zhang, P.; Sun, B. Distribution Characteristics of the Advance Abutment Pressure in a Deep Stope. *J. Geophys. Eng.* **2020**, *17*, 686–699. [CrossRef]
40. Zhu, Q.; Sui, L.; Yin, Y.; Liu, J.; Ouyang, Z.; Wang, D. Mechanism and Empirical Study of Rockburst in the Adjacent Area of a Fully Mechanized Top-Coal Caving Face Based on Microseismic Technology. *Appl. Sci.* **2023**, *13*, 6317. [CrossRef]
41. Zhao, Y.; Zhou, J.; Liu, W. Characteristics of Ground Pressure and Mechanism of Coal Burst in the Gob Side Roadway at Xinjie Deep Mining Area. *J. China Coal Soc.* **2020**, *45*, 1595–1606.
42. Dou, L.; He, X. *Theory and Technology of Rock Burst Prevention*; China University of Mining and Technology Press: Xuzhou, China, 2001.
43. He, J.; Dou, L.; Cai, W.; Li, Z.; Ding, Y. In Situ Test Study of Characteristics of Coal Mining Dynamic Load. *Shock Vib.* **2015**, *2015*, 121053. [CrossRef]

44. Cai, W.; Bai, X.; Si, G.; Cao, W.; Gong, S.; Dou, L. A Monitoring Investigation into Rock Burst Mechanism Based on the Coupled Theory of Static and Dynamic Stresses. *Rock Mech. Rock Eng.* **2020**, *53*, 5451–5471. [CrossRef]
45. Li, X.; Zhou, Z.; Lok, T.S.; Hong, L.; Yin, T. Innovative Testing Technique of Rock Subjected to Coupled Static and Dynamic Loads. *Int. J. Rock Mech. Min. Sci.* **2008**, *45*, 739–748. [CrossRef]
46. Chang, L.; Konietzky, H. Application of the Mohr-Coulomb Yield Criterion for Rocks with Multiple Joint Sets Using Fast Lagrangian Analysis of Continua 2D (FLAC2D) Software. *Energies* **2018**, *11*, 614. [CrossRef]
47. He, W.; He, F.; Zhao, Y. Field and Simulation Study of the Rational Coal Pillar Width in Extra-Thick Coal Seams. *Energy Sci. Eng.* **2020**, *8*, 627–646. [CrossRef]
48. He, X.; Zhang, C.; Han, P. Overburden Damage Degree-Based Optimization of High-Intensity Mining Parameters and Engineering Practices in China's Western Mining Area. *Geofluids* **2020**, *2020*, 8889663. [CrossRef]
49. Du, F.; Ma, J.; Guo, X.; Wang, T.; Dong, X.; Li, J.; He, S.; Nuerjuma, D. Rockburst Mechanism and the Law of Energy Accumulation and Release in Mining Roadway: A Case Study. *Int. J. Coal Sci. Technol.* **2022**, *9*, 67. [CrossRef]
50. Das, A.J.; Mandal, P.K.; Ghosh, N.; Singh, A.P.; Kumar, R.; Tewari, S.; Bhattacharjee, R. Evaluation of Energy Accumulation, Strain Burst Potential and Stability of Rock Mass during Underground Extraction of a Highly Stressed Coal Seam under Massive Strata-a Field Study. *Eng. Geol.* **2023**, *322*, 107178. [CrossRef]
51. Lu, J.; Yin, G.; Gao, H.; Li, X.; Zhang, D.; Deng, B.; Wu, M.; Li, M. True Triaxial Experimental Study of Disturbed Compound Dynamic Disaster in Deep Underground Coal Mine. *Rock Mech. Rock Eng.* **2020**, *53*, 2347–2364. [CrossRef]
52. Zhang, Y.; Okere, C.J.; Su, G. Effect of Loading Rates on Accurate In-Situ Stress Determination in Different Lithologies via Kaiser Effect. *Arab. J. Geosci.* **2021**, *14*, 1304. [CrossRef]
53. Miao, S.J.; Cai, M.F.; Guo, Q.F.; Huang, Z.J. Rock Burst Prediction Based on In-Situ Stress and Energy Accumulation Theory. *Int. J. Rock Mech. Min. Sci.* **2016**, *83*, 86–94. [CrossRef]
54. Jia, Z.; Li, C.; Zhang, R.; Wang, M.; Gao, M.; Zhang, Z.; Zhang, Z.; Ren, L.; Xie, J. Energy Evolution of Coal at Different Depths Under Unloading Conditions. *Rock Mech. Rock Eng.* **2019**, *52*, 4637–4649. [CrossRef]
55. Xu, J.; Jiang, J.; Xu, N.; Liu, Q.; Gao, Y. A New Energy Index for Evaluating the Tendency of Rockburst and Its Engineering Application. *Eng. Geol.* **2017**, *230*, 46–54. [CrossRef]
56. Hao, Q.; Zhang, R.; Gao, M.; Xie, J.; Ren, L.; Zhang, A. Characterization of Energy-Driven Damage Mechanism and Gas Seepage in Coal under Mining-Induced Stress Conditions. *Int. J. Rock Mech. Min. Sci.* **2024**, *181*, 105834. [CrossRef]
57. Cui, F.; Sun, J.; Lai, X.; Jia, C.; Zhang, S. Study on the Energy Release Law of Overburden Rock Breaking and Anti-Rockburst Technology in the Knife Handle Working Face of a Gently Inclined Coal Seam. *Appl. Sci.* **2023**, *13*, 11809. [CrossRef]
58. Yadav, A.; Behera, B.; Sahoo, S.K.; Singh, G.S.P.; Sharma, S.K. Numerical Analysis of the Gob Stress Distribution Using a Modified Elastic Model as the Gob Constitutive Model. *J. Inst. Eng. Ser. D* **2020**, *101*, 127–139. [CrossRef]
59. Zhang, G.; Chen, L.; Wen, Z.; Chen, M.; Tao, G.; Li, Y.; Zuo, H. Squeezing Failure Behavior of Roof-Coal Masses in a Gob-Side Entry Driven under Unstable Overlying Strata. *Energy Sci. Eng.* **2020**, *8*, 2443–2456. [CrossRef]

applied sciences

MDPI

Article

Slope Stability Analysis of Open-Pit Mine Considering Weathering Effects

Wei Liu [1], Gang Sheng [1], Xin Kang [1], Min Yang [2], Danqi Li [3] and Saisai Wu [2,*]

1 Ankang Branch of Shaanxi Provincial Land Engineering Construction Group Co., Ltd., Ankang 725000, China
2 School of Resources Engineering, Shanxi Key Laboratory of Geotechnical and Underground Space Engineering, XAUAT, Xi'an 710055, China
3 Minerals, Energy and Chemical Engineering, Curtin University, Kalgoorlie, WA 6430, Australia; danqi.li@curtin.edu.au
* Correspondence: saisai.wu@xauat.edu.cn

Abstract: Weathering processes gradually alter the physical and mechanical attributes of slope materials, weakening the structural integrity and stability of slopes. This paper presents an in-depth analysis of slope stability in an open-pit mine, emphasizing the pivotal role of weathering effects in determining slope stability. To accurately capture the impact of weathering on slope stability, a comprehensive analysis model was developed, incorporating field observations, laboratory testing, and numerical simulations. The effects of weathering on the mechanical properties of black shale were studied through extensive laboratory tests. The uniaxial compressive strength, shear strength, and modulus of elasticity significantly decreased with increasing weathering, indicating a heightened vulnerability to slope failure. The correlation function between mechanical parameters and weathering time was obtained, providing the basis for evaluating the stability of mine slopes. It was found that more severe weathering conditions were strongly correlated with elevated risks of slope failure, including landslides and collapses. Based on these findings, practical recommendations are provided for slope reinforcement and management strategies, aimed at mitigating slope failure risks and ensuring the safe and efficient operation of the mine. By incorporating weathering effects into slope stability analysis, mine operators can make informed decisions that account for the dynamic nature of slope materials and their susceptibility to weathering, thereby improving overall mine performance and safety.

Keywords: slope stability; weathering; black shale; mechanical properties

1. Introduction

Slope stability analysis is a pivotal aspect of mine planning and operations in open-pit mines, as it directly impacts the safety of mine workers and the efficiency of mining activities [1–3]. Over the years, numerous studies have been conducted to assess slope stability. With the advancing technology in open-pit mining, the scale and depth of excavations are constantly increasing. Slope stability is a crucial aspect of civil engineering and geotechnical studies. The following are the most influential parameters in slope stability: soil properties, geological structure, slope geometry, groundwater conditions, and external loads [4]. As the depth of mining increases, slope heights are also escalating annually, resulting in concave or deeply concave mining fields. In open-pit mines, every marginal increase in slope angle can significantly augment the stripping volume, emphasizing the importance of optimizing slope design and accurately evaluating slope stability [5,6]. Therefore, a comprehensive approach to slope stability analysis, considering various factors such as geological conditions and weathering effects is essential for ensuring the safety and sustainability of mining operations.

Slope stability is often influenced by a range of geological and environmental factors, with weathering effects being particularly significant contributors [7–10]. Rocks vary in composition and structure, leading to differences in stress–strain relationships and deformation characteristics. These properties are further influenced by factors such as temperature and humidity, which are integral parts of the weathering process [11–15]. Rock weathering is a natural process that involves the gradual breakdown and alteration of rocks due to their interactions with the environment over time [16–20]. This ongoing process leads to intricate changes in the petrographical, mineralogical, microstructural, and geomechanical properties of rocks. Typically, weathering reduces the strength of rocks by enhancing the deformability and degradation. Furthermore, weathering alters the mineral composition of rocks, causing certain minerals to undergo dissolution, oxidation, or hydrolysis through chemical reactions. These chemical alterations result in changes in the physical and mechanical properties, further compromising the tensile strength of rock and increasing its vulnerability to failure [21]. Given the adverse impact of weathering on strength and deformational characteristics of rocks, a thorough evaluation of the physical–mechanical behavior of rocks under weathering conditions is crucial in the context of underground mines [22,23]. Previous investigations into the geotechnical state of railway tunnels have revealed that weathering can pose significant challenges, leading to substantial economic losses [24].

Numerous studies have been conducted to evaluate the effects of weathering on the mechanical properties of granitoid rocks. Heidari and Momeni [25] observed a decrease in the elastic modulus of granodiorite as weathering time increased. Leveraging the correlation between weathering grades and changes in mechanical and petrographic properties, Momeni and Khanlari established a weathering classification system for granitoid rocks. Abad and Tugrul [26] delved into the characteristics of granitic rock masses across various weathering zones in tropical environments. Additionally, researchers have discovered that weathering reduces the elastic modulus of rocks by modifying their mineralogical characteristics. Specifically, for calcareous rocks, the elastic modulus has been shown to exhibit a negative exponential relationship with calcite content.

In open-pit mines, weathering can significantly reduce the strength and stability of slope materials, posing potential safety hazards [27–32]. Despite the crucial role of weathering effects in determining slope stability, many traditional analysis methods overlook or oversimplify these factors [33–38]. This approach can lead to inaccurate predictions of slope stability and potentially underestimate the risks of slope failure. Therefore, it is imperative to incorporate weathering effects into slope stability assessments to ensure the safe and efficient operation of mines. Additionally, a thorough understanding of the mechanical properties of rocks at a given time and location is crucial for studying related geotechnical engineering. The process of ore mining itself involves studying the mechanical properties of rocks. Gaining a comprehensive understanding of the physical and mechanical properties of surrounding rocks, particularly under the influence of weathering, holds significant theoretical and practical value. It is essential for ensuring the safe mining of open-pit mines and optimizing mining boundaries.

Although weathering effects play a crucial role in determining slope stability, few researchers have investigated the effects of weathering and the traditional analysis methods either overlook these factors or oversimplify them. In this study, a comprehensive analysis of slope stability in open-pit mines, taking into account the influence of weathering effects, was conducted. Laboratory tests were performed to investigate the impact of weathering on the mechanical properties of black-shale specimens. Numerical simulations were employed to evaluate slope stability and determine safety factors under various weathering conditions. By incorporating weathering effects into slope stability analysis, this study aimed to provide a more accurate and reliable assessment of slope stability in open-pit mines. The findings of this research could significantly contribute to enhancing mine safety, mitigating slope failure risks, and optimizing mine planning and operations.

2. Laboratory Tests of Weathering Effects

2.1. Testing Specimens

The engineering background of this study was one of the phosphate mines located in Yunnan Province, China. The predominant rock type surrounding the ore body in this mine is black shale. Given the low strength and rapid weathering characteristics of black shale in the mining area, the stability of the slopes, as well as the optimization of slope angles and boundaries, pose significant concerns. To address the concerns, black-shale samples collected from the mining areas were utilized as specimens to determine their fundamental mechanical properties under weathering effects. This particular phosphate mine was located at the northern edge of the Yunnan–Guizhou Plateau, where the terrain is complex with intense cutting and well-developed valleys, belonging to the deeply incised karst highland mountain landforms. The topography of the mining area is steep, with ore outcrops at altitudes ranging from 2600 to 3000 m. These topographical conditions have a significant impact on the preservation and enrichment depth of weathering products. The rock specimens were drilled perpendicular to the bedding direction at the depth of 600 m and processed into standard cylindrical specimens, which was in accordance with the rock mechanics test specification recommended by the International Society for Rock Mechanics (ISRM) [39]. Weathering may alter the mineral composition and physical properties of rocks, as well as their mechanical properties, but only the mechanical properties were focused on in this study. For the uniaxial compression test, the specimens had a diameter of 50 mm and a height-to-diameter ratio of 2. For the direct shear mechanical test, the height-to-diameter ratio was set to 1, while for the Brazilian splitting test it was 0.5. The specimens used for each test are shown in Figure 1.

Figure 1. Examples of the specimens: (**a**) Brazilian splitting test, (**b**) uniaxial compression test, and (**c**) direct shear test.

2.2. Experimental Methodologies

A comprehensive suite of tests, including uniaxial compressive strength tests, direct shear tests, and Brazilian splitting tests, were conducted on the specimens (Figure 2). For the uniaxial compressive strength tests the specimen was placed between the platens of a compressive testing machine. A gradually increasing compressive load at 0.5 MPa/s was applied to the specimen at a controlled rate. For direct shear tests, the specimen was placed between the upper and lower shear boxes. A predetermined normal load was applied to the specimen to simulate the in situ stress conditions. A gradually increasing horizontal load at 0.1 MPa/s was applied to cause shear deformation. For the Brazilian splitting tests, the specimen was placed between two loading platens. A compressive load at 0.5 MPa/s was applied along the diameter of the specimen. Prior to testing, the specimens were carefully placed in the actual mining environment to undergo natural weathering. The specimens were placed in the areas near the slope edge where significant diurnal temperature range co-occurs during day and night, which is a prominent feature of the plateau climate. To understand the evolution of physical and mechanical parameters during weathering, the specimens were periodically retrieved, and examined for relevant physical and mechanical parameters. The testing programs are outlined in Table 1. A total of 15 Brazilian splitting tests, 15 uniaxial compression tests, and 15 direct shear tests were conducted. During the tests, since the weathering process occurred in the actual mining environment, the temperature as well as the precipitation were different each day. It was recorded that the maximum surface temperature reached 75.2 °C, while the minimum surface temperature was 16.4 °C, resulting in a temperature difference of 55.8 °C. The large diurnal temperature variation not only affected the physical properties of rocks, such as causing micro-cracks on the surface of rocks, but also promoted the freeze–thaw weathering process of rocks. When the temperature dropped at night, moisture on the surface of rocks easily condensed into frost or ice, exerting a freezing force on the rocks, thereby accelerating the weathering process. The area also has abundant precipitation, and weathering was even more intense. It was recorded that during the 28 days of weathering, there were 9 days of rain.

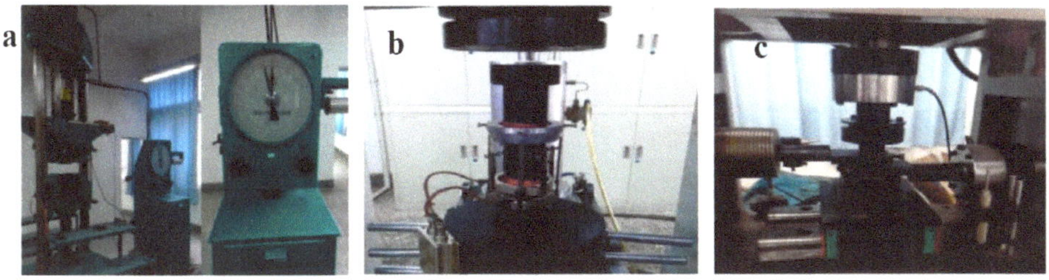

Figure 2. Testing equipment: (**a**) uniaxial compressive strength test, (**b**) direct shear test, and (**c**) Brazilian splitting test.

Table 1. The design of the experimental programs.

ID	Weathering (Days)	Number of Specimens		
		Brazilian Split Tests	Uniaxial Compression Tests	Direct Shear Tests
1	0	3	3	3
2	7	3	3	3
3	14	3	3	3
4	21	3	3	3
5	28	3	3	3

2.3. *Mechanical Properties Test Results*

The weathering effects exerted a profound influence on the experimental outcomes of the Brazilian splitting test. The failed specimens in the Brazilian splitting tests, uniaxial compression tests, and direct shear tests are displayed in Figure 3. As detailed in Table 2, the average tensile strengths of black shale decreased gradually with the increase in weathering time. The integrity of the rocks was compromised by weathering, leading to an increase in microcracks and porosity within the rock matrix. This, in turn, diminished the overall strength of the rock. In the case of unweathered rocks, the Brazilian splitting test typically produced relatively smooth and flat fracture surfaces. However, weathered rocks exhibited irregular fracture surfaces due to the presence of internal microcracks and pores, resulting in multiple fracture planes. This phenomenon contributed to the increased dispersion of test results.

Table 2. The average tensile strengths of black shale at different weathering times.

Weathering (Days)	Specimen		Tensile Strength
	Diameter (mm)	Height (mm)	
0	47.08	23.6	9.82
7	46.94	24.42	6.91
14	47.36	24.12	5.09
21	47.20	24.02	4.02
28	47.26	23.96	2.98

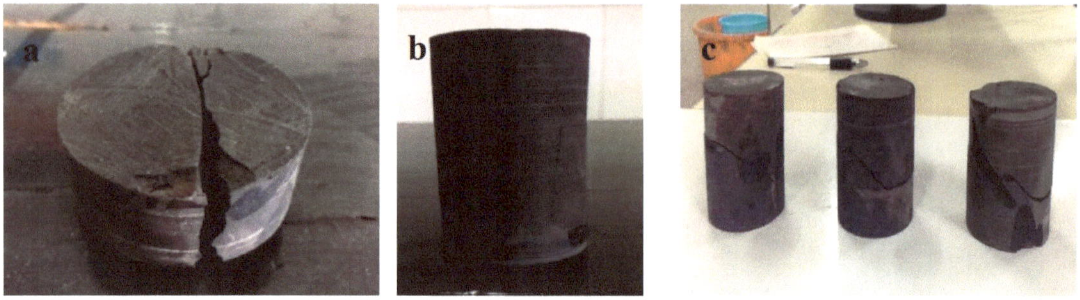

Figure 3. The failed specimens: (**a**) Brazilian splitting test, (**b**) uniaxial compression test, and (**c**) direct shear test.

The uniaxial compression tests revealed that weathering significantly affected the compression properties of the rocks (Table 3). As weathering progressed, the uniaxial compressive strength and elastic modulus of black shale gradually decreased, while the Poisson's ratio gradually increased. The decrease in elastic modulus indicated that the rock became more susceptible to deformation under external forces. This is attributed to the gradual development of transverse and longitudinal cracks within the rock during the weathering process. These cracks weaken the structural integrity of the rock, leading to a reduction in its compressive strength and stiffness. The longitudinal cracks directly contribute to the decrease in uniaxial compressive strength of the specimen, while the development of transverse cracks accelerates radial deformation during the loading process. Consequently, the Poisson's ratio gradually increases with the increase in weathering time. This decline in strength can be attributed to the breakdown of mineral grains and the formation of microcracks and pores within the rock matrix. Furthermore, weathering enhances the deformability of rocks. Rock specimens with higher degrees of weathering exhibit greater strain under compressive loads compared with their less weathered counterparts. This enhanced deformability is primarily due to the softening of the rock matrix and the development of microcracks and pores.

Table 3. The uniaxial compression mechanical parameters of black shale at different weathering times.

Weathering (Days)	Specimen		Uniaxial Compression Strength (MPa)	Elastic Modulus (GPa)	Poisson's Ratio
	Diameter (mm)	Height (mm)			
0	47.22	95.44	66.48	13.29	0.160
7	47.18	95.80	39.51	4.70	0.295
14	47.26	96.26	24.25	4.34	0.339
21	47.04	94.46	15.29	3.92	0.379
28	47.12	95.08	10.09	3.55	0.408

The mechanical properties of black shale obtained from direct shear tests at various weathering times are displayed in Figure 4. The test results demonstrate significant decreases in shear strength for the weathered rocks. During the direct shear tests, a linear correlation was observed between the deformation and the applied load, as summarized in Table 4. This linearity suggests that the rocks exhibited elastic behavior until they reached the point of failure. Based on this linear relationship, the cohesion and internal friction angle of black shale were calculated. Generally, as the weathering time increased, the cohesion and internal friction angle of the black shale tended to decrease gradually. This decrease in shear strength can be attributed to the weakening of the rock matrix due to weathering processes. Weathering leads to the breakdown of mineral grains, resulting in the formation of cracks and pores within the rock structure. The cracks and pores reduce the effective contact area between the grains, thereby decreasing the shear strength of the rock.

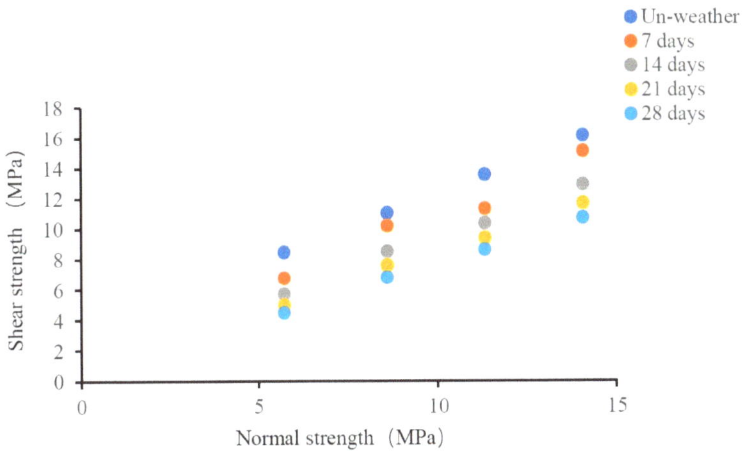

Figure 4. The direct shear testing parameters at different weathering times.

Table 4. Cohesion and internal friction angle at different weathering times.

Weathering (Days)	Fitting Results	Correlation Coefficient (R2)	Cohesion	Internal Friction Angle
0	y = 0.9086x + 3.2702	0.9999	3.28	42.22
7	y = 0.9298x + 1.6042	0.9611	1.60	42.92
14	y = 0.8398x + 1.0663	0.9941	1.07	40.02
21	y = 0.7849x + 0.6717	0.9946	0.67	38.10
28	y = 0.7368x + 0.3809	0.9977	0.38	36.41

A comprehensive set of physical and mechanical parameters was obtained through laboratory tests, which included uniaxial compressive strength, elastic modulus, Poisson's ratio, uniaxial tensile strength, cohesion, and internal friction angles, as detailed in Tables 5 and 6. Additionally, the determination coefficients for each correlation function are also provided. The uniaxial tensile strength, uniaxial compression strength, and cohesion of the rock specimens exhibited a negative exponential relationship with weathering time. This indicates that as the weathering time increased, these mechanical properties decreased The modulus of elasticity demonstrated a negative quadratic relationship with weathering time, suggesting a decrease in stiffness over time. Conversely, the Poisson's ratio exhibited a positive quadratic relationship with weathering time, indicating an increase in lateral deformation relative to axial deformation.

Table 5. Mechanical properties of black shale obtained from laboratory tests.

Weathering (Days)	Uniaxial Tensile Strength (MPa)	Uniaxial Compression Strength (MPa)	Elastic Modulus (GPa)	Poisson's Ratio	Internal Friction Angle	Cohesion (MPa)
0	9.82	66.48	13.29	0.160	42.22	3.28
7	6.91	39.51	4.70	0.295	42.92	1.60
14	5.09	24.25	4.34	0.339	40.02	1.07
21	4.02	15.29	3.92	0.379	38.10	0.67
28	2.98	10.09	3.55	0.408	36.41	0.38

Table 6. The correlation function between mechanical parameters and weathering time.

Parameters	Type of Function	Fitting Formula	Correlation Coefficient (R2)
Uniaxial tensile strength	Exponential	$y = 9.4964 \times 10^{-0.042}x$	0.9957
Uniaxial compression strength	Exponential	$y = 64.337 \times 10^{-0.067}x$	0.9982
Elastic modulus	Quadratic	$y = 0.0157 \times 2 - 0.7738x + 13.079$	0.9939
Poisson's ratio	Quadratic	$y = -0.0003 \times 2 + 0.0171x + 0.1693$	0.9797
Cohesion	Exponential	$y = 3.0276 \times 10^{-0.074}x$	0.9916

The correlation functions derived between weathering time and mechanical properties offer profound insights into deleterious effects of weathering on rocks. As the degree of weathering intensified, the elastic modulus of the rock diminished, indicating a heightened susceptibility to deformation under external forces. Weathering compromises the structural integrity of rocks, leading to the formation of microcracks and voids within the rock matrix. These microstructural defects not only weaken the tensile strength of the rock but also serve as stress concentration points, predisposing the rock to failure at lower stress levels. Weathering weakens the structural planes within rocks, leading to a decrease in cohesion and internal friction angles. This reduction in the strength of structural planes contributes to the overall decline in the rock's tensile strength, making it more susceptible to failure under tensile loads. The observed reduction in strength, augmentation of deformability, and enhancement of permeability of weathered rocks carry profound implications for a range of engineering applications. Consequently, it is imperative to meticulously consider the impact of weathering during the design and analysis of rock engineering projects. It is worth noting that weathering is a highly intricate process influenced by numerous factors. Therefore, when conducting experiments and analyzing data, it is essential to comprehensively account for various factors to ensure the accuracy and reliability of the findings.

3. Numerical Simulation of Slope Stability Analysis

3.1. Numerical Simulation Methodologies

To incorporate the influences of weathering into slope stability analysis, Fast Lagrangian Analysis of Continua (FLAC3D, Version 5.00) was utilized, which was developed by the Itasca Company, USA. FLAC3D is a simulation and calculation software that specializes in simulating and analyzing the plastic flow behavior of three-dimensional structures composed of soil, rock, and other materials. The software can accurately simulate the plastic failure and flow characteristics of materials, making it particularly suitable for simulating large deformation problems and nonlinear stress–strain relationships. The physical and mechanical parameters of the rock mass and various rock types within the mining area were obtained through site investigation. The method for obtaining the overburden thickness and mechanical properties of the open-pit mine was through drilling measurement technology. The cores extracted from the boreholes were analyzed for their physical and mechanical properties. A total of 12 boreholes were drilled, with the obtained results displayed in Table 7.

Table 7. The physical and mechanical parameters of other rock types in mining areas.

Rock Properties	Uniaxial Compressive Strength/MPa	Elastic Modulus/GPa	Tensile Strength/MPa	Poisson's Ratio	Cohesive Strength/MPa	Internal Friction Angle/°
Phosphate ore layer	50.27	15.80	5.57	0.29	12.78	41.30
Light gray dolomite	48.13	12.05	3.68	0.34	7.48	31.30
Dark gray dolomite	59.10	16.26	6.18	0.28	13.05	47.23

During the modeling process, a reasonable degree of simplification was applied to the geological prototype of the slope to ensure computational efficiency and accuracy. For the boundary conditions, the exposed surface of the slope was designated as the free boundary, while the bottom and both sides were constrained with normal displacement boundaries. The initial stress field was primarily determined by considering the self-weight stress of the rock mass. Initially, the stress field of the model was set to achieve equilibrium, reflecting the undisturbed geological conditions. The constitutive model and material division within the mining areas are clearly depicted in Figure 5. The model was configured with the X-direction aligned with the slope inclination and extending 1000 m in length, while the Y-direction represented the vertical direction with a height of 500 m. Taking into account the geological environment conditions, topography, and lithology of the open-pit slope in the mining areas, two models were specifically designed with final slope angles of 45° and 47°. These models were employed to simulate the gradual development of the open-pit slope. To delve deeper into consequences of weathering on slope stability, the mechanical parameters of black shale were adjusted to reflect various degrees of weathering, as outlined in Table 5. By incorporating the weathered properties, the model was able to simulate the altered behavior of the slope under the influence of weathering.

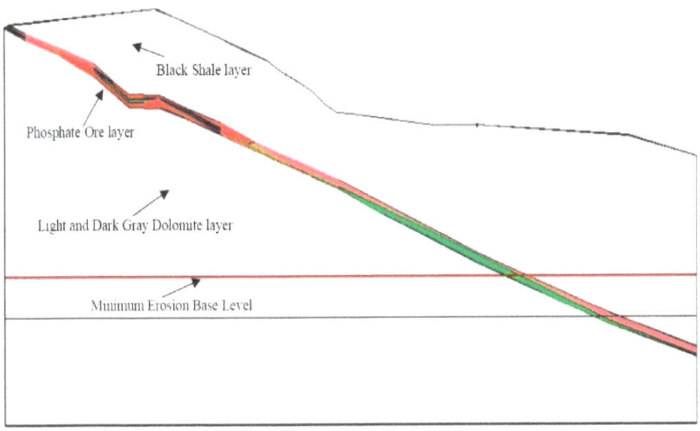

Figure 5. Constitutive model and material division in the mining areas.

3.2. Weathering Assessment Results

The computational model established for slope stability analysis was tailored to reflect the specific geological conditions. The model dimensions were approximately 900 m × 360 m, and it was discretized into 170 × 66 grids. In the real mining operations, the open-pit slope is excavated down to the current level of 2080 m, resulting in a slope height of 50 m. Subsequently, the slope is further excavated to the next stage, increasing its height to 90 m. This process continues, with the slope advancing to subsequent stages and ultimately reaching a height of 120 m. Therefore, to assess the slope stability under varying conditions, simulations and calculations were conducted using slope heights of 50 m, 90 m, and 120 m. These calculations were performed for slope angles of both 45° and 47°, taking into account the mechanical properties of the rock mass at different weathering degrees. For each step height, a detailed analysis was conducted to evaluate the slope stability at both slope angles, incorporating the varying mechanical parameters associated with different weathering degrees. As an illustrative example, the calculated results for a slope height of 50 m, slope angles of 45° and 47°, and a weathering time of 28 days are presented in Figures 6 and 7.

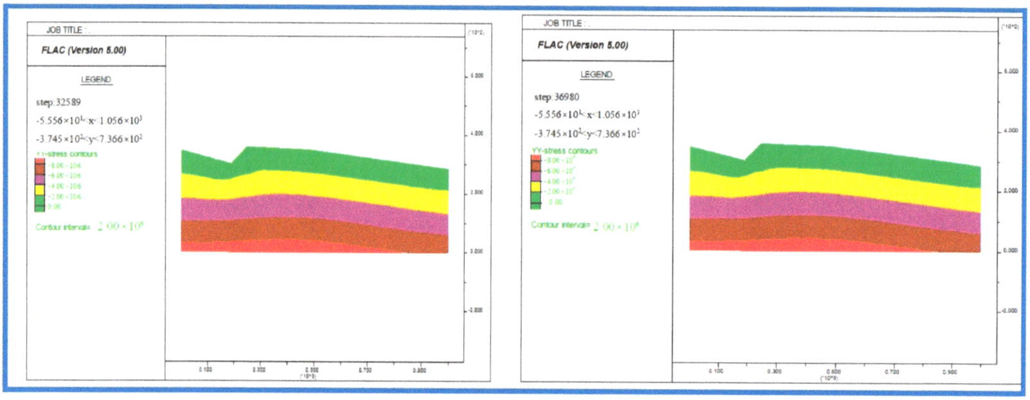

a: Vertical stress distribution, Unweathered (left), at weathering time of 28 days (right)

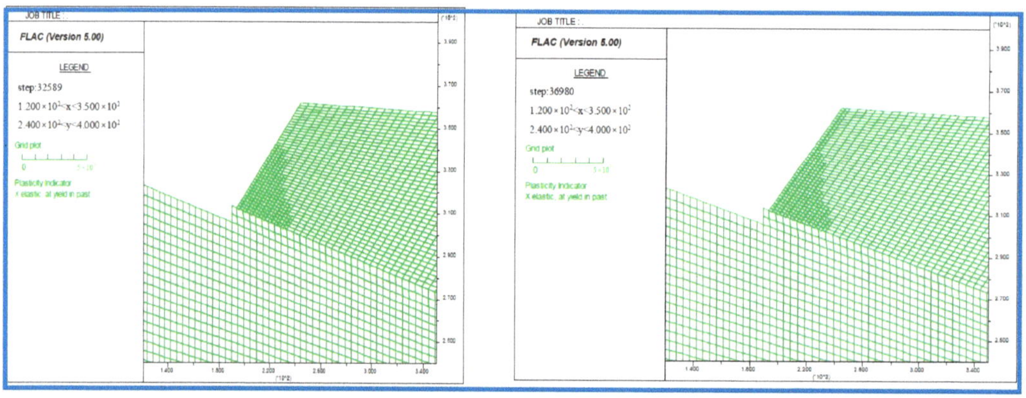

b: Plastic-zone distribution, Unweathered (left), at weathering time of 28 days (right)

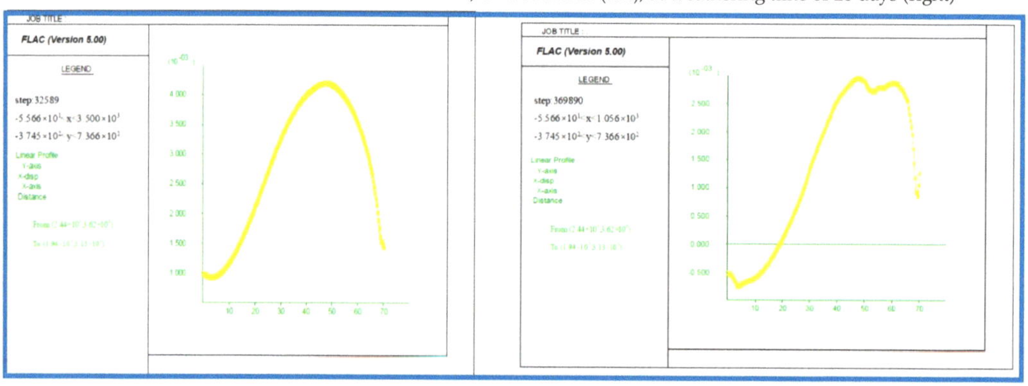

c: Horizontal-displacement distribution curve of slope surface, Unweathered (left), at weathering time of 28 days (right)

Figure 6. Calculated results for slope height of 50 m at angle of 45°.

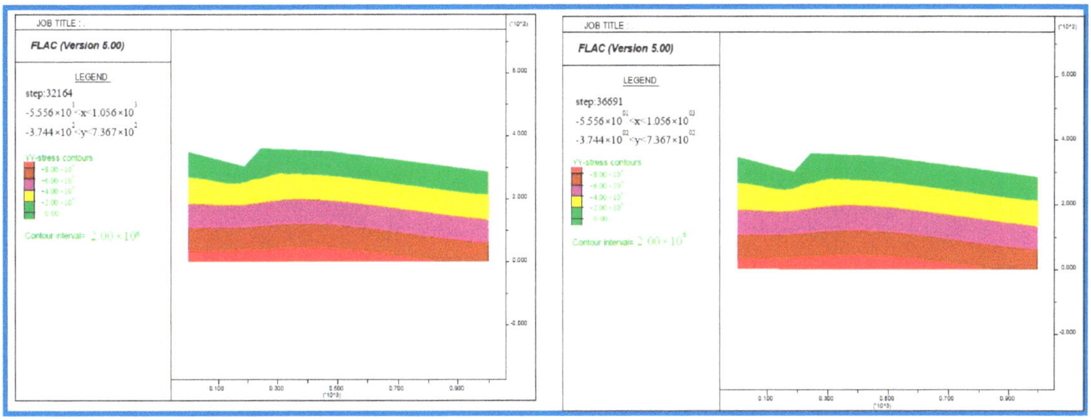

a: Vertical stress distribution, Unweathered (left), at weathering time of 28 days (right)

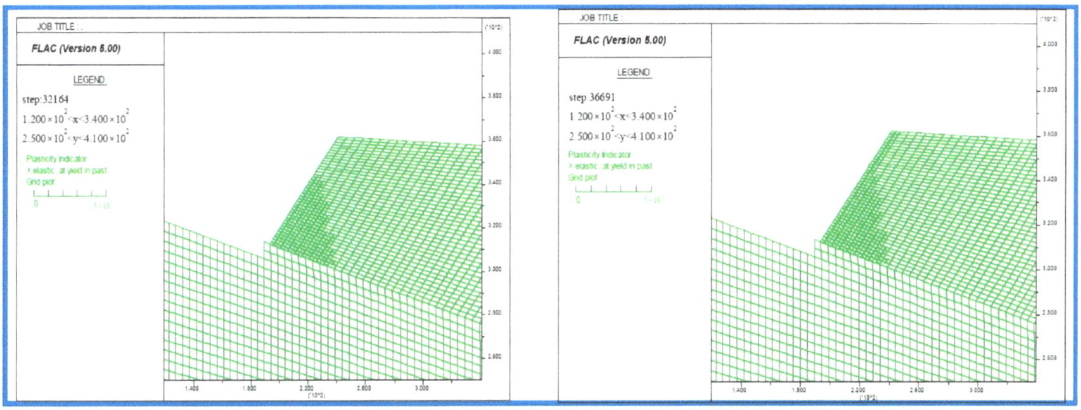

b: Plastic-zone distribution, Unweathered (left), at weathering time of 28 days (right)

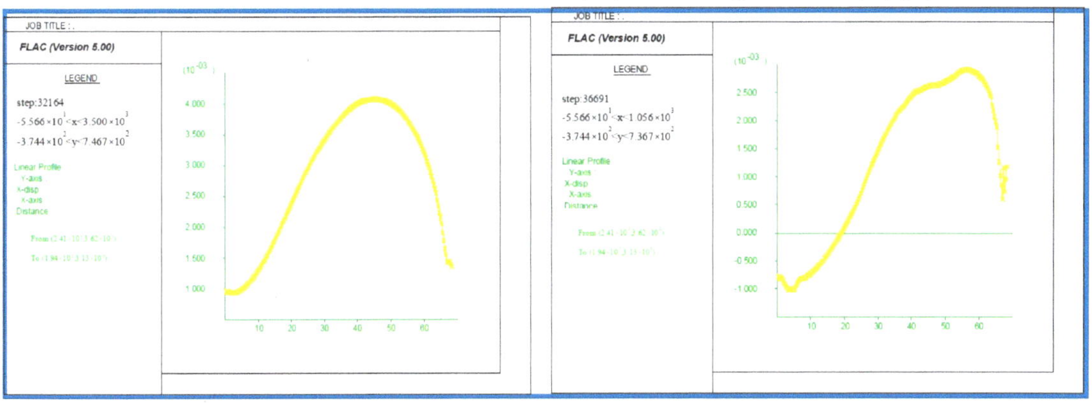

c: Horizontal-displacement distribution curve of slope surface, Unweathered (left), at weathering time of 28 days (right)

Figure 7. Calculated results for slope height of 50 m at angle of 47°.

Regarding the horizontal displacement of the slope surface, a gradual decrease was observed. At the slope's apex, the horizontal displacement was directed positively along the

x-axis. However, as the slope extended downwards, the influence of excavation activities was equal to the displacement, resulting in negative values. This indicated a higher degree of deformation and damage at the slope's base. When comparing slopes with angles of 45° and 47°, an increase in the final slope angle led to an elevation in horizontal stress on the slope surface. To assess the slope's stability, the distribution of plastic zones was analyzed to identify failure-prone areas. It was evident that in the initial excavation stages, prior to weathering of the black shale, there was no discernible presence of plastic zones on the slope surface or within its interior. However, as the weathering process advanced, the extent of plastic zones both on the slope surface and within its depths began to gradually expand. Consequently, the areas susceptible to shear and tensile failures within the slope also increased in scope.

Regarding the impact of weathering effects, it was observed that in the initial stages of weathering, a distinct range of shear and tensile failure areas had already emerged within the slope. At slope angle of 45°, during the early weathering, there was no evidence of plastic failure either on the slope surface or within its interior. However, at the weathering duration of 28 days, extensive plastic zones had developed within the slope, and the slope also exhibited widespread plastic deformation. This underscores that in the later stages of black-shale weathering, the material in its weathered state is susceptible to plastic failure. Comparing the plastic-zone distributions of slopes with angles of 45 and 47, it was evident that as the slope angle increased, the extent of the plastic zone on the slope surface gradually enlarged. The areas of stress, strain, shear failure, and tensile failure within the slope progressively expanded, potentially forming a contiguous zone across the slope surface and interior. Given that the strength of the black shale was significantly reduced after 28 days of weathering, it is advisable to implement appropriate support measures for the exposed slope.

4. Slope Stability Evaluation and Discussion

Utilizing the FLAC finite software (version 5.00), initial findings revealed that when the black-shale slope underwent 28 days of weathering at the slope angle of 47°, the slope's stability was significantly compromised. To further evaluate the potential for slope failure under different scenarios, a slope stability analysis was performed using the limit equilibrium method [40]. This method involves applying static equilibrium conditions to determine the safety factors, which represent the ratio of resisting forces to the driving forces acting on the potential sliding mass. The static equilibrium conditions are applied in both horizontal and vertical directions to the sliding mass. The resisting forces on the sliding surface are calculated based on the shear strength parameters, including cohesion and internal friction angle. When applying the limit equilibrium method to analyze slope stability in open-pit mines, the slope within the mining area is systematically divided into a sequence of vertical slices with uniform width. For each slice, static equilibrium equations are established in both horizontal and vertical directions. By considering any individual slice, the balance of forces in the X- and Y-directions, as well as the moment balance, are calculated. The Swedish arc method is then utilized to derive the expression for calculating the engineering stability coefficient, as outlined in Equation (1). Additionally, when the pore pressure ratio is known, a specific formula is employed to compute the stability coefficient, as detailed in Equation (2). This approach enables a precise assessment of slope stability in open-pit mines, taking into account various factors that influence stability.

$$F = \frac{\sum\limits_{i=1}^{n} \left[\bar{c} b_i \sec\theta_i + (\gamma h_i - \gamma_w h_{iw}) b_i \cos\theta_i \tan\overline{\varphi}\right]}{\sum\limits_{i=1}^{n} (W_i \sin\theta_i + Q_i\,/R)} \tag{1}$$

$$F = \frac{\sum_{i=1}^{n} [\bar{c}b_i \sec\theta_i + (1 - r_u)W_i \cos\theta_i \tan\overline{\varphi}]}{\sum_{i=1}^{n} (W_i \sin\theta_i + Q_i / R)} \tag{2}$$

where C represent effective stress shear strength index of soil cohesion, KPa; φ is the effective stress shear strength index of internal friction angle; α represents the moment arm from the center of gravity of the strip to the center of the sliding arc; R is the radius of the sliding arc, m; Q represents the seismic inertial force (horizontal direction), KN; W is the weight of the strip of soil, KN; θ represents the inclination angle of the sliding surface of the strip; and r_u is the ratio of pore pressure, which is the ratio of the weight of the soil column to the weight of the water column.

The resisting forces on the slope surface are calculated, taking into account the shear strength parameters specific to the black shale as well as the normal stress exerted on each individual slice. The safety factor, a crucial metric, is defined as the ratio between the resisting forces and the driving forces acting on the potential sliding mass. To accurately assess the slope safety factor during various weathering periods, the physical and mechanical parameters are carefully assigned, drawing upon the results of rigorous laboratory tests. Stability calculations are conducted for the slope at different weathering stages, considering two slope angles of 45 and 47, with the aim of exploring the optimal final slope angle under diverse conditions.

Tables 8 and 9 present the calculated safety factors for slopes, considering various combinations of slope angles, heights, and weathering times. For slopes with an angle of 45, it is observed that under identical slope height conditions, the strength of the slope surface rock diminishes gradually as weathering time increases. Consequently, the safety factor of the slope decreases progressively. Additionally, when comparing slopes with the same weathering time, an increase in slope height also contributes to a decrease in the safety factor. Overall, it is evident that the safety factor of the slope decreases with both increasing weathering time and slope height. Furthermore, the safety factors for slopes of 50 m and 90 m are found to comply with the specified requirements. However, for a 120 m slope, as the weathering time gradually increases to 28 days, the safety factor diminishes to 1.085, falling below the minimum safety factor stipulated by the specification.

Table 8. The safety factors under different heights and weathering times at angle of 45°.

Slope Height (m)	0 Days	7 Days	14 Days	21 Days	28 Days
50	1.788	1.773	1.753	1.538	1.130
90	1.598	1.590	1.578	1.567	1.122
120	1.287	1.282	1.276	1.269	1.085

Table 9. The safety factors under different heights and weathering times at angle of 47°.

Slope Height (m)	0 Days	7 Days	14 Days	21 Days	28 Days
50	1.720	1.705	1.683	1.599	1.094
90	1.353	1.346	1.336	1.325	1.050
120	1.230	1.225	1.219	1.231	1.031

For slopes with an angle of 47, it is apparent that an augmentation in the final slope angle leads to a commensurate reduction in the safety factor. After 28 days of weathering, the safety factors for slopes of varying heights are recorded as 1.094, 1.050, and 1.031, all of which are below the minimum safety factor mandated by relevant standards. While these safety factors remain marginally above 1, there is a discernible trend towards instability, particularly for slopes subject to continuous weathering effects [41,42]. This trend is detrimental to slope stability and may signal a potential risk of failure. Consequently, for slopes exceeding 90 m in height, it is prudent to exercise caution when considering a final slope angle of 47°. If the intended service life surpasses one month, it is generally advisable to avoid such steep slope angles, as they may jeopardize the long-term stability and safety of the slope.

It should be noted that the results obtained through the Fast Lagrangian Analysis of Continua modelling as well as the limit equilibrium methods were based on a reasonable degree of simplification of the models as well as empirical analysis. For example, the boundary conditions and initial stress field were primarily determined by considering the self-weight stress of the rock mass. The obtained numerical results will be also compared with the recorded data in the mine sites in the future. Therefore, the obtained results help us to further understand the effects of weathering on the slope stability, but caution must be taken when using the numerical results as a source of reliable results. The numerical results should be combined with the geological environment conditions, topography, and lithology of the open-pit slope in the mining areas. It is also recommended that further research be conducted in the future and that analysis be undertaken to establish a more explicit relationship between the weathering parameters and the mechanical properties of rock layering at different depths.

5. Conclusions

The impacts of weathering on the stability of open-pit mine slopes were rigorously examined through laboratory tests. The correlation functions between weathering time and mechanical properties were obtained, providing valuable insights into the detrimental effects of weathering on rocks. The finite element method serves as a robust tool to simulate and analyze slope stability under varying weathering conditions. The weathering processes lead to a loosening of the rock structure and a corresponding decrease in its strength and stability. The limit equilibrium methods were used for slope stability analysis, and revealed that an increase in weathering time gradually weakens the slope surface rock, ultimately resulting in a reduction in the slope safety factor. For slopes exceeding 90 m in height, it is advisable to avoid a final slope angle of 47 degrees if the intended service life exceeds one month. This study underscores the crucial importance of considering weathering effects in the design and maintenance of open-pit mine slopes which also help to ensure the long-term stability and safety of the slope structures.

Author Contributions: Conceptualization, W.L. and S.W.; methodology, D.L. and X.K; software, M.Y.; validation, W.L., X.K. and G.S.; formal analysis, M.Y.; resources, W.L., X.K. and G.S.; data curation, W.L., X.K. and G.S.; writing—original draft preparation, S.W. and D.L.; writing—review and editing, S.W. and D.L; project administration, S.W.; funding acquisition, W.L., X.K. and G.S. All authors have read and agreed to the published version of the manuscript.

Funding: This paper is funded by National Natural Science Foundation of China (52474144, 52004196), the Shaanxi Innovation Capability Support Program, the Science and Technology Innovation team (2023CX-TD-12), the Shaanxi Province Department of Education Key Projects of Scientific Research Program (23JS036), and the Ministry of Education, "Chunhui Plan" (HZKY20220534).

Institutional Review Board Statement: Not applicable.

Informed Consent Statement: Not applicable.

Data Availability Statement: The data presented in this study are available on request from the corresponding author.

Conflicts of Interest: Author Wei Liu, Gang Sheng and Xin Kang were employed by the company Shaanxi Provincial Land Engineering Construction Group. The remaining authors declare that the research was conducted in the absence of any commercial or financial relationships that could be construed as a potential conflict of interest.

References

1. Ferrero, A.M. The shear strength of reinforced rock joints. *Int. J. Rock Mech. Min. Sci.* **1995**, *32*, 595–605. [CrossRef]
2. Bieniawski, Z.T. Mechanism of brittle fracture of rock: Part II—Experimental studies. *Int. J. Rock Mech. Min. Sci. Geomech. Abstr.* **1967**, *4*, 407–423. [CrossRef]
3. Hoek, E.; Bieniawski, Z.T. Brittle fracture propagation in rock under compression. *Int. J. Fract. Mech.* **1965**, *1*, 137–155. [CrossRef]
4. Li, Y.C.; Tang, C.A.; Li, D.Q.; Wu, C.Z. A New Shear Strength Criterion of Three-Dimensional Rock Joints. *Rock Mech. Rock Eng.* **2020**, *53*, 1477–1483. [CrossRef]
5. Wu, S.; Ma, X.; Zhang, X.; Chen, J.; Yao, Y.; Li, D. Investigation into hydrogen induced fracture of cable bolts under deep stress corrosion coupling conditions. *Tunn. Undergr. Space Technol.* **2024**, *147*, 105729. [CrossRef]
6. Windsor, C.R. Rock reinforcing systems. *Int. J. Rock Mech. Min. Sci.* **1997**, *34*, 919–951. [CrossRef]
7. Ghobadi, M.H.; Momeni, A.A. Assessment of granitic rocks degradability susceptive to acid solutions in urban area. *Environ. Earth Sci.* **2011**, *64*, 753–760. [CrossRef]
8. Tapponnier, P.; Brace, W.F. Development of stress-induced microcracks in Westerly Granite. *Int. J. Rock Mech. Min. Sci. Geomech. Abstr.* **1976**, *13*, 103–112. [CrossRef]
9. Wu, S.; Hao, W.Q.; Yao, Y.; Li, D.Q. Investigation into durability degradation and fracture of cable bolts through laboratorial tests and hydrogeochemical modelling in underground conditions. *Tunn. Undergr. Space Technol.* **2023**, *138*, 105198. [CrossRef]
10. Gong, B.; Jiang, Y.J.; Chen, L.J. Feasibility investigation of the mechanical behavior of methane hydrate-bearing specimens using the multiple failure method. *J. Nat. Gas Sci. Eng.* **2019**, *69*, 13. [CrossRef]
11. Wang, S.R.; Wu, X.G.; Zhao, Y.H.; Hagan, P.; Cao, C. Evolution Characteristics of Composite Pressure-Arch in Thin Bedrock of Overlying Strata During Shallow Coal Mining. *Int. J. Appl. Mech.* **2019**, *11*, 20. [CrossRef]
12. Gong, B.; Jiang, Y.J.; Yan, P.; Zhang, S.H. Discrete element numerical simulation of mechanical properties of methane hydrate-bearing specimen considering deposit angles. *J. Nat. Gas Sci. Eng.* **2020**, *76*, 17. [CrossRef]
13. Wu, S.S.; Chen, H.H.; Craig, P.; Ramandi, H.L.; Timms, W.; Hagan, P.C.; Crosky, A.; Hebblewhite, B.; Saydam, S. An experimental framework for simulating stress corrosion cracking in cable bolts. *Tunn. Undergr. Space Technol.* **2018**, *76*, 121–132. [CrossRef]
14. Gong, J.; He, M.; Zhang, J.; Liang, W.; Wang, S. Dynamic impact mechanical properties of red sandstone based on digital image correlation method. *Int. J. Min. Reclam. Environ.* **2024**, 1–16. [CrossRef]
15. Gökceoğlu, C.; Ulusay, R.; Sönmez, H. Factors affecting the durability of selected weak and clay-bearing rocks from Turkey, with particular emphasis on the influence of the number of drying and wetting cycles. *Eng. Geol.* **2000**, *57*, 215–237. [CrossRef]
16. Liu, J.Q.; Chen, W.Z.; Liu, T.G.; Yu, J.X.; Dong, J.L.; Nie, W. Effects of Initial Porosity and Water Pressure on Seepage-Erosion Properties of Water Inrush in Completely Weathered Granite. *Geofluids* **2018**, *2018*, 4103645.
17. Wang, S.R.; Xiao, H.G.; Zou, Z.S.; Cao, C.; Wang, Y.H.; Wang, Z.L. Mechanical Performances of Transverse Rib Bar During Pull-Out Test. *Int. J. Appl. Mech.* **2019**, *11*, 15. [CrossRef]
18. Wu, S.S.; Li, J.P.; Guo, J.P.; Shi, G.B.; Gu, Q.H.; Lu, C.W. Stress corrosion cracking fracture mechanism of cold-drawn high-carbon cable bolts. *Mater. Sci. Eng. A-Struct. Mater. Prop. Microstruct. Process.* **2020**, *769*, 10. [CrossRef]
19. Li, Y.C.; Wu, C.Z.; Jang, B.A. Effect of Bedding Plane on the Permeability Evolution of Typical Sedimentary Rocks Under Triaxial Compression. *Rock Mech. Rock Eng.* **2014**, *53*, 5283–5291. [CrossRef]
20. Basu, A.; TCelestino, B.; Bortolucci, A.A. Evaluation of rock mechanical behaviors under uniaxial compression with reference to assessed weathering grades. *Rock Mech. Rock Eng.* **2009**, *42*, 73–93. [CrossRef]
21. Arias, D.; Pando, L.; López-Fernández, C.; Díaz-Díaz, L.M.; Rubio-Ordóñez, Á. Deep weathering of granitic rocks: A case of tunnelling in NW Spain. *Catena* **2016**, *137*, 572–580. [CrossRef]
22. Heidari, M.; Khanlari, G.; Momeni, A.; Jafargholizadeh, H. The relationship between geomechanical properties and weathering indices of granitic rocks, Hamedan, Iran. *Geomech. Geoengin.* **2011**, *6*, 59–68. [CrossRef]
23. Ceryan, S.; Zorlu, K.; Gokceoglu, C.; Temel, A. The use of cation packing index for characterizing the weathering degree of granitic rocks. *Eng. Geol.* **2008**, *98*, 60–74. [CrossRef]
24. Ceryan, S. New weathering indices for evaluating durability and weathering characterization of crystalline rock material: A case study from NE Turkey. *J. Afr. Earth Sci.* **2015**, *103*, 54–64. [CrossRef]
25. Heidari, M.; Momeni, A.A.; Naseri, F. New weathering classifications for granitic rocks based on geomechanical parameters. *Eng. Geol.* **2013**, *166*, 65–73. [CrossRef]
26. Momeni, A.A.; Khanlari, G.R.; Heidari, M.; Sepahi, A.A.; Bazvand, E. New engineering geological weathering classifications for granitoid rocks. *Eng. Geol.* **2015**, *185*, 43–51. [CrossRef]
27. Abad, S.A.N.K.; Tugrul, A.; Gokceoglu, C.; Armaghani, D.J. Characteristics of weathering zones of granitic rocks in Malaysia for geotechnical engineering design. *Eng. Geol.* **2016**, *200*, 94–103. [CrossRef]
28. Wu, S.; Northover, M.; Craig, P.; Canbulat, I.; Hagan, P.C.; Saydam, S. Environmental influence on mesh corrosion in underground coal mines. *Int. J. Min. Reclam. Environ.* **2018**, *32*, 519–535. [CrossRef]
29. Qi, S.; Yue, Z.Q.; Liu, C.; Zhou, Y. Significance of outward dipping strata in argillaceous limestones in the area of the Three Gorges reservoir, China. *Bull. Eng. Geol. Environ.* **2009**, *68*, 195–200. [CrossRef]
30. Wu, S.S.; Ramandi, H.L.; Chen, H.H.; Crosky, A.; Hagan, P.; Saydam, S. Mineralogically influenced stress corrosion cracking of rockbolts and cable bolts in underground mines. *Int. J. Rock Mech. Min. Sci.* **2019**, *119*, 109–116. [CrossRef]

31. Wu, S.; Zhang, X.; Li, J.; Wang, Z. Investigation for influences of seepage on mechanical properties of rocks using acoustic emission technique. *Geofluids* **2020**, *2020*, 6693920. [CrossRef]
32. Momeni, A.; Hashemi, S.S.; Khanlari, G.R.; Heidari, M. The effect of weathering on durability and deformability properties of granitoid rocks. *Bull. Eng. Geol. Environ.* **2017**, *76*, 1037–1049. [CrossRef]
33. Germanovich, L.N.; Salganik, R.L.; Dyskin, A.V.; Lee, K.K. Mechanisms of brittle fracture of rock with pre-existing cracks in compression. *Pure Appl. Geophys.* **1994**, *143*, 117–149. [CrossRef]
34. Wu, S.; Chen, H.; Ramandi, H.L.; Hagan, P.C.; Hebblewhite, B.; Crosky, A.; Saydam, S. Investigation of cable bolts for stress corrosion cracking failure. *Constr. Build. Mater.* **2018**, *187*, 1224–1231. [CrossRef]
35. Erarslan, N.; Williams, D.J. Mixed-Mode Fracturing of Rocks Under Static and Cyclic Loading. *Rock Mech. Rock Eng.* **2013**, *46*, 1035–1052. [CrossRef]
36. Wu, S.; Zhang, Z.R.; Chen, J.H.; Yao, Y.; Li, D.Q. Characterisation of stress corrosion durability and time-dependent performance of cable bolts in underground mine environments. *Eng. Fail. Anal.* **2023**, *150*, 107292. [CrossRef]
37. Zhu, D.; Yu, B.; Wang, D.; Zhang, Y. Fusion of finite element and machine learning methods to predict rock shear strength parameters. *J. Geophys. Eng.* **2024**, *21*, 1183–1193. [CrossRef]
38. Wu, S.; Guo, J.; Shi, G.; Li, J.; Lu, C. Laboratory-Based Investigation into Stress Corrosion Cracking of Cable Bolts. *Materials* **2019**, *12*, 16. [CrossRef]
39. ISRM Society. *Suggested Methods for Determining the Uniaxial Compressive Strength and Deformability of Rock Materials*; International Society for Rock Mechanics: Lisbon, Portugal, 1978; pp. 137–156.
40. Liu, L.-L.; Wang, Y. Quantification of stratigraphic boundary uncertainty from limited boreholes and its effect on slope stability analysis. *Eng. Geol.* **2022**, *306*, 106770. [CrossRef]
41. Rezaei, M.; Seyed, S.Z. Mousavi, Slope stability analysis of an open pit mine with considering the weathering agent: Field, laboratory and numerical studies. *Eng. Geol.* **2024**, *333*, 107503. [CrossRef]
42. Xu, S.; Wang, B.; Wang, D.; Zhang, J. A practical stability/instability chart analysis for slope large deformations using the material point method. *Eng. Geol.* **2024**, *338*, 107611. [CrossRef]

applied sciences

MDPI

Article

Deformation Characteristics and Response Factors of Rock Bolt Body in Roadway with Layered Composite Roof

Ziyue Wang, Shangxin Fang and Cun Zhang *

School of Energy & Mining Engineering, China University of Mining and Technology-Beijing, Beijing 100083, China; wangziyue@tdkcsj.com (Z.W.); fliex0105@163.com (S.F.)
* Correspondence: cumt-zc@cumtb.edu.cn

Abstract: Layered composite roofs are characterized by developed bedding fissures, resulting in severe deformation and damage of rock bolts at the top of the roadway, as well as a poor roadway support effect. Increasing pretension force is an effective way to enhance the stiffness of the rock bolt support system. To clarify the influence and mechanism of the pretension force on the support effect of rock bolts in the layered roof, a roadway model of the layered roof was established using the interface unit of FLAC^{3D}, and the simulation rock bolts were constructed using the pile unit, which can simulate the mechanical behaviors of rock bolts, such as tension, shear, bending, fracture, and anchor failure, and the pretension force was applied. On this basis, the deformation and failure characteristics and influencing factors of rock bolts in the layered roadway roof under different surrounding rock conditions were simulated and analyzed. The research shows the following: ① Field measurements showed minor shear deformation in the rock bolts at the center of the roadway roof, with lateral displacements of 5.7 cm and 5.3 cm. Significant shear deformation occurred in the rock bolts at the roof corners, with lateral displacements of 18.2 cm and 17.6 cm. ② Simulations of rock bolt deformation characteristics matched the field measurements closely, confirming the reliability of the simulation method, parameter selection, and calculation sequence. ③ The primary factors affecting rock bolt deformation are the structural plane strength and surrounding rock strength. Rock bolts are most susceptible to lateral displacement when the structural plane strength is low, the strength difference between rock layers is large, and the weaker layer is below the structural plane. The presented research can provide a reference for the instability mechanism and support treatment of the layered composite roof roadway.

Keywords: layered roof; force of rock bolt; deformation of rock bolt; interface strength; surrounding rock structure

1. Introduction

Rock bolt support, as an economical and effective active support method, has been widely applied in engineering supports, such as tunnels, slopes, water conservancy and hydropower, and underground mines [1,2]. Rock bolt support is the main support form in Chinese coal mine roadways. In the excavation of about 8000 km of coal mine roadways every year, 80% adopt rock bolt support [3]. The factors that affect the stability of the roadway surrounding rock and the effect of rock bolt support usually include water corrosion [4], the width of the coal pillar reserved [5], as well as the diameter and angle of the rock bolt installation, rock strength, grouting material, rock bolt material, drilling diameter, pre-tightening force of the rock bolt, normal stress, surface roughness of the joint, expansion angle of the joint, etc. [6,7]. Many scholars at home and abroad have studied the mechanism and failure mechanism of rock bolts under different conditions through three aspects: laboratory tests, numerical simulations, and theoretical analyses.

In terms of laboratory tests, Nourizadeh et al. [8] studied the influence of the intermediate principal stress on the strength characteristics of rock bolts through true triaxial tests. The results show that as the confining pressure gradually increases from zero, the behavior of rock bolts before failure changes from linear to nonlinear. Entezam et al. [9] revealed the influence of adding fly ash in grouting on the axial bearing capacity of fully grouted rock bolts through rock bolt pull-out tests and uniaxial compression tests. The results show that after the addition of fly ash, the axial bearing capacity of rock bolts can be effectively improved. Yu et al. [10] applied confining pressure and pull-out load in indoor tests and conducted tests on rock bolts using ultrasonic guided waves. The propagation law of ultrasonic guided waves and the bonding quality of rock bolts under the action of pull-out load and confining pressure were studied. Jodeiri Shokri et al. [11] revealed the influence of the restraint diameter on the axial bearing capacity of rock bolts through grouting solidification and rock bolt pull-out tests. The results show that increasing the restraint diameter, the bearing capacity of rock bolts increases by approximately 32% and 45% on the 7th and 28th days, respectively. Nourizadeh et al. [12] prepared a rock bolt heating pull-out test device and studied the mechanical properties of resin rock bolts under high-temperature conditions. The results show that when the ambient temperature reaches 75 °C and 150 °C, the bonding capacity of resin rock bolts decreases by 6.6% and 31.3%, respectively.

In the field of numerical simulation, He et al. [13,14] conducted a large number of anchorage shear laboratory tests on 2G-NPR rock bolts. A three-dimensional refined numerical simulation method considering the elastic–plastic mechanical behavior of rock mass, the tensile–shear fracture criterion of rock bolts, and the damage and deterioration of the grouting body was proposed. Tahmasebinia et al. [15,16] analyzed the influence of different parameters of the anchor bolt on its ultimate bearing capacity through the finite element analysis software ABAQUS/Explicit. It is indicated that in the environment where the roadway is in a layered composite roof and under impact ground pressure, flexible bolts can play an important role due to their strong ability to resist dynamic loads. Chen et al. [17] used the PFC simulation method to study the macroscopic shear behavior and microscopic failure characteristics of a single rock bolt under different types of rock joints. Jiang et al. [18], based on the failure theory of rock bolts, modified the mechanical model of the pile element in the FLAC^{3D} program and revealed the distribution and evolution characteristics of the axial force and shear resistance of rock bolts.

From the perspective of theoretical analysis, Zhan et al. [19] proposed an effective characterization method for the stress concentration of the surrounding rock in coal mine roadways, replacing the traditional measurement method of the stress of the surrounding rock in the roadway. Kong et al. [20] studied the influence of the loss of the pre-tightening force of high-strength rock bolts on the basic mechanical properties of the structure. The results show that the loss of the pre-tightening force of high-strength rock bolts reduces the connection strength and structural stiffness, but the influence on the ultimate bearing capacity of the structure is very limited. Yi et al. [21] pointed out that the hardness of the surrounding rock of the roadway and the integrity of the rock mass are important factors affecting the anchoring performance of rock bolts in jointed rock. Therefore, a rock bolt pull-out test was designed, and the two influencing factors of rock hardness and integrity were comprehensively considered. The calculation formulas for the ultimate pull-out load and ultimate displacement of mortar rock bolts in jointed rock were established and proposed. Peng [22] introduced the underground rock bolt support mode in American coal mines, indicating that if the roof strata are weak, especially affected by weathering, the rock beams between the rock bolts will be damaged and fall off over time.

Engineering rock masses are often affected by frequent geological tectonic movements and complex stress fields, often resulting in many weak structural planes, such as fractures, joints, and faults. These weak structural planes seriously damage the integrity and continuity of the surrounding rock of the roadway [23]. Therefore, most coal mine roadways belong to layered composite roof roadways. The layered composite roof generally refers to the roof rock mass being several layers of heterogeneous soft and hard interlayered strata, which often contain weak interlayers such as thin coal and generally have poor mechanical properties and interlayer cementation. Therefore, the surrounding rock of the roadway roof is prone to large deformations and failures, such as dislocation and separation along the weak structural planes, which is also the decisive factor leading to the failure of rock bolt support [24]. Both Yang and Ma et al. [25,26] indicated that the composite roof poses a great threat to the safe production of the working face and the stability of the surrounding rock of the roadway. According to relevant data statistics, more than one-third of the coal roadway roofs in China belong to the layered composite roof, and the layered composite roof roadways have typical characteristics such as large surrounding rock deformation, sudden roof collapse, and great support difficulty during the mining process, which greatly weakens the anchoring effect of roadway rock bolts. Due to the significant difference in the mechanical properties of the weak interlayers, the deformation of the rock bolts is very serious, thereby causing difficulties in roadway support and severely restricting the safe production of the mine. According to statistics, the number of accidents caused by composite-roof-type roadways in the coal mine production process accounts for 2/3 of all roadway roof collapse accidents, but at present, scholars at home and abroad have conducted less research on the deformation of rock bolts and its influencing factors in the layered composite-roof-type roadways. In response to the above problems, in this paper, a "break-tension-suture" method for applying the pretension force of the pile unit simulation rock bolt was put forward. The weak surface of the layered composite roof structure was simulated through the use of the interface. A reasonable calculation sequence of the stress field of the surrounding rock before and after the rock bolt was devised. On this basis, by employing the aforesaid simulation method, the deformation and failure characteristics of the anchor bolts in the roadway with a layered composite roof and their influencing factors under different surrounding rock conditions were analyzed.

This paper selected six rock bolts from typical layered composite roof roadways and analyzed their force and deformation characteristics. Through FLAC^{3D}, a unified model of rock bolt deformation and stress evolution in the layered composite roof roadway was constructed. The force and deformation characteristics of the on-site rock bolts were inversely simulated. The influence of the strength of the surrounding rock and the structural plane on the shear and bending deformation of the rock bolts was further simulated and analyzed.

2. Field Sampling Analysis of Roadway Rock Bolts

2.1. Engineering Background

The coal seam mined at the 41072 working face of Changping Mine is the main mined $3^{\#}$ coal seam of the mine. The average burial depth is 500 m. The average dip angle is 6°, belonging to the nearly horizontal coal seam. The thickness of the coal seam ranges from 4.60 to 6.83 m, with an average thickness of 5.86 m. The length of the 41072 working face is 300 m, and the designed mining height is 5.86 m. Fully mechanized mining with full-seam extraction is adopted. The average advancing speed is 2.54 m/d. The treatment method of the goaf is the total caving method. Furthermore, according to the geological data of the mine, the hydrogeological conditions of the mine are simple: the coal seam is not prone to spontaneous combustion, and the coal dust has no explosiveness. The mine belongs to a low-impact tendency mine. However, due to the large mining height and burial depth of the working face, the periodic weighting of the mine is intense. The 41072 roadway in Changping Mine was originally the return airway of the 4307 working face. After the mining of 4307 was completed, the roadway was retained to serve the 4107 working face. The layout of roadway 41072 is shown in Figure 1. The 4107 working face mainly mines the $3^{\#}$ coal seam, with an average thickness of 5.75 m. The average uniaxial compressive strength of the coal body is distributed between 12.8 and 15.4 MPa. The immediate roof is sandy mudstone, with a thickness of 1.57 m. The fissures are developed, the rock is soft, and the strength is mainly concentrated between 27.6 and 32.6 MPa. The main roof is medium sandstone, with a thickness of 9.26 m. The rock is hard, the strength is between 70.0 and 80.0 MPa, and the uniaxial compressive strength is large. The immediate floor is sandy mudstone, with a thickness of 6.89 m. The strength is between 25.8 and 34.9 MPa, and the uniaxial compressive strength is small. The old floor is sandstone, with a thickness of 2.20 m, hard texture, and an average strength of 70.5 MPa.

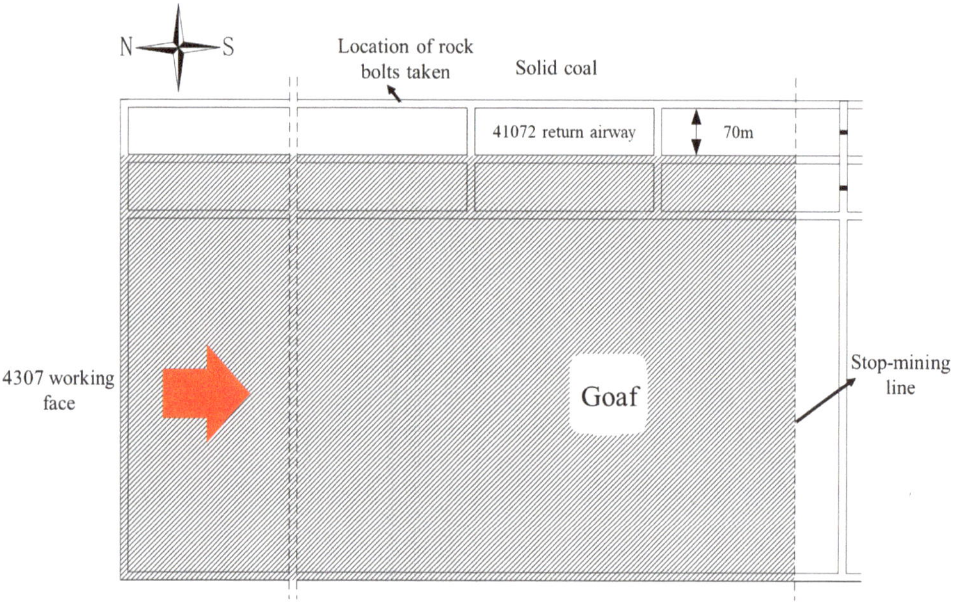

Figure 1. Layout of 41072 Roadway.

The measurement results of in situ stress in the fourth panel area of Changping Mine show that the maximum σ_H is 10.81 MPa and the minimum is 8.85 MPa. In terms of magnitude, the magnitude of the in situ stress field in the fourth panel area is moderately low. The lateral pressure coefficient σ_H/σ_V is greater than 1. The in situ stress in the panel area is dominated by horizontal stress and belongs to the type of tectonic stress field. The maximum horizontal principal stress is generally in the north-east-east direction. The geological and hydrological conditions in the fourth panel area are relatively simple.

The 41072 roadway is excavated along the floor, with 2 m thick top coal at the top and a layer of gangue 500 mm below the immediate roof. The 41072 roadway adopts the combined support of rock bolts and cables. The rock bolts have a diameter of 22 mm and a length of 2400 mm. The steel grade is BHRB500, where 'B' is the abbreviation of the manufacturer, 'HRB' is the abbreviation of 'Hot Rolled Ribbed Bars', and '500' indicates a yield strength of 500 MPa. The thread at the tail of the bolt is M24, where 'M' represents the common thread and '24' indicates that the major diameter (nominal diameter) of the thread is 24 mm. It is often used for fastening mechanical connection components. There are six rock bolts in each row, with a row spacing of 900 mm and a spacing of 1000 mm. The top corner rock bolts are 300 mm away from the roadway side and are all set perpendicularly to the roof. The resin-extended anchorage uses two anchoring agents, one with a specification of MSK2335 and the other with a specification of MSZ2360. Among them, 'MS' represents 'Resin Anchoring Agent', '23' is the diameter of the anchoring agent, and both are 23 mm. The length of the anchoring agent is denoted as '60', which is 350 mm and 600 mm, respectively. The drilling diameter is 30 mm, and the theoretical anchoring length is 1208 mm. The roadway support of 41072 is shown in Figure 2.

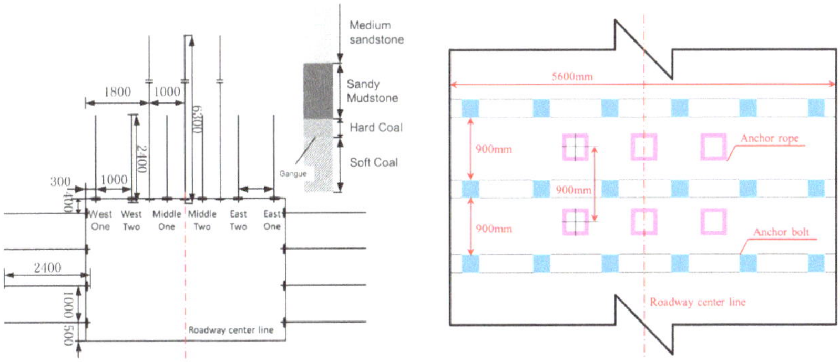

Figure 2. Support Schematic Diagram of 41072 Roadway.

According to the on-site borehole peeping results of the surrounding rock, above the coal seam roof of the 41072 roadway in Changping Mine is a typical layered composite soft rock roof of mudstone and sandy mudstone. The inner layered joints and fissures of this mudstone or sandy mudstone are developed, and the macroscopic bedding or joint spacing is generally 20–100 mm. This makes the roadway roof more fragmented, with poor integrity, low strength, and stiffness. Figure 3 shows the composite roof of roadway exposed on site. In addition, due to the influence of roadway excavation and mining, the integrity, strength and stiffness of the fissured layered composite soft rock roof are further reduced, causing the deformation of the surrounding rock of the roadway roof to increase rapidly, and the roof subsidence is severe. The roof subsidence is between 0.5 and 1 m, and the subsidence in individual sections is nearly 2 m. Moreover, the axial force of the rock bolts is constantly increasing. Field measurements show that many rock bolts and even cables have broken. The stability of the surrounding rock of the roadway can no longer meet the mining needs of the 4107 working face. Simple roof-picking treatment is carried out on site, which may eventually induce serious roof caving accidents. The subsidence of roadway roof in the field is shown in Figure 4.

Figure 3. Composite Roof of Roadway Exposed On Site.

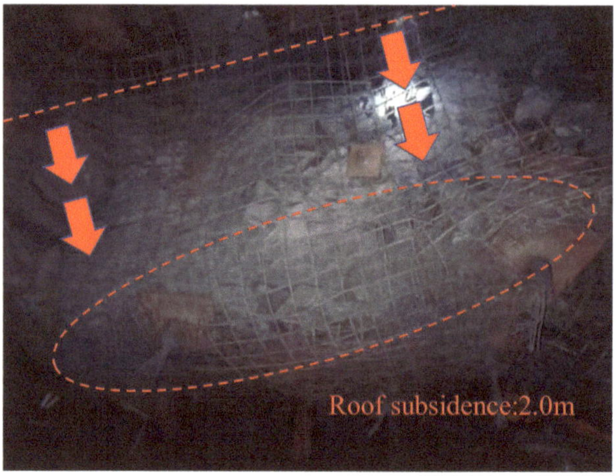

Figure 4. Subsidence of Roadway Roof.

2.2. Analysis of Deformation Characteristics of Roof Rock Bolts in 41072 Roadway of Changping Mine

In order to analyze the deformation characteristics of the rock bolts on the roadway roof, six rock bolts were selected on site to analyze their deformation characteristics. The bolts in Figure 5 are directly removed during the process of roof brushing in the severely deformed area of the roadway. The deformation characteristics of roof bolt of roadway 41072 are shown in Figure 6. The $1^{\#}$ and $6^{\#}$ rock bolts are taken from the top corner positions. Obvious shear deformation occurs at the interface between the roof and the coal seam. The rock bolts in the fine sandstone section of the roof only bend within a range of 5 cm near the bedding plane. The rock bolts in the coal seam above the entire gangue are obviously bent, and the shear dislocation direction is towards the middle of the roadway. The lateral shear dislocation of the $1^{\#}$ rock bolt reaches 16.2 cm, and the lateral shear dislocation of the $6^{\#}$ rock bolt is 15.6 cm. No obvious deformation occurs in the rock bolt body below the gangue surface.

Figure 5. Rock Bolts' Removal Position.

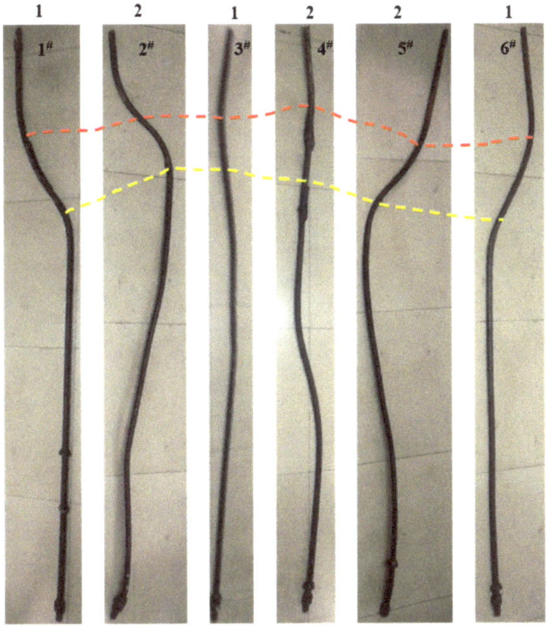

Figure 6. Deformation Characteristics of Roof Rock Bolts in 41072 Roadway.

The 2# and 5# rock bolts are taken from the west two and east two positions (Figure 2). Obvious shear deformation occurs at the interface between the roof and the coal seam. The deformation form is similar to that of the top corner rock bolts. The lateral shear dislocations of the 2# and 5# rock bolts are 21.5 cm and 13.4 cm, respectively. Continuous bending deformation occurs in the rock bolt body below the gangue surface, and the bending angles are 10° and 7°, respectively.

The 3# and 4# rock bolts are taken from the middle of the roadway. Compared with the deformation of the rock bolts at other positions, the bending and shear deformation of the rock bolts in the middle is relatively small. Slight shear deformation occurs at the interface between the roof and the coal seam. The lateral dislocations are 3.7 cm and 3.3 cm, respectively. Bending deformation also occurs in the lower part of the rock bolt body below the gangue surface.

3. Model Construction of Roadway with Layered Composite Roof

3.1. Numerical Simulation of Structural Plane

According to the borehole peeping results, there are eight obvious bedding planes in the roof, which are 0.2 m, 0.4 m, 0.6 m, 0.8 m, 1.0 m, 1.3 m, and 1.8 m away from the top surface of the roadway, respectively; there is one bedding plane in the floor, which is the interface between the floor coal and rock. Based on this, a FLAC^{3D} numerical model containing bedding was constructed, and the interface was used to simulate the bedding plane. Eight interfaces were set in the roof and one interface was set in the floor. The strength and stiffness of the interface were set much lower than that of the intact coal and rock mass to simulate the weak structural plane. At the same time, the "update" parameter of the interface was set to off to prevent the search for new contacts after the movement on the structural plane and to achieve the separation of nodes and simulate the discontinuous deformation such as the slip and opening of the weak structural plane. The mechanical performance parameters of the structural plane are shown in Table 1.

Table 1. Mechanical Performance Parameters of Structural Planes.

Bedding Plane Number	Height/m	Normal Stiffness/GPa/m	Tangential Stiffness/GPa/m	Tensile Strength/MPa	Cohesion/kPa	Friction Angle/°	Dilatancy Angle/°
1	6	1	1	0.1	10	20	6
2	10.1	1	1	0.1	10	20	6
3	10.3	0.8	0.8	0.08	8	16	4.8
4	10.5	1	1	0.1	10	20	6
5	10.7	1	1	0.1	10	20	6
6	10.9	0.8	0.8	0.08	8	16	4.8
7	11.1	1	1	0.1	10	20	6
8	11.4	1	1	0.1	10	20	6
9	11.9	0.8	0.8	0.08	8	16	4.8

3.2. *Numerical Model of Roadway with Layered Composite Roof*

The model size is 40 × 1 × 17.9 m^3, and a total of 20,580 elements and 30,090 nodes are divided. The thickness of the model in the Y direction is 1 m. The pile unit (the parameters of the rock bolt and the anchorage section are shown in Table 2, and each rock bolt is divided into 24 pile units) is used to simulate the rock bolt model. One row of rock bolts is arranged at Y = 0.5 m on the roadway roof, with a spacing of 800 mm, totaling six bolts, numbered from left to right as $1^{\#}$ to $6^{\#}$, and the anchorage length is 800 mm. The maximum horizontal principal stress σ_H = 10 MPa, the minimum horizontal principal stress σ_h = 5 MPa, the vertical stress σ_V = 10 MPa, the axial direction of the roadway is perpendicular to the direction of the maximum principal stress, the gravitational acceleration is 10 m/s^2, and the model adopts the large deformation calculation mode. A total of nine interfaces are built within the anchorage range to simulate the bedding plane. The constitutive model of the surrounding rock adopts the Mohr–Coulomb model. The boundary constraints of the model are as follows: the front and rear, left and right of the model are hinged, the bottom surface is fixed, and the corresponding stress constraints are applied on all surfaces of the model. The grid within the range of rock bolt support is refined, and the cell specification is 0.1 m × 0.1 m × 0.1 m. The model diagram is shown in Figures 7 and 8.

Table 2. Parameters of Rock Bolts and Anchoring Agents.

Rod Diameter d_b/mm	Rod Length L/m	Elastic Modulus of Rod E_b/GPa	Yield Strength of Rod σ_s/MPa	Drilling Diameter/mm	Elastic Modulus of Resin Anchoring Agent E_g/GPa	Shear Strength of Anchoring Interface S_p/MPa	Poisson's Ratio of Resin Anchoring Agent v_g
22	2.4	200	500	30	16	30	0.4

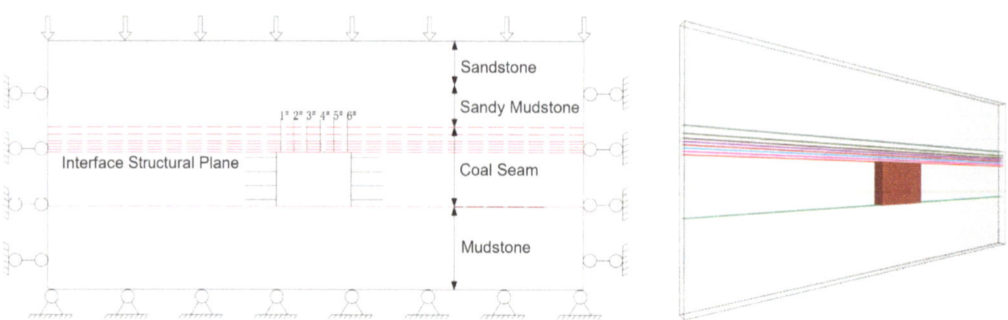

Figure 7. Model Schematic Diagram.

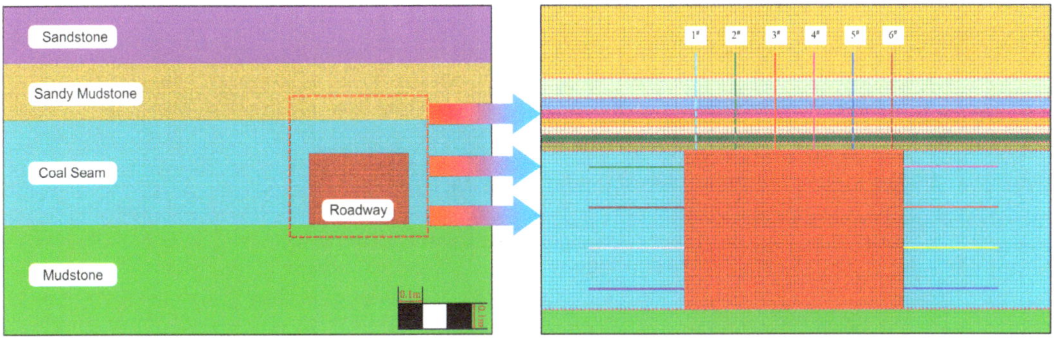

Figure 8. Schematic Diagram of Numerical Model.

4. Simulation Realization of Rock Bolt Pile Unit

4.1. Numerical Model of Roadway with Layered Composite Roof

In FLAC^{3D}, the cable unit is generally used to simulate rock bolts, but the cable unit can only bear axial loads and cannot reflect shear loads, while the shear resistance of rock bolts has a great influence on the support effect. In order to better simulate the comprehensive effect of the tensile and shear resistance of rock bolts, the pile unit is selected: The pile unit combines the beam element and the anchor cable element. Besides being able to bear the axial pulling load, it also has the ability to resist bending and shear; The pile unit can also calculate the stress softening between the pile and the grid and the rupture of the pile, thereby simulating the rupture of the rock bolt and the decoupling of the anchorage interface. The use of the pile unit can better simulate the mechanical behaviors, such as tension, bending, shear, fracture, and anchor failure of the rock bolt. In this paper, the HBRB500 high-strength rock bolt with a diameter of 22 mm is uniformly used for simulation. The drilling diameter is 30 mm. The parameters of the rock bolt and the anchoring agent are listed in Table 2.

In the pile unit, "cs-scoh" refers to the peak bond force, which determines when the bonding interface decouples and detaches from the anchor; "cs-sk" refers to the bond stiffness, which determines the resistance-increasing speed of the anchorage body; cs-scoh and cs-sk jointly determine the mechanical properties of the anchorage body and have a very significant influence on the support effect. To provide a basis for determining cs-scoh and cs-sk of the numerical rock bolt, a pull-out test was carried out in the laboratory. A steel pipe with a length of 125 mm and an inner diameter of 30 mm was used to simulate the drilling hole, and a left-handed high-strength threaded steel with a diameter of 22 mm was selected for the rock bolt. After repeated tests, when the anchorage section length was 125 mm, the average maximum pull-out force was 140 kN, and the corresponding average displacement was 13 mm. Calculate cs-scoh = 1.12×10^6 N/m. Considering that the on-site rock bolts are anchored in the coal and rock mass, the peak strength of the bonding interface should be lower than the laboratory test results. Zhu et al. [27] took advantage of the repeatability and scalability of numerical simulation calculations and proposed calibrating the mechanical parameters of the coal and rock mass through the repetitive trial-and-error method. In this paper, the above method was adopted to obtain cs-scoh = 4.37×10^5 N/m, cs-sk = cs-scoh/13 mm = 3.3×10^7 N/m. The parameters of pile Units are shown in Table 3.

Table 3. Parameters of Pile Units.

Elastic Modulus/Pa	Poisson's Ratio	Cross-Sectional Area/m^2	Perimeter/m	Yield Force/N	Moment of Inertia I_y/m^4	Moment of Inertia I_z/m^4	Polar Moment of Inertia/m^4	Bond Force (cs-scoh)/(N/m)	Bond Stiffness (cs-sk)/(N/m^2)
200×10^9	0.25	3.8×10^{-4}	0.094	2.47×10^5	1.15×10^{-8}	1.15×10^{-8}	2.3×10^{-8}	4.37×10^5	3.3×10^7

4.2. Method of Applying Pretension Force of Pile Unit

In $FLAC^{3D}$, there is no direct command to apply the pretension force to the rock bolts simulated by the pile unit. Through the built-in fish script of $FLAC^{3D}$, this paper proposes the method of "break–tension–suture", realizing the application of the pretension force to the rock bolts simulated by the pile unit.

The rock bolt is 2.4 m long and is simulated using 24 pile units. The anchorage section is 0.8 m long, and the free section is 1.6 m long. Two sections of pile units are created at the position where the rock bolt is set. The shorter section contains eight pile units and simulates the anchorage section. The parameters are set according to the parameters of the anchorage section pile unit in Table 3. The longer section contains 17 pile units and simulates the free section of the rock bolt. The cs-scoh and cs-sk of the pile unit in the free section are both set to 0, and other parameters are the same as those of the anchorage section. A rigid connection is established between the tail of the pile unit in the free section and the surrounding grid to simulate the rock bolt tray (Figure 9a).

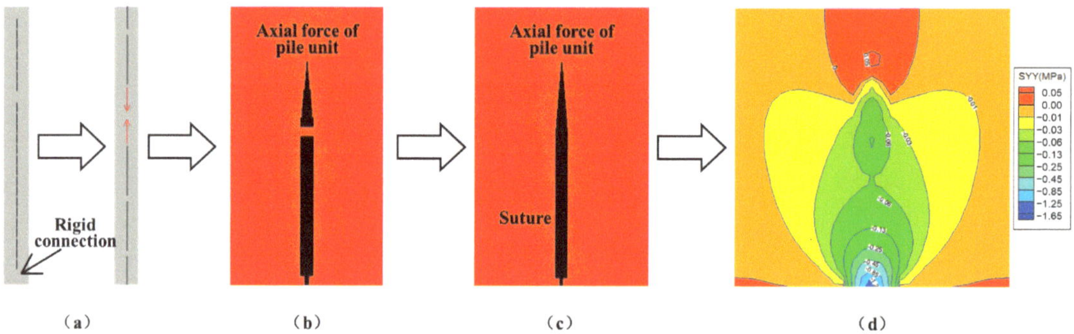

Figure 9. Steps of Applying Pretension Force of Pile Unit: (**a**) break; (**b**) tension; (**c**) suture; (**d**) Prestress diffusion.

Tension loads are applied to the nodes at the openings of the anchorage section and the free section of the rock bolt. The magnitude of the tension load is the preset pretension force of the rock bolt. The pile unit is stretched, and the operation is balanced to simulate the application process of the pretension force of the rock bolt (Figure 9b,c). Taking the two nodes at the openings of the anchorage section and the free section as the starting and ending points, a new pile unit is established. The parameters are set according to the pile unit in the free section. The opening of the rock bolt is "sutured", and the tension load applied to the nodes is deleted. The operation is balanced to complete the pretension of the rock bolt. The grid is released, and the operation is carried out to realize the diffusion of the prestress in the surrounding rock (Figure 9d). It should be noted that due to the calculation of flexible structures and ultimate bearing capacity involved in the simulation, a large deformation calculation mode is adopted in the simulation.

The shape of the support stress field after the pretension of the rock bolt by the above method is observed. The overall compressive stress zone is in a "pomegranate shape". The magnitude and shape are quite consistent with the support stress field shapes in the similar simulation by Lin [28] and the numerical simulation by Zhang [29]. This proves the effectiveness of the "break-tension-suture" method.

5. Simulation Analysis of Deformation and Stress Characteristics of Rock Bolts

5.1. Calculation Sequence

The pressure of the surrounding rock in the roadway ahead of the tunneling face begins to be relieved. After the roadway is excavated, the stress in the shallow part of the surrounding rock is rapidly released. After the rock bolts are installed and pretensioned, the diffusion of the pretension force and the unloading of the in situ stress of the original rock occur simultaneously. The surrounding rock continuously migrates under the combined action of the in situ stress field of the original rock, the mining-induced stress field and the stress field of the rock bolt support until it reaches equilibrium.

In order to correspond to the above stress change process, the calculation sequence is set as follows: Initialize the numerical model, apply the in situ stress and calculate the equilibrium; excavate the roadway, apply the in situ stress on the periphery (roof, floor, and two sides) and gradually reduce it to simulate the real release process of the surrounding rock stress after the roadway excavation. The power exponential equation is used for the release, that is, the release is fast at the beginning and slow in the later stage; when the stress is released to 30% of the in situ stress [30], fix the grid, install the rock bolts and pretension them using the "break-tension-suture" method; release the grid, and the surrounding rock continues to migrate under the combined action of the in situ stress field of the original rock, the mining-induced stress field and the stress field of the rock bolt support until it reaches equilibrium. The calculation sequence is shown in Figure 10.

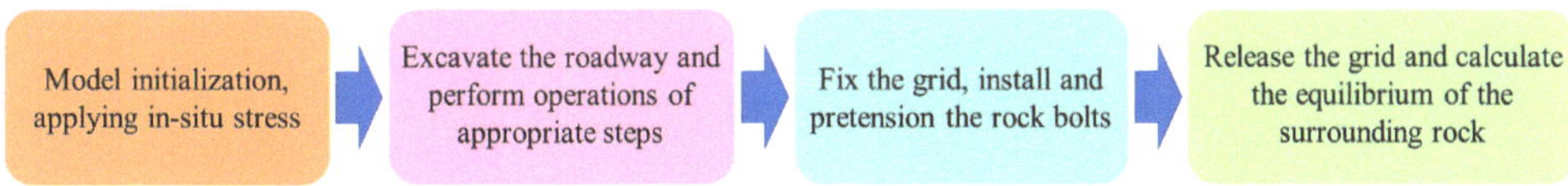

Figure 10. Calculation Sequence.

5.2. Simulation Analysis of Rock Bolt Deformation

The horizontal stress of the roadway surrounding rock is an important factor affecting the deformation of rock bolts. With the increase in horizontal stress, the deformation of rock bolts increases significantly. At this time, the subsidence of the roadway roof, floor heave, and displacements of the two sides increase significantly. The angle between the axial direction of the roadway and the maximum horizontal stress is different, and the degree of its influence on the deformation of rock bolts also varies. When the axial direction of the roadway is parallel to the direction of the maximum horizontal stress, the deformation of the rock bolts is the smallest. When the axial direction forms an acute angle with the direction of the maximum horizontal stress, the deformation of the rock bolts is biased towards one side of the roadway. When the axial direction is perpendicular to the maximum horizontal stress, the influence is the greatest, and obvious deformation occurs in the rock bolts. The deformation of the rock bolts obtained according to the simulation calculation method is shown in Figure 11. The $1^{\#}$ rock bolt shows obvious shear deformation at the interface between the roof and the coal seam. The shear dislocation direction is towards the middle of the roadway, and the dislocation distance is 13.3 cm. The $2^{\#}$ rock bolt shows obvious shear deformation at the interface between the roof and the coal seam. The deformation form is similar to that of the $1^{\#}$ rock bolt, and its lateral dislocation distance is 12.3 cm. Continuous bending deformation occurs in the lower part of the gangue, and the bending angle is 4.5°. The $3^{\#}$ rock bolt shows shear deformation at the interface between the roof and the hard coal, and the lateral dislocation distance is 4.1 cm. Compared with the $1^{\#}$ and $2^{\#}$ rock bolts, its bending deformation degree is smaller.

The deformation form of the rock bolts in the field and the simulation is consistent, both deforming and expanding towards the interior of the roadway along the interface of the roof coal seam with the center line of the roadway as the axis. In terms of the degree of deformation, the lateral dislocations of the $1^{\#}$ and $3^{\#}$ rock bolts in the field and the simulation are 16.2 cm, 13.3 cm, 3.7 cm, and 4.1 cm, respectively. The errors are 0.179 and 0.098. To sum up, compared with the actual deformation of the roof rock bolts in the 41072 roadway of Changping Mine (Figure 6), the simulation results of the rock bolt deformation are relatively consistent in terms of the deformation form and degree, indicating that the model parameter selection is reasonable, the calculation sequence is relatively consistent with the actual excavation process, and the model can be used to analyze the deformation of the bolt body.

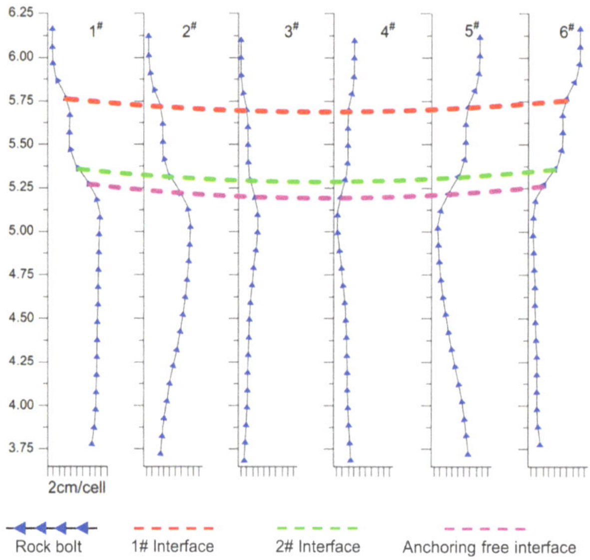

Figure 11. Simulated Deformation Results of Roof Rock Bolts in 41072 Roadway.

5.3. Simulation Analysis of Rock Bolt Force

The force conditions of the rock bolts are shown in Figures 12 and 13. The axial force of the $1^{\#}$ rock bolt is the smallest, which is 84.5 kN, indicating that the subsidence of the surrounding rock within its support range is small. The axial force of the anchorage section increases greatly at the interface between the roof and the coal seam, indicating that there is a separation layer at this interface. The axial force of the $2^{\#}$ rock bolt is the largest, which is 204.8 kN, indicating that the surrounding rock within its support range has a large subsidence and obvious separation layer. The axial force of the $3^{\#}$ rock bolt is slightly lower than that of the $2^{\#}$ rock bolt, which is 180.3 kN. There is no obvious change in the axial force of the anchorage section at the interface between the roof and the coal seam, indicating that the rock layers at this location have coordinated subsidence and the separation layer at the interface is small. The $1^{\#}$, $2^{\#}$, and $3^{\#}$ rock bolts are all subject to a large shear force at the interface between the roof and the coal seam. The shear force of the $1^{\#}$ rock bolt is the largest, which is 105.3 kN. It is not subject to shear force at the gangue surface and the interface between the anchorage and the free section, indicating that the coal seams above and below the gangue surface deform coordinately and there is no dislocation. Therefore, the force conditions of the $1^{\#}$, $2^{\#}$, and $3^{\#}$ rock bolts correspond to the deformation characteristics of the bolt bodies. In addition, according to the technical specifications of rock bolt support for coal mine roadways, the maximum bearing capacity of the roof rock bolts in the roadways of Changping Mine is 125 kN. Therefore, according to the simulated actual force conditions of the rock bolts, the safety factors of $1^{\#}$, $2^{\#}$, and $3^{\#}$ rock bolts during operation are calculated to be 1.48, 0.61, and 0.69, respectively. Under normal circumstances, the safety factor of rock bolts should be between 1.5 and 2.5. Within this range, the stability requirements of coal mine roadways can be met. Therefore, the above-mentioned rock bolts all undergo obvious deformation and lose their bearing capacity during actual operation, which is consistent with the on-site monitoring results of rock bolt deformation, once again verifying the reliability of the simulation method in this paper.

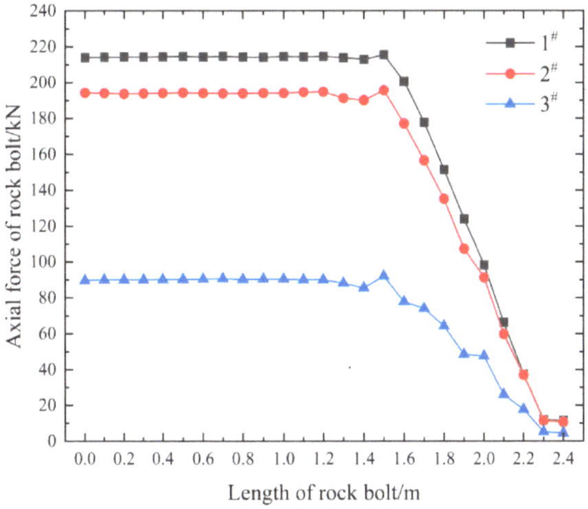

Figure 12. Axial Force of Roof Rock Bolts.

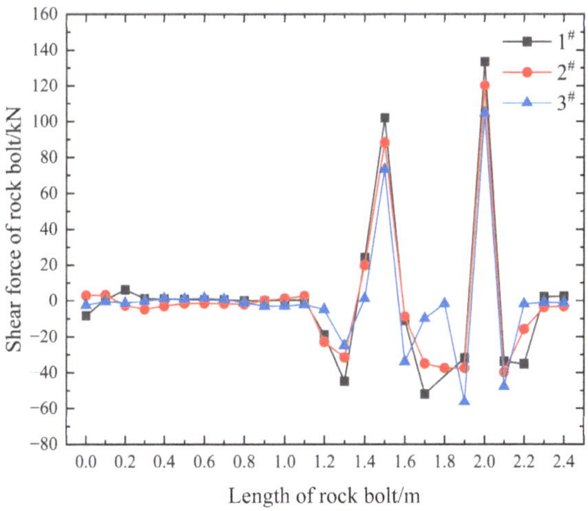

Figure 13. Shear Force of Roof Rock Bolts.

6. Analysis of Influencing Factors of Rock Bolt Body Deformation

6.1. Simulation Schemes of Different Surrounding Rock and Structural Plane Strengths

With the continuous improvement in the quality of the steel of the rock bolt body, the tensile strength of the rock bolt body continues to increase. Pure tensile fractures of rock bolts rarely occur at the site of roadway support. The fracture form of rock bolts is generally tensile–shear fracture. Therefore, the influencing factors of the shear and bending deformation of rock bolts should be analyzed. Based on the simulation of the 41072 roadway in Changping Mine in this section, four groups of comparative simulations were conducted to analyze the influence of the strength of the surrounding rock and the strength of the structural plane on the shear and bending deformation of the rock bolts. The details are shown in Table 4.

Table 4. Mechanical Performance Parameters of Surrounding Rock of Structural Planes.

Test Number	I	II	III	IV
1# Interface strength	High	Low	Low	High
Coal seam 1	High	High	Low	Low
2# Interface strength	Low	High	High	Low
Coal seam 2	Low	Low	Low	High

6.2. Analysis of Simulation Results

6.2.1. The Deformation Characteristics of Rock Bolts

Based on the roadway model with layered composite roof constructed in the above text, the deformation characteristics of the anchor bolt body under different factors were simulated. Figure 14 shows the simulation results of the deformation characteristics of the anchor bolt under different strengths of the surrounding rock and structural planes.

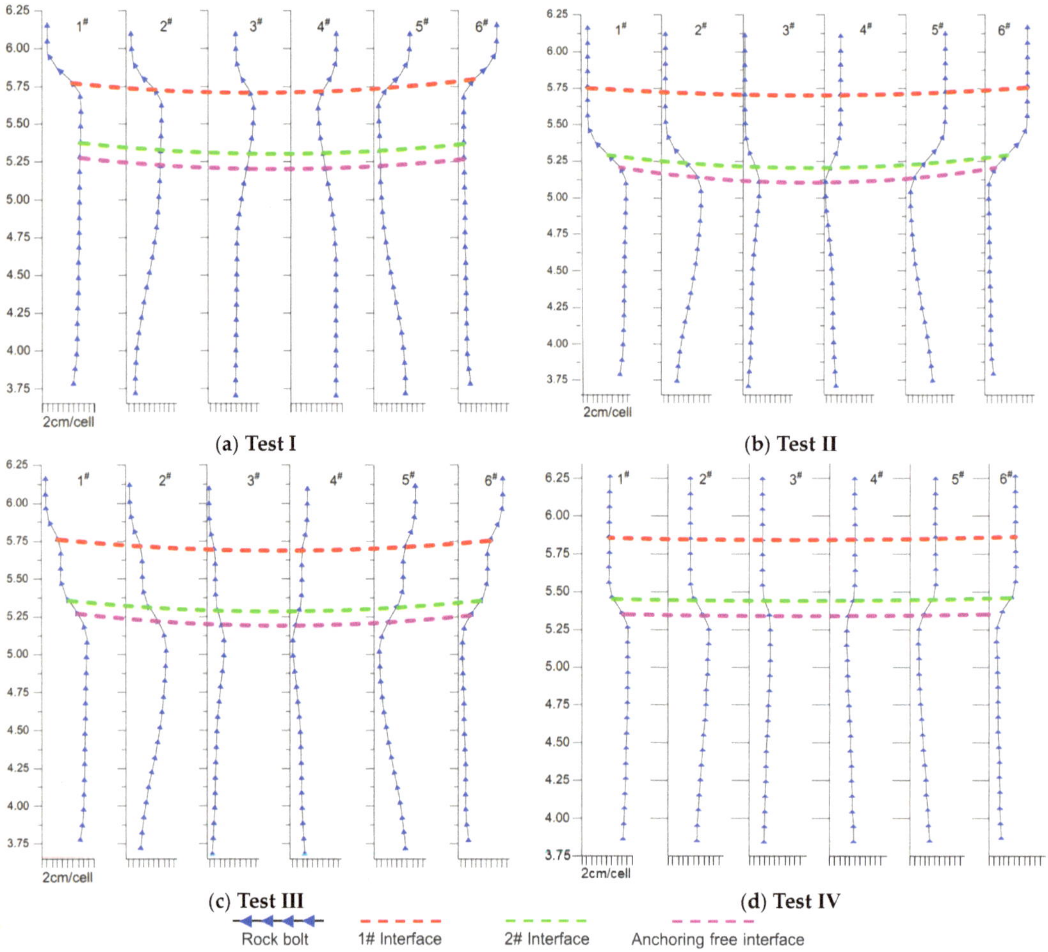

Figure 14. Deformation Characteristics of Rock Bolts with Different Surrounding Rock and Structural Plane Strengths.

It can be concluded from Figure 14 that in Test I, no obvious deformation occurred at the 1# interface of the bolt body, and the relative dislocation was not obvious, but obvious shear deformation occurred at the 2# interface of the anchor bolt, and there was a large relative dislocation of the rock strata above and below the 2# interface. In Test II, shear deformation occurred at both the 1# and 2# interfaces of the anchor bolt, and the relative dislocation of the upper and lower rock strata was obvious, but the deformation degree of the anchor bolt near the 1# interface was smaller than that near the 2# interface. In Test III, obvious shear deformation occurred at the 1# interface, and the relative dislocation was obvious, while no obvious deformation occurred at the 2# interface, and the relative dislocation was not obvious. In Test IV, shear deformation occurred at both the 1# and 2# interfaces of the anchor bolt, and the relative dislocation of the upper and lower rock strata was obvious. Comparing the deformation of the anchor bolt near the 2# interface of Test I and the 1# interface of Test II, it can also be shown that under the condition of the same structural plane strength, the higher the strength of the surrounding rock, the smaller the shear and bending deformation of the anchor bolt. Comparing the deformation of the anchor bolt near the 2# interface of Test I and the 2# interface of Test IV, the strength of both interfaces was small, and the difference in the strength of the rock strata above and below the interface was also large. The weak coal seam of the 2# interface in Test I was below the interface, and the weak coal seam in Test IV was above the 2# interface. This indicates that under the condition of the same structural plane strength, when the weak rock mass is below the structural plane, the structural plane is more prone to transverse dislocation.

6.2.2. The Force Analysis of Rock Bolts

During the simulation process, the axial force and shear force at each position of the anchor bolt body were monitored in real time to reveal the deformation and force characteristics and influencing factors of the anchor bolt body under the action of the layered composite roof. Figures 15 and 16, respectively, show the axial force and shear force characteristics of the anchor bolts with different strengths of the surrounding rock and structural planes.

It can be concluded from Figures 15 and 16 that in Test I, the axial force and shear force of the bolt body increased significantly at the 2# interface, so the bolt body underwent obvious deformation at the 2# interface, while no obvious deformation occurred at the 1# interface, and the shear force condition also indicated that the bolt body was not subjected to shear at this point. In Test II, the axial force of the bolt body increased significantly at the 1# interface, and the shear force of the bolt body also indicated that the bolt body was subjected to obvious shear at the 1# interface. The 2# interface was located in the non-anchored section of the anchor bolt, and it was impossible to determine whether the axial force increased nearby, but the shear force condition of the bolt body indicated that the bolt body was subjected to shear force at this point. Comparing the bending deformation conditions of the anchor bolts near the 1# and 2# interfaces, the deformation degree of the anchor bolt near the 1# interface was smaller than that near the 2# interface, but the shear force of the anchor bolt near the 1# interface was higher than that near the 2# interface, indicating that when the surrounding rock strength was higher, the anchor bolt could mobilize a greater deformation resistance with a smaller deformation. In Test III, the axial force of the bolt body increased significantly at the 1# interface, and the shear force of the bolt body also indicated that the bolt body was subjected to obvious shear at the 1# interface. No obvious deformation occurred at the 2# interface, and the shear force condition also indicated that the bolt body was not subjected to shear at this point. In Test IV, the axial force of the bolt body did not increase significantly at the 1# interface, and the shear force of the bolt body indicated that the bolt body was not subjected to obvious shear at the 1# interface. However, the shear force of the bolt body at the 2# interface increased significantly, and the bolt body also underwent obvious shear deformation.

Analyzing the above four tests, when the strength of the structural plane was small, the rock strata near the structural plane all underwent lateral dislocation, indicating that the strength of the structural plane was the primary factor controlling the lateral dislocation of the structural plane. Secondly, when the strength of the surrounding rock above and below the structure differed greatly, the structural plane was also prone to lateral dislocation, which was caused by the uncoordinated deformation of the rock strata above and below the structural plane. When the weak rock layer was located below the structural plane, the lateral dislocation of the structural plane was more likely to occur.

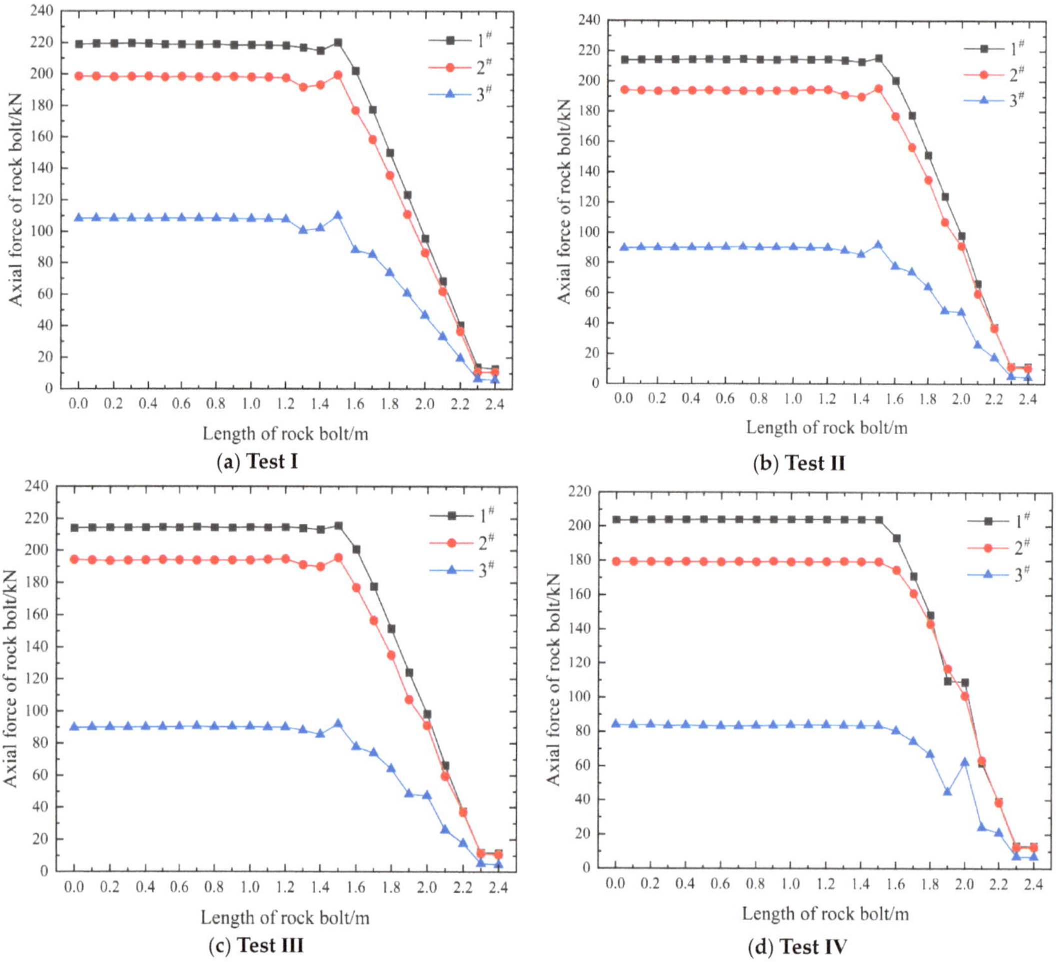

Figure 15. Axial Force Characteristics of Rock Bolts with Different Surrounding Rock and Structural Plane Strengths.

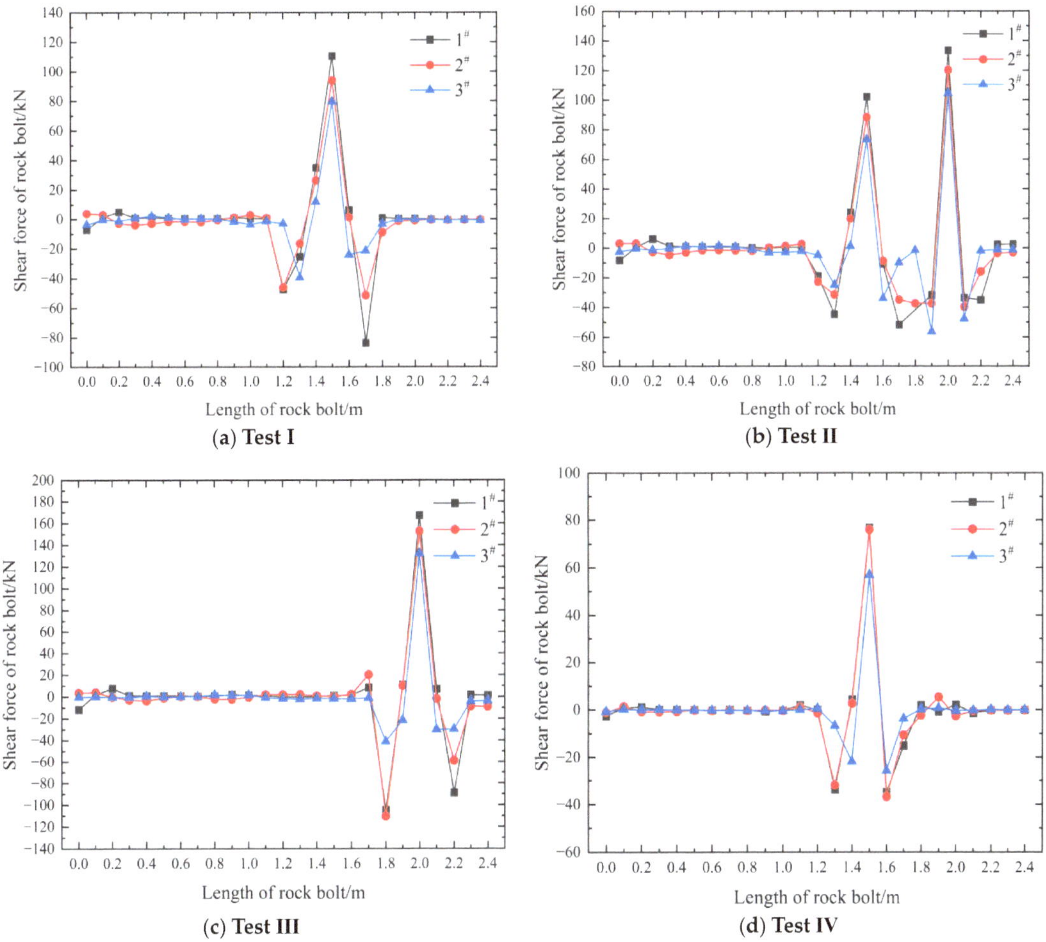

Figure 16. Shear Force Characteristics of Rock Bolts with Different Surrounding Rock and Structural Plane Strengths.

To sum up, the rock bolt and the roadway surrounding rock have an interactive relationship. In addition to the strength of the surrounding rock itself, the strength of the bedding of the roadway surrounding rock structural plane is also an important factor affecting the force, deformation and failure of the rock bolt. Cai et al. [31] used a linear drop model once to analyze the stress distribution characteristics of the rock bolt under the opening of the bedding rock mass. It shows that the opening of the bedding applies tensile stress to the rock bolt. Li and Stillborg et al. [32] believe that when the bedding strength of the surrounding rock is small, it generates a large axial tensile force on both sides of the rock bolt, which leads to decoupling of the anchoring interface and gradually expands to both sides. Hyett et al. [33] believe that the shear stress of the anchoring interface has a linear relationship with the deformation of the rock bolt and the surrounding rock and further analyzed the influence of the anchoring interface and the bedding strength on the force and deformation of the rock bolt. The above research results are consistent with this paper. When the bedding strength of the structural plane is low, the strength of the upper and lower rock strata differs greatly, and the weak rock layer is located below the structural plane, the rock bolt is most prone to lateral dislocation.

7. Conclusions

(1) Six rock bolts at the roadway roof were selected on site to analyze their deformation characteristics. It was found that the bending and shear deformation of the rock bolts in the middle of the roadway roof was relatively small, and slight shear deformation occurred at the interface between the roof and the coal seam, with the lateral shear dislocation being 3.7 cm and 3.3 cm, respectively. However, the rock bolts at the two top corners of the roadway showed obvious shear deformation at the interface between the roof and the coal seam, with the lateral shear dislocation being 16.2 cm and 15.6 cm, respectively.

(2) A "break-tension-suture" method for applying the pretension force of the pile unit simulation rock bolt was proposed. The interface was used to simulate the weak surface of the layered composite roof structure, and a reasonable calculation sequence of the stress field of the surrounding rock before and after the rock bolt was invented. Using the above simulation method to invert the deformation characteristics of the on-site rock bolts, it was found that the simulated deformation form and degree of the rock bolts were relatively consistent with the on-site situation. The lateral dislocations of $1^{\#}$ and $3^{\#}$ rock bolts on site and in the simulation were 16.2 cm, 13.3 cm, 3.7 cm, and 4.1 cm, respectively, indicating that the simulation method was reliable, the parameter selection was reasonable, and the calculation sequence was consistent with the actual excavation process.

(3) A unified model of deformation and stress evolution of rock bolts in the layered composite roof roadway was constructed. The influences of the strength of the surrounding rock and the strength of the structural plane on the shear and bending deformation of the rock bolts were analyzed. It was found that the influencing factors of the rock bolt deformation were the strength of the structural plane > the strength of the surrounding rock. When the strength of the structural plane was low, the strength difference between the upper and lower rock layers was large, and the weak rock layer was located below the structural plane, the rock bolts were most prone to lateral dislocation.

(4) This study found that the deformation of the top corner bolt was the largest, and on-site measurements were conducted to strengthen the top corner bolt monitoring. Furthermore, regarding the impact of structural deformation on bolts, grouting reinforcement can be adopted on site to ensure the integrity of the roadway roof. This article did not change the pretension force of the bolts in the simulation. However, on-site practice has shown that the active restraining force provided by high pretensioning bolts can not only limit the opening of weak surfaces in the structure, but also increase the shear strength of the structural surface and suppress its activity. Thus, future research will focus on the influence of pretension on bolt deformation.

Author Contributions: Z.W. and C.Z. conceived and designed the study. C.Z. and S.F. were involved in the numerical simulation and analysis. Z.W. was involved in the field measurement and prepared all figures. All authors have read and agreed to the published version of the manuscript.

Funding: Financial support for this work is provided by the National Key R&D Program of China (2023YFC2907602), the Science and Technology Innovation and Entrepreneurship Fund of Tiandi Science and Technology Co., Ltd. (2023-TD-ZD011-001), the National Natural Science Foundation of China (52104155), and the Fundamental Research Funds for the Central Universities (2024ZKPYNY01).

Data Availability Statement: All data used to support the findings of this study are available from the corresponding author upon request.

Conflicts of Interest: The authors declare no conflicts of interest.

References

1. Jodeiri Shokri, B.; Mirzaghorbanali, A.; Nourizadeh, H.; McDougall, K.; Karunasena, W.; Aziz, N.; Entezam, S.; Entezam, A. Axial Load Transfer Mechanism in Fully Grouted Rock Bolting System: A Systematic Review. *Appl. Sci.* **2024**, *14*, 2076–3417. [CrossRef]
2. Chen, J.; Liu, P.; Zhao, H.; Zhang, C.; Zhang, J. Analytical studying the axial performance of fully encapsulated rock bolts. *Eng. Fail. Anal.* **2021**, *128*, 105580. [CrossRef]

3. Zhang, K.; Zhang, G.; Hou, R.; Wu, Y.; Zhou, H. Stress evolution in roadway rock bolts during mining in a fully mechanized longwall face, and an evaluation of rock bolt support design. *Rock Mech. Rock Eng.* **2015**, *48*, 333–344. [CrossRef]
4. Zhang, C.; Bai, Q.; Han, P. A review of water rock interaction in underground coal mining: Problems and analysis. *Bull. Eng. Geol. Environ.* **2023**, *82*, 157. [CrossRef]
5. Zhang, C.; Zhao, Y.; Han, P.; Bai, Q. Coal pillar failure analysis and instability evaluation methods: A short review and prospect. *Eng. Fail. Anal.* **2022**, *138*, 106344. [CrossRef]
6. Chen, Y.; Xiao, H. State-of-the-art on the anchorage performance of rock bolts subjected to shear load. *Int. J. Coal Sci. Technol.* **2024**, *11*, 9. [CrossRef]
7. Gu, S.C.; Wang, P.; Yang, C.F. Mechanical characteristics and stability analysis of surrounding rock reinforcement in rectangular roadway. *Sci. Rep.* **2022**, *12*, 22234. [CrossRef]
8. Nourizadeh, H.; Mirzaghorbanali, A.; Serati, M.; Mutaz, E.; McDougall, K.; Aziz, N. Failure characterization of fully grouted rock bolts under triaxial testing. *J. Rock Mech. Geotech. Eng.* **2024**, *16*, 778–789. [CrossRef]
9. Entezam, S.; Shokri, B.J.; Nourizadeh, H.; Motallebiyan, A.; Mirzaghorbanali, A.; McDougall, K.; Aziz, N.; Karunasena, K. *Investigation of the Effect of Using Fly Ash in the Grout Mixture on Performing the Fully Grouted Rock Bolt Systems*; University of Wollongong: Wollongong, NSW, Australia, 2023.
10. Yu, S.; Niu, L.; Chen, J. Experimental and Numerical Studies on Bond Quality of Fully Grouted Rockbolt under Confining Pressure and Pull-Out Load. *Shock Vib.* **2022**, *2022*, 7012510. [CrossRef]
11. Jodeiri Shokri, B.J.; Entezam, S.; Nourizadeh, H.; Motallebiyan, A.; Mirzaghorbanali, A.; McDougall, K.; Aziz, N.; Karunasena, K. The effect of changing confinement diameter on axial load transfer mechanisms of fully grouted rock bolts. In Proceedings of the 2023 Resource Operators Conference, University of Wollongong, Wollongong, NSW, Australia, 9–10 February 2023; pp. 290–295.
12. Nourizadeh, H.; Mirzaghorbanali, A.; McDougall, K.; Jeewantha, L.; Craig, P.; Motallebiyan, A.; Shokri, B.J.; Rastegarmanesh, A.; Aziz, N. Characterization of mechanical and bonding properties of anchoring resins under elevated temperature. *Int. J. Rock Mech. Min. Sci.* **2023**, *170*, 105506. [CrossRef]
13. He, M.; Ren, S.; Guo, L.; Lin, W.; Zhang, T.; Tao, Z. Experimental study on influence of host rock strength on shear performance of Micro-NPR steel bolted rock joints. *Int. J. Rock Mech. Min. Sci.* **2022**, *159*, 105236. [CrossRef]
14. He, M.; Ren, S.; Xu, H.; Luo, S.; Tao, Z.; Zhu, C. Experimental study on the shear performance of quasi-NPR steel bolted rock joints. *J. Rock Mech. Geotech. Eng.* **2023**, *15*, 350–362. [CrossRef]
15. Tahmasebinia, F.; Yang, A.; Feghali, P.; Skrzypkowski, K. Structural evaluation of cable bolts under static loading. *Appl. Sci.* **2023**, *13*, 1326. [CrossRef]
16. Tahmasebinia, F.; Yang, A.; Feghali, P.; Skrzypkowski, K. A numerical investigation to calculate ultimate limit state capacity of cable bolts subjected to impact loading. *Appl. Sci.* **2022**, *13*, 15. [CrossRef]
17. Chen, Y.; Lin, H.; Xie, S.; Ding, X.; He, D.; Yong, W.; Gao, F. Effect of joint microcharacteristics on macroshear behavior of single-bolted rock joints by the numerical modelling with PFC. *Environ. Earth Sci.* **2022**, *81*, 276. [CrossRef]
18. Jiang, Y.; Zhang, S.; Luan, H.; Wang, C.; Wang, G.; Han, W. Numerical modelling of the performance of bolted rough joint subjected to shear load. *Geomech. Geophys. Geo-Energy Geo-Resour.* **2022**, *8*, 140. [CrossRef]
19. Zhan, K.; Wen, X.; Wang, X.; Kong, C. A method for characterization of stress concentration degree of coal mine roadway surrounding rock. *J. Geophys. Eng.* **2023**, *20*, 699–711. [CrossRef]
20. Kong, Q.; Li, Y.; Wang, S.; Yuan, C.; Sang, X. The influence of high-strength bolt preload loss on structural mechanical properties. *Eng. Struct.* **2022**, *271*, 114955. [CrossRef]
21. Yi, W.; Wang, M.; Zhao, S.; Tong, J.; Liu, C. The effect of rock hardness and integrity on the failure mechanism of mortar bolt composite structure in a jointed rock mass. *Eng. Fail. Anal.* **2023**, *143*, 106831. [CrossRef]
22. Peng, S. Roof bolting and underground roof falls. *Geohazard Mech.* **2023**, *1*, 32–37. [CrossRef]
23. Ren, S.; Tao, Z.; He, M.; Li, M.; Sui, Q. Numerical simulation study on shear resistance of anchorage joints considering tensile–shear fracture criterion of 2G-NPR bolt. *Int. J. Coal Sci. Technol.* **2023**, *10*, 58. [CrossRef]
24. Zhang, Y.; Jiang, Y.; Wang, Z.; Yin, Q.; Chen, M. Anchorage effect of bolt on en-echelon fractures: A comparison between energy-absorbing bolt and conventional rigid bolt. *Eng. Fail. Anal.* **2022**, *137*, 106256. [CrossRef]
25. Yang, W.; Xia, X. Study on mining failure law of the weak and weathered composite roof in a thin bedrock working face. *J. Geophys. Eng.* **2018**, *15*, 2370–2377. [CrossRef]
26. Ma, S.; Wang, Y.; Wang, H.; Gao, L. Stability mechanism and control technology of large inclined working face roof in large-scale caving area. *J. Geophys. Eng.* **2023**, *20*, 91–102. [CrossRef]
27. Zhu, D.; Yu, B.; Wang, D.; Zhang, Y. Fusion of finite element and machine learning methods to predict rock shear strength parameters. *J. Geophys. Eng.* **2024**, *21*, gxae064. [CrossRef]
28. Lin, J.; Shi, Y.; Sun, Z.Y.; Wang, Z.S.; Cai, J.F. Large scale model test on the distribution characteristics of the prestressed field of end-anchored bolts. *Chin. J. Rock Mech. Eng.* **2016**, *35*, 2237–2247.
29. Zhang, Z.; Kang, H.P.; Wang, J.H. Pre-tensioned stress coordination function analysis of bolt-cable anchor support in coal roadway. *J. China Coal Soc.* **2010**, *35*, 881–886.
30. Kang, H.P.; Jiang, P.F.; Gao, F.Q. Analysis on stability of rock surrounding heading faces and technical approaches for rapid heading. *J. China Coal Soc.* **2021**, *46*, 2023–2045.
31. Cai, Y.; Esaki, T.; Jiang, Y. A rock bolt and rock mass interaction model. *Int. J. Rock Mech. Min. Sci.* **2004**, *41*, 1055–1067. [CrossRef]

32. Li, C.; Stillborg, B. Analytical models for rock bolts. *Int. J. Rock Mech. Min. Sci.* **1999**, *36*, 1013–1029. [CrossRef]
33. Hyett, A.J.; Moosavi, M.; Bawden, W.F. Load distribution along fully grouted bolts, with emphasis on cable bolt reinforcement. *Int. J. Numer. Anal. Methods Geomech.* **1996**, *20*, 517–544. [CrossRef]

applied sciences

MDPI

Article

Theoretical Discrimination Method of Water-Flowing Fractured Zone Development Height Based on Thin Plate Theory

Fengfeng Wu [1,2], Zhiqiang Gao [1,2,*], Huaidong Liu [1,2], Xin Yu [1,2] and Haoyuan Gu [1,2]

[1] Key Laboratory of Deep Coal Resource Mining, Ministry of Education, China University of Mining and Technology, Xuzhou 221116, China; wufengfeng@cumt.edu.cn (F.W.); liuhuaidong@cumt.edu.cn (H.L.)
[2] School of Mines, China University of Mining and Technology, Xuzhou 221116, China
* Correspondence: ts22020013a31tm@cumt.edu.cn

Abstract: The water-flowing fractured zone development height (WFZDH) is of great importance for water prevention and control in coal mines. The purpose of this research is to obtain a WFZDH prediction method of the first mining face based on thin plate theory, considering the rock stratum as a thin plate. By analyzing the thin plate, we expect to derive formulas for deflection, thus further analyzing the deformation of the rock formation. Existing methods tend to analyze the rock stratum as if they were beams, and their results are errors from reality. The proposed method is more realistic in analyzing the rock stratum as a plate. The theoretical discrimination method for the WFZDH based on thin-plate theory was investigated using theoretical analysis, numerical simulation, and field measurements. A mechanical model of the key stratum (a hard and thick rock stratum that controls the activity of all rock formations overlying a mining site, either locally or up to the surface) as a thin plate was established. The formulae for the deflection of the key stratum and the critical span for fracture were obtained from this model. The failure of the key stratum must meet two conditions: the key stratum's suspended span exceeds the critical span at which key strata first fracture, and the free space height below the key stratum is greater than its maximum deflection. Based on the above demarcation basis and key stratum failure conditions, the method of discriminating the WFZDH and its applicable conditions are proposed. In accordance with Yeping Coal Mine's geological background, the method was applied to discriminate the WFZDH, and the WFZDH was calculated to be 54 m. The results of the numerical simulation show that WFZDH is 55 m, and the measured results using the double-end water plugging device observation method and the Borehole TV method are 55.3 m~58.9 m. By comparing and analyzing the results obtained via various methods, the results show that the WFZDH analyzed using thin-plate theory is similar to those measured in the field and obtained through numerical simulation, verifying the appropriateness and practicability of the WFZDH discrimination method based on thin-plate theory. This research obtained the WFZDH of Yeping Coal Mine, which ensured its safe mining and provided guidance for the estimation of WFZDH in other mines with similar conditions.

Keywords: thin plate theory; water-flowing fractured zone development height; key stratum; mining covered rock; maximum deflection

1. Introduction

China has diversified coal resources with abundant reserves, but the hydrogeological conditions are complex and diverse. Most mining areas are plagued by mine water hazards, and there is a constant occurrence of water intrusion accidents [1–3]. Pang Mingkun [4] studied the flow state of the groundwater system and analyzed the mechanism of water emergencies in coal mines. The existence of bed separations above the coal seam can gather a large amount of water or act as a channel to direct water to the working face, leading to water ingress accidents. The study of the generation and distribution of bed separations is essential for preventing water inrush accidents in mines [5–7]. In the western part of

China, there are more aquifers, and the problem of large water surges in mines can occur at any time. Also, due to the thicker coal seams in the western mine, the fissures and bed separations above them are more fully developed, which is more likely to lead to sudden water accidents [8–10]. The roof bed separations act as possible water storage spaces and water conduits, with a high probability of triggering water damage in mines [11,12]. The water-flowing fractured zone development height (WFZDH) is closely related to coal mine water control work. It is of great significance to understand the WFZDH under various geological environments in order to achieve safe, economical, and environmentally sustainable mine development [13,14].

In light of the findings on the deformation and failure of rock formation above the working face, Chinese scholars have conducted extensive research in this field. In 1981, Qian Minggao made a creative proposal for the "masonry beam" mechanical model for the overlying strata [15]. Since then, Qian Minggao has enriched and developed this theory and proposed the "key stratum" theory [16,17]. This theory posits that the position of the main key stratum exerts a significant influence on the WFZDH. Once this critical height is reached, the main key stratum will begin to fracture. These fractures will then propagate, or spread and extend, through the entire thickness of the main key stratum. The fractures do not remain confined within the stratum but rather continue to grow and expand upwards towards the surface. When the main key stratum has not yet exceeded its critical height, the fracture is impeded at the base of the sub-key stratum in proximity to its critical height. Xu Jialin [18], Gong Peilin [19], Yu Bin [20], and others conducted studies into the fracturing and movement laws of overlying strata under conditions of large mining depths and the presence of a competent, rigid roof layer. They supplemented and improved the key stratum theory. Gong Peilin [19] and Yi Yongjie [21] applied the key stratum theory in analyzing the WFZDH and verified the key stratum theory with field measurements. In studying the overlying layer fracture mechanism in mining faces, the thin plate theory has been applied to a certain extent. Chen Zhonghui, Xie Heping [22], Zhong Yang, Yin Jianhua [23], and Song Yanqi [24] established the mechanical model of thin plates and analyzed roof deformation using the corresponding numerical calculation method. Yu Hui [25], Huang Changguo [26], and Liu Zhengchun et al. [27] established corresponding thin plate mechanical models for different working face conditions and obtained the calculation formula for roof fracture distance. For roof load calculation, Liu Xiaobo [28], Zhang Yidong [29], Lin Haifei [30], WU Fengfeng [31], and others conducted corresponding research work and obtained the critical load for roof strata failure through different methods. In addition to the research on the calculation of roof deformation, roof load, and roof step distance, Wang Xinfeng [32] used thin plate bending theory to establish roof mechanical models at different mining stages and revealed the roof 'O-X'-type breaking mechanism; Tu Hongsheng [33] established the steep-face roof panel model and studied the deflection deformation characteristics of the steeply inclined working face. Yang Shengli [34] analyzed the key stratum-breaking forms under different thickness conditions. Chang Xikun [35] studied the characteristics of WFZDH in shallow, thick coal seam underwater. He Xiang [36] used multiple regression analysis to establish a non-linear model for predicting the WFZDH by combining factors such as mining height, percentage of hard rock, mining depth, and extraction rate. Guo Wenbing [37] studied the damage transfer process of upper rock under intensive mining and proposed a new theoretical method based on the overburden failure transfer process to predict the WFZDH. Xu [38] developed a trapezoidal fracture model to explain water intrusion during coal mining based on overburden strata movement induced by coal mining.

Currently, there are more studies on the failure characteristics of mine roofs based on elastic thin plate theory. However, there is relatively less research on WFZDH using the thin plate theory. Therefore, this paper analyzes the deflection and the key stratum span in the process of bending and deformation to breakage by establishing a thin-plate mechanical model. It studies the links between key stratum failure and the WFZDH, proposes a method for discriminating the WFZDH based on thin plate theory and applicable conditions, and verifies the reasonableness of this method by combining it with the actual geological conditions of the 1301 face in Yeping Coal Mine.

2. Mechanical Modeling of Key Stratum and Its Fracture Analysis

2.1. Mechanical Modeling of Key Stratum Sheet

In elastic mechanics, a geometric object surrounded by two horizontal planes and four vertical planes is called a plate. The distance between the two horizontal planes, denoted as h, is referred to as the plate thickness. If h is less than one-fifth of the shorter side, the plate is considered a thin plate [39]. The elastic thin plate model is illustrated in Figure 1.

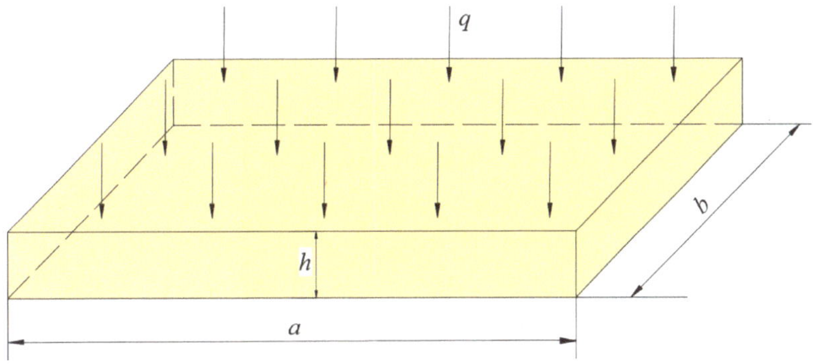

Figure 1. Thin Plate Mechanics Model.

The key stratum can be regarded as a continuous elastomer, and the load borne by the thin plate can be decomposed into vertical load and horizontal load, while the vertical load causes the plate to undergo bending deformation. Since the 1301 working face is the first face mined in this area, it can be considered that the key stratum is surrounded by solid support conditions, and the effect of horizontal load on the thin plate is ignored.

The small deflection bending theory postulates that the deflection is significantly less than its thickness, owing to its certain bending stiffness. Based on Kirchhoff's plate theory, the following basic assumptions are made for the overlying strata: ① The vertical strain caused by vertical stress $\varepsilon_z = 0$; ② The vertical strain components ε_z, γ_{zx}, and γ_{zy} can be ignored; ③ The horizontal displacement is zero.

2.2. Key Stratum Breakage Analysis

In the case where the collapsed zone with loose and fragmented rocks is unable to fully fill the goaf, the key stratum determines the division of the WFZ [40]. The WFZDH is affected by the key stratum; whether the key stratum experiences fracturing determines if the overlying rock fissures can continue to propagate upwards. When the key strata remain intact, the fissures are impeded at its bottom. As the working face continues to advance, if the key stratum remains unbroken, it can be considered that the WFZ has developed to the bottom of the key strata at this point.

The relationship between the maximum subsidence deformation and the height of free space underneath determines whether the rock strata will be damaged [41]. Meanwhile, it is only when the working face has advanced a sufficient distance, causing the rock strata below the key strata to fracture, that the suspended span of the key strata reaches its fracturing distance, at which point the key strata will fracture. Therefore, the key stratum fracture needs to meet two conditions: (1). In the direction of advancement, the suspended span of the key strata is greater than the critical span at which the key strata first fracture.

(2). The free space height below the key stratum is greater than its maximum deflection, and the expression is shown in Equation (1):

$$\begin{cases} a_j > a_{jT} \\ w_{\max} < \Delta = M - \sum\limits_{i=1}^{j-1} h_i(k_i - 1), \end{cases} \quad (1)$$

where a_j is the suspended span of key strata, m; a_{jT} is the critical span at which the key strata first fracture, m; Δ is the free space height below the key strata, m; M is the coal seam thickness, m; k_i is the residual bulking coefficient of the jth stratum, the average coefficient of the caved strata is related to the distance from the coal seam [37].

The deflection differential equation of the thin plate is as follows:

$$D\nabla^4 w = q \quad (2)$$

$$D = \frac{Eh^3}{12(1-\mu^2)}, \quad (3)$$

where D is the bending stiffness, N/m; ∇ is the Laplacian operator; q is the vertical load on the thin plate. MPa; μ is the Poisson ratio; h is the thin plate thickness, m; E is the elastic modulus of the thin plate, MPa.

3. Theoretical Method for Determining the WFZDH

Based on the previous analysis, the key factors in determining WFZDH are the working face advancement distance and whether the key strata experience fracturing. Therefore, this section will analyze the key strata as a thin plate and specifically examine the conditions for key strata fracturing. Theoretical analyses of this issue can help to understand the essence and inner mechanism of things in depth and provide guidance and predictions for practical applications. In addition, theoretical analysis provides the basic assumptions and model construction for numerical simulation.

3.1. Key Stratum Fracture Analysis

For the first mining face, the surrounding area is unmined solid coal. Before the initial fracturing of the key stratum, it can be regarded as a thin plate with four edges fixed, subjected to a uniformly distributed vertical load q. Assume that the suspended span of the key stratum at the moment of initial fracturing is $2a \times 2b$. The coordinate axes are taken as shown in Figure 2, and the boundary conditions are as shown in Equation (4):

$$\begin{cases} (w)_{x=\pm a} = 0 \\ (w)_{y=\pm b} = 0 \\ \left(\frac{\partial w}{\partial x}\right)_{x=\pm a} = 0 \\ \left(\frac{\partial w}{\partial y}\right)_{y=\pm b} = 0. \end{cases} \quad (4)$$

The Galerkin method is used to solve the small deflection problem of thin plates. The Galerkin variational equation is expressed as Equation (5), and the deflection expression of the thin plate is given using Equation (6):

$$\begin{cases} \int_V \left(\frac{\partial \sigma_z}{\partial z} + \frac{\partial \tau_{zx}}{\partial x} + \frac{\partial_{zy}}{\partial y} + f_z\right) w_m dV = 0 \\ \int \int_A D(\nabla^4 w) w_m dxdy = \int \int_A q w_m dxdy \end{cases} \quad (5)$$

$$w = \sum_m C_m w_m = \left(x^2 - a^2\right)^2 \left(y^2 - b^2\right)^2 \left(C_1 + C_2 x^2 + C_3 y^2 + \ldots\right), \tag{6}$$

where C_m are m undetermined coefficients that are independent of each other, and w_m is the setting function that satisfies the displacement boundary condition.

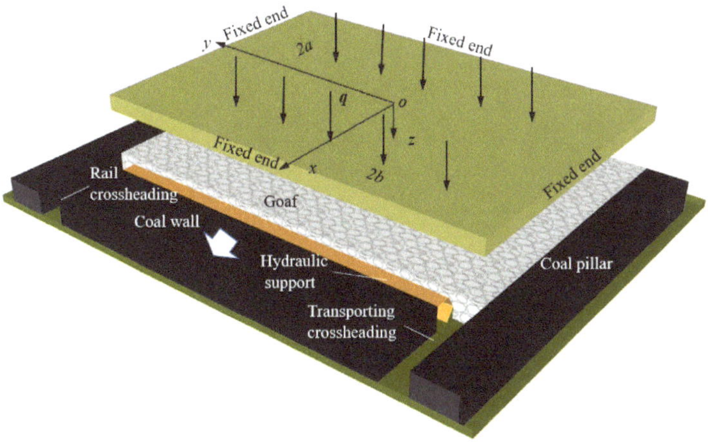

Figure 2. Four edges clamped into a thin plate.

Suppose that only one coefficient is considered in Equation (6):

$$w_1 = C_1 w_1 = C_1 \left(x^2 - a^2\right)^2 \left(y^2 - b^2\right)^2. \tag{7}$$

So we get

$$w_m = w_1 = \left(x^2 - a^2\right)^2 \left(y^2 - b^2\right)^2 \tag{8}$$

$$\begin{aligned} \nabla^4 w &= \tfrac{\partial^4 w}{\partial x^4} + \tfrac{\partial^4 w}{\partial y^4} + 2\tfrac{\partial^4 w}{\partial x^2 \partial y^2} \\ &= 8C_1\left[3\left(y^2 - b^2\right)^2 + 3\left(x^2 - a^2\right)^2 + 4\left(3x^2 - a^2\right)\left(3y^2 - b^2\right)\right] \end{aligned}. \tag{9}$$

By substituting Equation (5), we get

$$4D\int_0^a \int_0^b \nabla^4 w \left(x^2 - a^2\right)^2 \left(y^2 - b^2\right)^2 dxdy = 4q\int_0^a \int_0^b \left(x^2 - a^2\right)^2 \left(y^2 - b^2\right)^2 dxdy. \tag{10}$$

After the integration of Equation (10), C_1 is solved and then substituted into Equation (6), resulting in the deflection expression of the key stratum thin plate:

$$w = \frac{128q\left(x^2 - a^2\right)^2\left(y^2 - b^2\right)^2}{\left(a^4 + b^4 + 0.57a^2b^2\right)D}. \tag{11}$$

In Equation (11), it can be seen that when $x = 0$ and $y = 0$, the thin plate is allowed to deflect to its maximum as follows:

$$w_{\max} = (w)_{x=y=0} = \frac{128qa^4b^4}{D\left(a^4 + b^4 + 0.57a^2b^2\right)}. \tag{12}$$

Especially when the thin plate is square, with $b = a$, the deflection of the key stratum thin plate is as follows:

$$w = \frac{128q\left(x^2 - a^2\right)^2\left(y^2 - b^2\right)^2}{2.57a^4D}. \tag{13}$$

The maximum deflection is

$$w_{\max} = (w)_{x=y=0} = \frac{128qa^4}{2.57D}. \tag{14}$$

3.2. Analysis of Key Stratum Thin Plate Collapse Distance

After coal extraction, the layers above the goaf lose support and become suspended. The suspended span increases as the working face progresses. The relationship between the suspended span a_j and the working face advance distance l_j is shown in Figure 3.

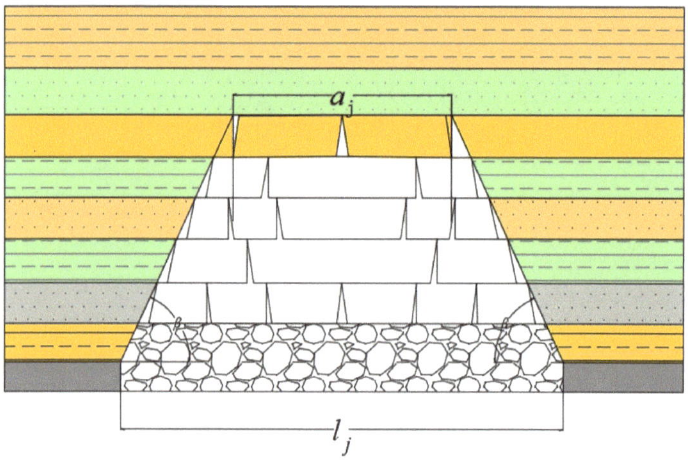

Figure 3. Key stratum fracture diagram.

In the case where all four edges are fixed, Figure 4 shows the bending moment distribution of the key stratum thin plate [15]. The maximum principal bending moment M_a is at the center of the long side, forming a positive bending moment zone in the center. According to the Marcus modified solution shown below

$$|M_{\mathrm{a}}| = \frac{(1-\mu^2)(1+\mu\lambda^2)}{12(1+\lambda^4)} \cdot qa^2, \tag{15}$$

where μ is the Poisson's ratio of the key stratum; q is the load carried by the key stratum, MPa; λ is the geometric shape coefficient of the key stratum, where $\lambda_1 = a/b$; a is the distance from the key stratum to the dew, m; b is the suspended span, m.

Consider the relation between stress and bending moment:

$$M_a = \frac{h^2\sigma_t}{6}, \tag{16}$$

where h is the key stratum thickness, m; σ_t is the tensile strength of the key stratum, MPa.

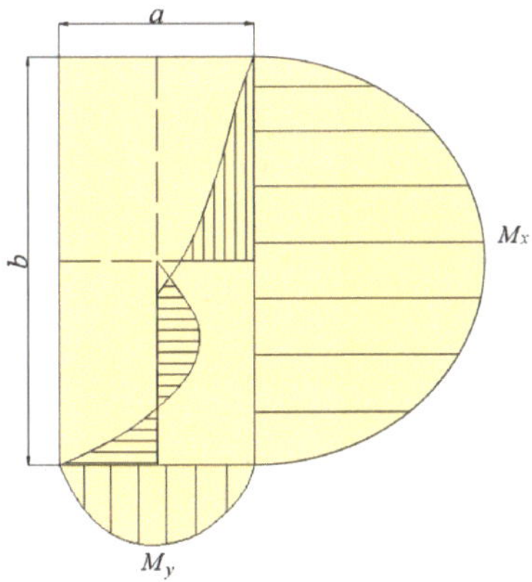

Figure 4. Bending moment distribution of four-edge clamped thin plate.

Substituting Equation (16) into Equation (15), the suspended span in the strike direction of the key stratum at the time of its initial fracture under four-edge fixed boundary condition is obtained as follows:

$$a_{jT} = \frac{h}{1-\mu^2}\sqrt{\frac{2\sigma_t}{q}\cdot\frac{1+\lambda^4}{1+\mu\lambda^2}}. \tag{17}$$

When the key stratum undergoes initial fracture, $a_j = a_{jT}$. The connection between the suspended span of the key stratum at the initial fracture and the working face advance distance can be derived from Figure 4, as shown in Equation (18).

$$a_{jT} = l_j - 2H_j \cot\beta \tag{18}$$

where H_j is the distance from the coal seam to the bottom of the key stratum, and β is the overburden fracture angle, which is generally 60°.

Substituting Equation (18) into Equation (17), the expression for the working face advance distance when the key stratum's initial fracture is obtained, as shown in Equation (19):

$$l_{jT} = \frac{h_j}{1-{\mu_j}^2}\sqrt{\frac{2\sigma_{tj}}{q_j}\cdot\frac{1+\lambda_j^4}{1+\mu_j\lambda_j^2}} + 2H_j\cot\beta. \tag{19}$$

As mining continues, the corresponding jth key stratum also has a sufficient suspended span to fracture. Due to the fracturing of the fractured rock mass, the free space height below the overlying key stratum decreases continuously as the fractured rock mass caves. When the free space height becomes less than the maximum deflection of the key stratum, the free space height below the key stratum will be insufficient for it to fracture, and the overlying strata fracturing will be obstructed by this key stratum. Thereafter, even if the working face advancement distance continues to increase, the rock stratum above this key stratum will no longer fracture [42].

3.3. Method for Determining the WFZDH

By analyzing the maximum bending subsidence deformation within the key stratum mechanical model and the critical span at the initial fracture of the key stratum, a WFZDH determination method is proposed:

Step 1: According to the rock stratum assignment conditions, combined with the key stratum discrimination method [15], the key strata positions above the coal seam can be obtained. Meanwhile, the load borne by each key stratum and their respective stratigraphic positions were also calculated. The key strata were then sequentially numbered from bottom to top as key stratum j (j = 1, 2, . . .).

Step 2: By considering the information on the working face dimensions and advance distance, in combination with the parameters of the key stratum, such as tensile strength, thickness, and the load it bears, the relationship between the suspended span and its initial fracture distance was determined. If the suspended span of a key stratum exceeds its initial fracture distance, the key stratum can undergo fracturing, enabling the next step of the evaluation to proceed.

Step 3: By sequentially substituting the bearing load, thickness, and Poisson's ratio of each key stratum into Equation (14), the maximum bending subsidence deformation of each key stratum was calculated. Meanwhile, the height of the free space below each key stratum was calculated using Equation (1). If the free space height below a key stratum is greater than its maximum deformation, it indicates that the key stratum has not fractured, and the WFZ will continue to develop upwards. Conversely, if the free space height is less than the maximum deformation, the key stratum will not fracture, and the WFZDH will be at the bottom of that key stratum.

There are two applicable conditions for the application of this discriminative method: Firstly, this method is primarily intended for determining the WFZDH in the working face with boundary conditions of four-sided fixed support. Secondly, it requires a comprehensive understanding of the thickness, lithology, and mechanical parameters of the strata. The WFZDH determination procedure is shown in Figure 5.

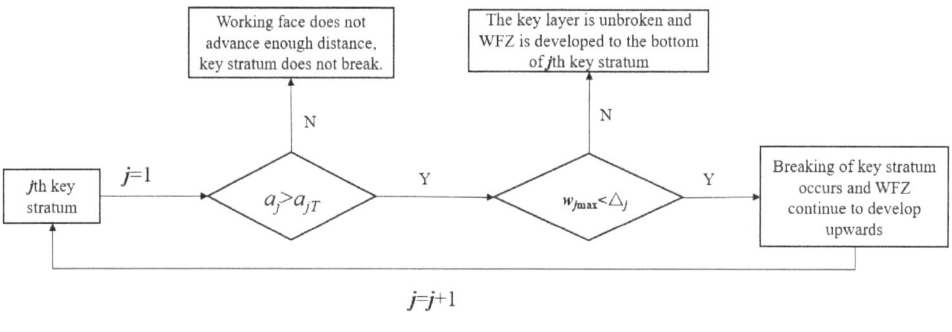

Figure 5. Flowchart for determining the WFZDH.

4. Example Analysis Validation

4.1. Overview

The 3# coal seam of Yeping Mine is a near-horizontal coal seam with a thickness of about 0~4 m and an average thickness of 3 m. It is buried at a depth of around 600 m. The 1301 working face is the initial mining face within this mining area. The inclined length is 180 m, and the mineable length is 1321 m. The overlying strata above the 3# coal seam contain a relatively large number of aquifers. Notably, the aquifer of the Luohe Formation, located approximately 120 m above the 3# coal seam, exhibits strong water abundance. The working face layout and the lithological characteristics of the overlying rocks are shown in Figure 6.

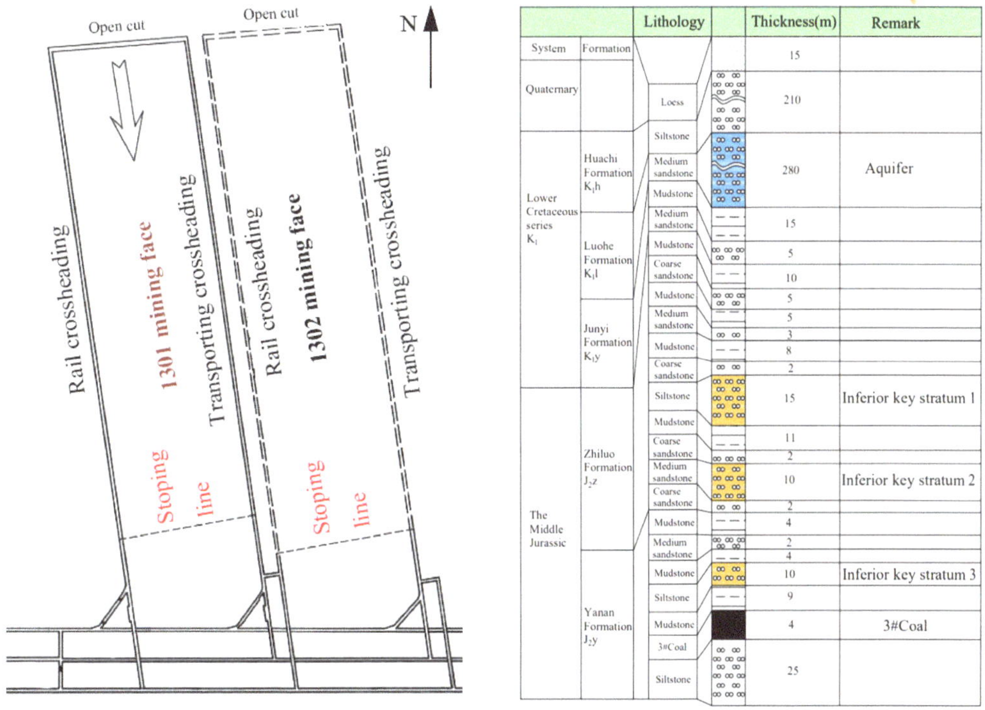

(**a**) (**b**)

Figure 6. (**a**) Working face layout; (**b**) stratigraphic section above the 3# coal seam.

4.2. Theoretical Calculation of the WFZDH

Using the key stratum theory [15], there are three inferior key strata between the aquifer and the 3# coal seam, which are the second stratum of siltstone, the seventh stratum of medium sandstone, and the tenth stratum of siltstone. Distances from the 3# coal seam are 9 m, 31 m, and 54 m, respectively. The thin plate theory was used to analyze the WFZDH. Table 1 shows the specific calculation results.

Table 1. Analysis and calculation results (a_j is the suspended span; a_{jT} is the ultimate span at the initial failure of key strata; $w_{\max}$ is the maximum deflection of key strata.).

Lithology	Thickness /m	Distance to Coal Seam/m	a_j /m	a_{jT} /m	$w_{\max}$ /m	△ /m	Break or Not
siltstone	10	9	34	45.23	2.47	3.82	Yes
medium sandstone	10	31	72	108.45	1.73	3.38	Yes
siltstone	15	54	94	156.78	5.5	2.92	No

From the results in Table 1, the maximum deflection of the inferior key stratum 3 is 5.5 m, which exceeds the free space height of 2.92 m underneath it. At this point, the void below inferior key stratum 3 has been filled, and no further bending deformation of inferior key stratum 3 is allowed. Therefore, inferior key stratum 3 will not be damaged, and the water-conducting fissures will be impeded by inferior key stratum 3. Consequently, inferior key stratum 3 is the lower boundary layer of the bending subsidence zone, indicating that the WFZDH in the 1301 working face is 54 m.

4.3. Numerical Simulation Analysis of WFZDH

4.3.1. Modeling

For extending the UDEC two-dimensional discrete medium description to three dimensions, the 3DEC (v.7.0) procedure is used. It can describe the basic constituent objects of a rock body, i.e., rock masses and structural surfaces. Also, it has a rich library of structural material models for geotechnical bodies, which is especially suitable for studying the motion, deformation, and damage characteristics of discrete media. To study the destruction of overburdened rock and the development of fractures as the working face advances, we used 3DEC to carry out simulation and analysis. The numerical simulation method is more complex and comprehensive, providing richer information and better visualization of results, as well as allowing for comparison between theoretical analysis and field measurement parts of this study.

In order to facilitate numerical model calculations based on the geological conditions of the Yeping Coal Mine, thinner rock layers were merged and simplified. Using 3DEC simulation software, a calculation model of 400 × 280 × 120 m (strike × tendency × height) was established, in which the working face size was 300 × 180 × 4 m (strike × tendency × height), as shown in Figure 7. The numerical model selected was a Mohr–Coulomb model for the block and the contact Coulomb slip model for the joint. Based on field drilling data and indoor test results, physical and mechanical parameters were determined, as shown in Table 2.

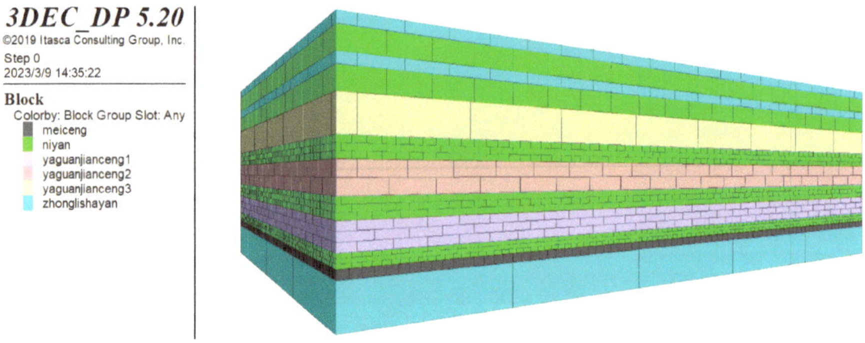

Figure 7. Numerical model diagram.

Table 2. Coal rock mechanics parameter table. D is the density; K is the bulk modulus; G is the shear modulus; C_m is the cohesion; σ_{tm} is the tensile strength; φ_m is the friction angle.

Lithology	Thickness/m	D/(kg·m^{-3})	K/GPa	G/GPa	φ_m/(°)	C_m/MPa	σ_{tm}/MPa
Medium sandstone	5	2.4	26.7	7	35	3	4
Mudstone	9	2.5	18.9	5	25	1.5	3
Medium sandstone	5	2.4	26.7	7	35	3	4
Mudstone	8	2.5	18.9	5	25	1.5	3
inferior key stratum 3	14	2.5	26.7	7	35	3	4
Mudstone	12	2.5	18	5	25	1.5	3
inferior key stratum 2	12	2.4	26.7	7	35	3	4
Mudstone	12	2.5	18	5	25	1.5	3
inferior key stratum 1	10	2.6	26.7	7	15	4	4
Mudstone	9	2.5	20	5	25	1.5	3
3#coal seam	4	1.5	13.2	10	20	1.2	2.6
Medium sandstone	20	2.5	18.9	7	35	3	4

To minimize boundary effects, 50 m coal pillars were left around each side of the 1301 working face. Each step of excavation was 5 m, with a mining height of 4 m, for a total of 60 steps of excavation. When the rock formation in the model breaks and gradually moves to a steady state, the model can be considered to have converged at this point.

Horizontal displacement constraints were applied around the perimeter boundary of the numerical model while constraining vertical displacements at the bottom boundary. The upper boundary was an unconstrained boundary. The numerical model height is only 120 m, while the depth of cover is 600 m, so it is necessary to apply vertical stress. The stress calculation expression is given in Equation (20):

$$P = \gamma H, \tag{20}$$

where γ is the average bulk density of the overlying rock layers, and H is the overlying rock thickness (m).

The result calculated according to Equation (20) is 12.96 MPa.

4.3.2. Simulation Result Analysis

The development of the overlying fissures under different advancing distances is shown in Figure 8.

According to Figure 8, when advancing to 40 m, inferior key stratum 1 reaches the first weighting distance and experiences its first weighting. The fractures continue to propagate upwards. As the working face advances to 60 m, inferior key stratum 1 undergoes periodic weighting. When the working face reaches 110 m, inferior key stratum 2 starts to experience its first weighting, and the overlying strata controlled by inferior key stratum 2 move synchronously. As advancing, inferior key stratum 2 undergoes periodic weighting, causing the fractures in the overlying soft rock to continuously expand. Until the working face reaches 300 m, inferior key stratum 3 does not experience weighting. At this point, it can be assumed that the fractures in the overlying strata no longer propagate upwards. The WFZ has developed up to the bottom of inferior key stratum 3, with an estimated height of 55 m.

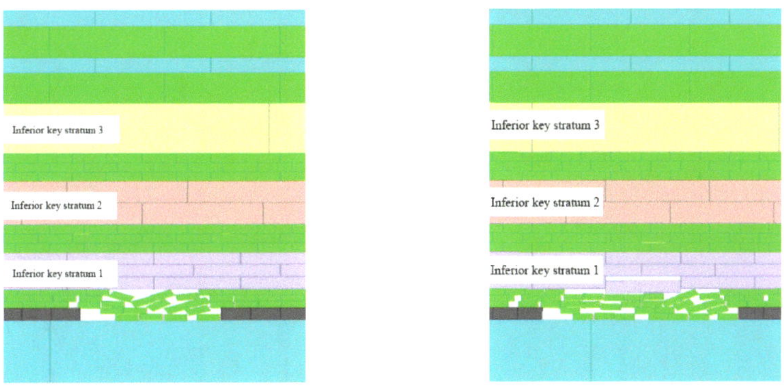

(a) (b)

Figure 8. *Cont.*

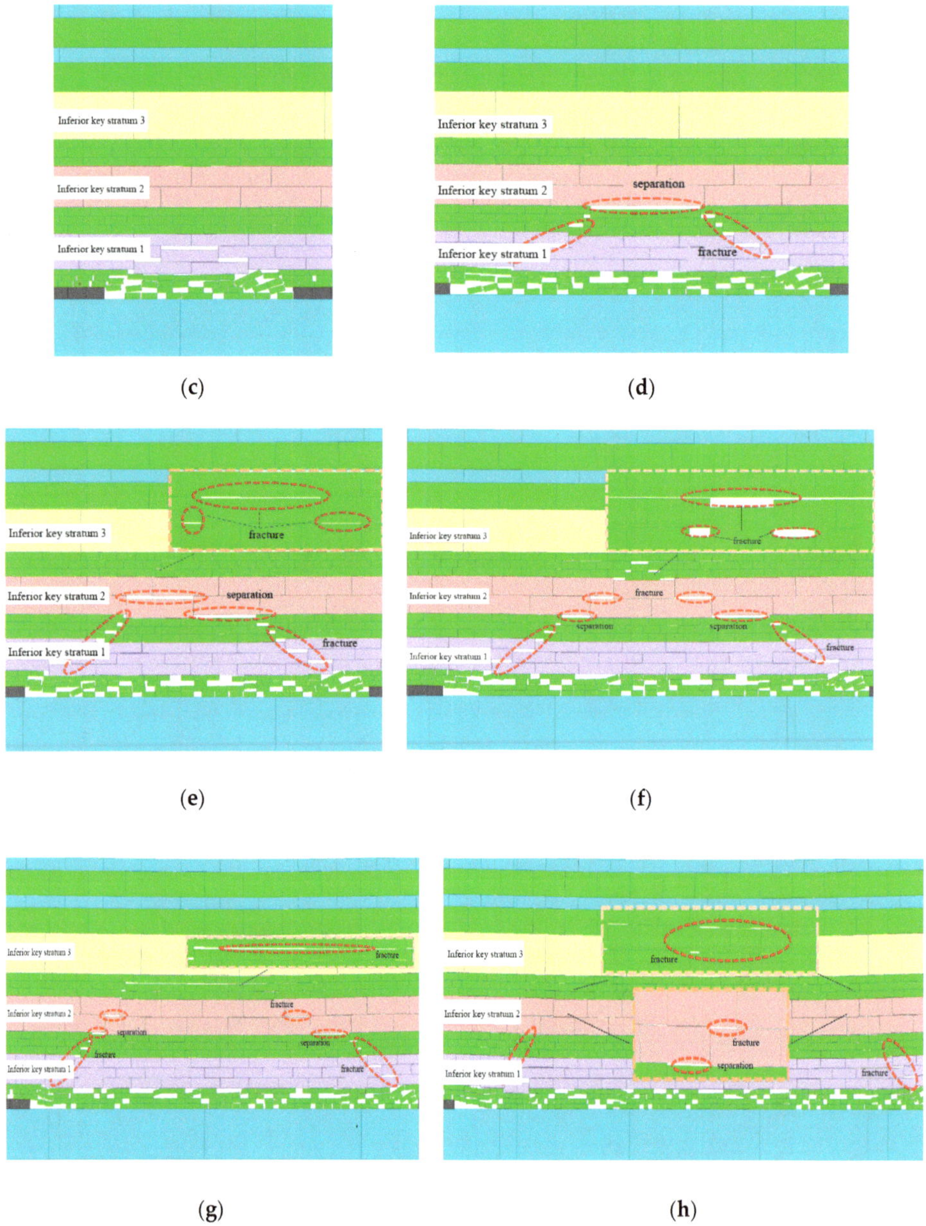

Figure 8. Fracture development of overburden rock: (**a**) advancing 20 m, (**b**) advancing 40 m, (**c**) advancing 60 m, (**d**) advancing 105 m, (**e**) advancing 110 m, (**f**) advancing 150 m, (**g**) advancing 200 m, and (**h**) advancing 300 m.

4.4. Field Measurements

4.4.1. Observation Program Design

This field measurement adopts the double-end water plugging device observation method and the Borehole TV method. The principles of these field detection methods are shown in Figure 9, and the detection equipment is shown in Figure 10. Field measurements can directly reflect the actual situation, and the data are highly credible, which is conducive to an in-depth understanding of the complexity of the problem. The measured data can be used to verify the accuracy of other methods. The observation locations were selected 200 m ahead of the 1302 rail crossheading. Three monitoring stations were set up, with each station having a 90 mm diameter, 75 m-deep observation borehole drilled using a drilling rig. To facilitate the movement of the drilling rig, the spacing between the three monitoring stations was set at 10 m. The borehole design parameters are shown in Table 3, and the drilling design diagram is shown in Figure 11.

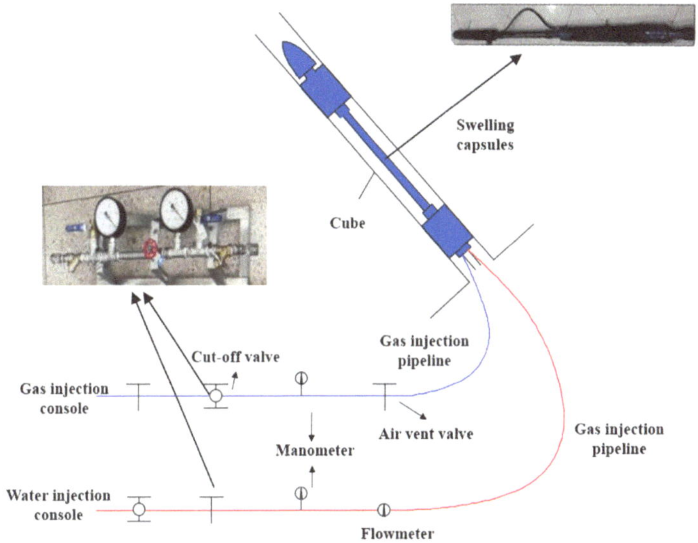

Figure 9. Detection system layout.

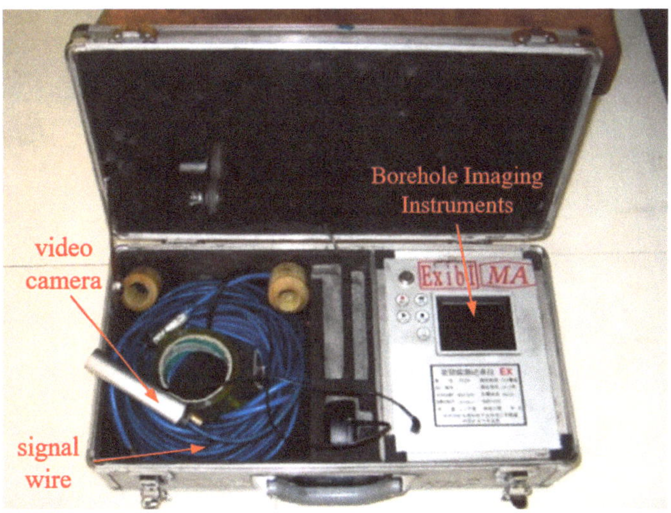

Figure 10. Detection equipment of borehole TV.

Table 3. Borehole design parameter table.

Borehole	Aperture/mm	Hole Length/m	Vertical Height/m	Angle of Elevation (°)	Axial Relationship with Roadway
1	90	75	60	55	It is perpendicular to the axial direction of the roadway and faces the goaf of 1301 working face.
2	90	75	60	60	
3	90	75	60	65	

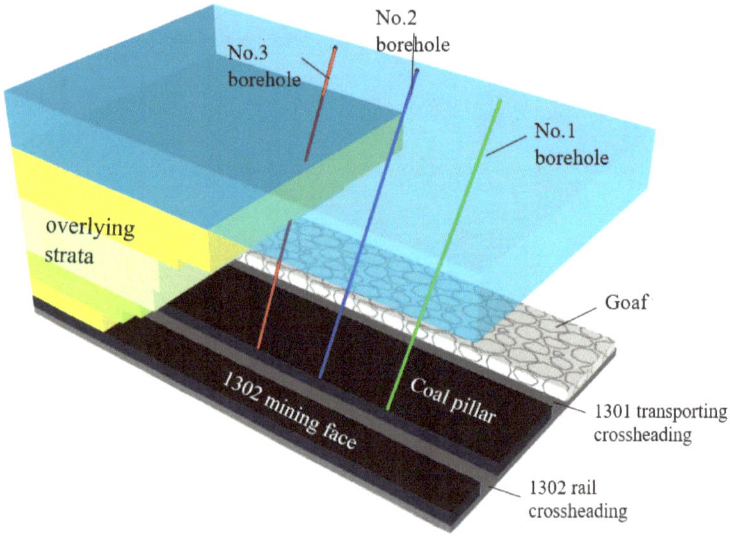

Figure 11. Borehole layout.

In addition to the quantitative analysis of overburden deformation using the double-ended plugger, direct observation and photographic documentation of overburden fracture development were conducted using borehole TV detection equipment (Zhongkuang Huatai Technology Development Co., LTD, Xuzhou, China).

4.4.2. Results and Analyses of Observations

The specific measurement process is shown in Figure 12.

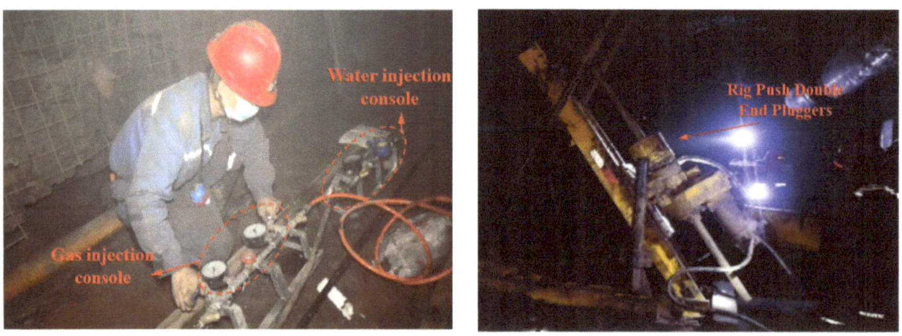

Figure 12. Control process of a specific measurement.

The observation results of the boreholes are shown in Figure 13. The variation in leakage across the three boreholes exhibits similar trends. Therefore, this analysis will focus on the variation in leakage and the reasons in Borehole 1 as a representative case. In the borehole section with a depth of 18~38 m, the overall leakage was slightly small, generally not exceeding 10 L/min, indicating that secondary fissures were not well-developed within this zone and that the strata were less affected by mining activities, maintaining relatively high integrity. The predominant fractures in this interval were the original, inherent fractures in the rock mass. However, in the borehole interval of 40–68 m in depth, the leakage during drilling increased sharply to 13–15.8 L/min. This significant increase in leakage suggests that this zone had been heavily impacted by the mining-induced deformation. The enhanced interconnectivity of these secondary fractures led to the formation of a highly permeable fracture network within this interval of the overburden strata. Therefore, the drilling data indicates that this deeper zone had been penetrated by the borehole, entering the WFZ.

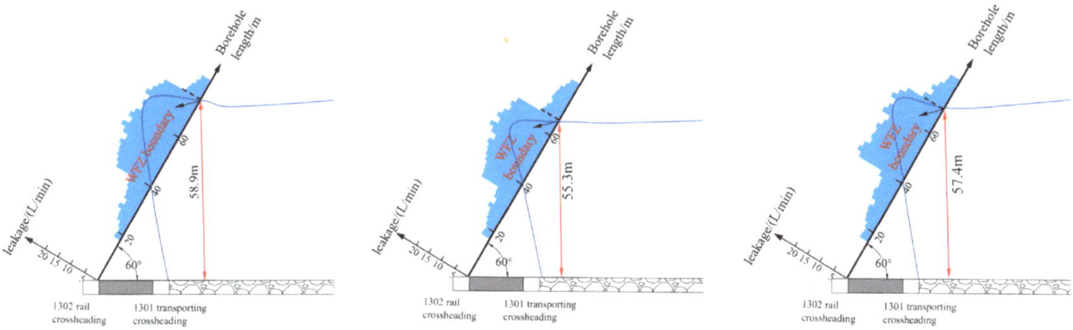

(**a**) (**b**) (**c**)

Figure 13. Changes in water leakage in different areas of boreholes: (**a**) Borehole 1, (**b**) Borehole 2, and (**c**) Borehole 3.

After 68 m, the leakage was 5.2~7.4 L/min, indicating that the overall permeability of the strata in this zone was relatively low. This suggests that the fractures were reduced in this interval compared to the zone above. Therefore, it can be considered that the boundary of the WFZ was at a depth of 68 m in the drill hole, making the WFZDH 58.9 m. The leakage in Borehole 2 and Borehole 3 is similar to that of Borehole 1, and the WFZDHs can be obtained from the analysis as 55.3 m and 57.4 m.

Figure 14 shows observations of the mining overburden fissures through Borehole 1. According to the pictures from the field observations, no fissures were present in the overburden until the hole depth was 30 m, and slight fissures started to appear at 30 m. At 36 m and 45 m, the rock stratum was seriously damaged. From 63 m onwards, the damage to the rock layer began to weaken, and the fissures decreased until 72 m when no obvious fissures appeared in the rock mass. It can be assumed that the WFZ develops to a depth of 63 m in the borehole, at which the WFZDH is 54.5 m.

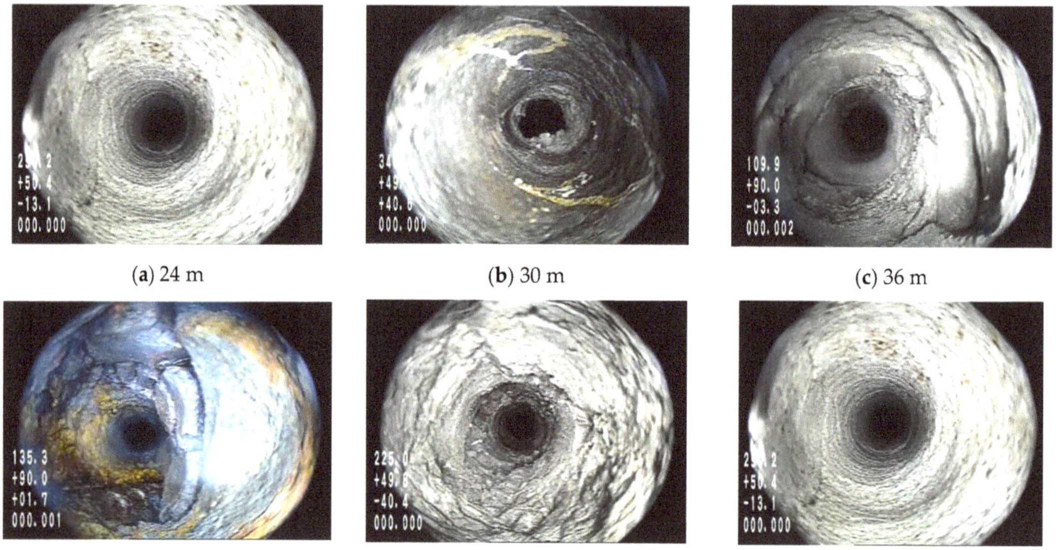

(a) 24 m (b) 30 m (c) 36 m

(d) 45 m (e) 63 m (f) 72 m

Figure 14. Borehole 1 overburden fracture development: (**a**) borehole depth 24 m; (**b**) borehole depth 30 m; (**c**) borehole depth 36 m; (**d**) borehole depth 45 m; (**e**) borehole depth 63 m; (**f**) borehole depth 72 m.

4.5. Comprehensive Analysis

The WFZDH was simulated using 3DEC at 55 m. Through the field measurement, the WFZDH of the 1301 working face is 55.3 m~58.9 m. It can be found that the results of theoretical analysis (54 m) and numerical simulation (55 m) are close to the measured results (55.3 m~58.9 m), which verifies the reasonableness and practicability of the analysis method of the WFZDH based on thin plate theory.

Through further analysis, it can be found that the result of theoretical analysis is smaller than the measured result. However, we found that in the literature [28–30], the results obtained from the theoretical analysis are greater than field measurement results. There are two main reasons for this: First, in conducting the theoretical analysis, we considered the boundary conditions at the working face to be four-sided fixed support. Under this condition, the theoretical analysis results may be underestimated. Second, field measurements were taken two months after the working face had passed through that location, by which time the WFZDH had already reached its maximum value. Therefore, the results from the theoretical calculations would be smaller than the field measurement results.

5. Conclusions

A theoretical discrimination method for the WFZDH based on thin-plate theory was proposed following a study of the 1301 working face at Yeping Coal Mine. The following conclusions were drawn:

(1) Key stratum failure needs to satisfy two conditions: the suspended span of the key stratum must be greater than the critical span at which the key strata first fracture, and the free space height below the key stratum must be greater than its maximum deflection. The deflection variation of the key stratum before its initial failure was analyzed, as well as the relationship between the suspended span of the key stratum and the advancement distance of the working face. Based on thin-plate theory, the theoretical discrimination method for the WFZDH was proposed.

(2) According to the geological information at Yeping Coal Mine, the proposed theoretical discrimination method was applied to calculate the WFZDH development in the 1301 working face. The field measurement results are similar to theoretical discrimination method, which verifies the reasonableness and practicability of the discrimination method.
(3) The method only requires knowledge of the geological information above the working face and a simple calculation can be used to obtain the WFZDH. However, this method does not consider the influence of factors such as possible pre-existing fissures in the overlying rock layers on the WFZDH, and there will be a certain degree of error between its calculation results and the actual values. At the same time, this research has some limitations, and it is mainly for the first mining or a single working face. Further research is needed on the WFZDHs for subsequent working faces and after multiple working faces have been mined.

Author Contributions: F.W. and Z.G. conceived and designed the study. H.L. and X.Y. were involved in theoretical modeling and analysis. H.G. was involved in the Field measurement and prepared all figures. All authors have read and agreed to the published version of the manuscript.

Funding: This work was supported by the National Natural Science Foundations of China (Grant No. 52174138).

Institutional Review Board Statement: Not applicable.

Informed Consent Statement: Not applicable.

Data Availability Statement: All data used to support the findings of this study are available from the corresponding author upon request.

Acknowledgments: The authors acknowledge the support of Wei Hu and Haoliang Jiang, who collaborated in the research.

Conflicts of Interest: The authors declare no conflicts of interest.

References

1. Fan, K.; Li, W.; Wang, Q.; Liu, S.; Xue, S.; Xie, C.; Wang, Z. Formation mechanism and prediction method of water inrush from separated layers within coal seam mining: A case study in the Shilawusu mining area. *Eng. Fail. Anal.* **2019**, *103*, 158–172. [CrossRef]
2. Gui, H.; Lin, M.; Song, X. Identification and application of roof bed separation (water) in coal mines. *J. Mine Water Environ.* **2018**, *37*, 376–384. [CrossRef]
3. Ma, D.; Cai, X.; Li, Q.; Duan, H. In-situ and numerical investigation of groundwater inrush hazard from grouted karst collapse pillar in longwall mining. *J. Water.* **2018**, *10*, 1187. [CrossRef]
4. Pang, M.; Zhang, T.; Guo, Y.; Zhang, L. Re-crushing process and non-Darcian seepage characteristics of broken coal medium in coal mine water inrush. *Sci. Rep.* **2021**, *11*, 11380. [CrossRef] [PubMed]
5. Zhao, Z.-G.; Zhang, Y.-J.; Li, C.; Wan, Z.; Li, Y.-N.; Wang, K.; Xu, J.-F. Monitoring of coal mine roadway roof separation based on fiber Bragg grating displacement sensors. *Int. J. Rock Mech. Min. Sci.* **2015**, *74*, 128–132. [CrossRef]
6. Yao, X.L.; Reddish, D.J.; Whittaker, B.N. Non-linear finite element analysis of surface subsidence arising from inclined seam extraction. *Int. J. Rock Mech. Min. Sci. Geomech. Abstr.* **1993**, *30*, 431–441. [CrossRef]
7. Chen, S.G.; Guo, H. Numerical simulation of bed separation development and grout injecting into separations. *Geotech. Geol. Eng.* **2008**, *26*, 375–385. [CrossRef]
8. Nie, W.J.; Zhao, X.G.; Du, H.D.; Cao, W.C.; Song, S.J.; Tian, D. Effects of different water conducting fissures on surface soil properties in Yushenfu mining area. *J. Xi'an Univ. Sci. Technol.* **2021**, *41*, 464–473.
9. Wang, S.M.; Shen, Y.J.; Sun, Q.; Hou, E.K. Scientific problems and technical prospect of geological support for coal loss reduction mining in ecologically fragile areas of Western China. *J. Min. Rock Strat. Control Eng.* **2020**, *2*, 5–19.
10. Guo, X. Study on Laws of Movement and Damage of Overlying Strata and Development of Water-Flowing Fractured Zone in Yushenfu Mining Area. Master's Thesis, China University of Geosciences, Beijing, China, 2020.
11. Lin, Q.; Qiao, W. Water prevention and control technology of roof bed separation in Cuimu Mine. *Coal Sci. Technol.* **2016**, *44*, 129–134.

12. Jiang, J.Q.; Wang, P.; Wu, Q.L.; Zhang, P.P. Evolution laws and prediction of separated stratum space under overlying high position magmatic rocks. *Chin. J. Geotech. Eng.* **2015**, *37*, 1769–1779.
13. Zhang, X.C.; Pan, X.; Liu, J.T.; Zhou, P. EW416 full mechanized face roof abscission layer water genetic analysis in Daming Coalmine, Tiefa Coalfield. *Coal Geol. China* **2018**, *30*, 48–51.
14. Sun, Y.J.; Zuo, J.P.; Li, Y.B.; Liu, C.H.; Li, Y.H.; Shi, Y. Micro-seismic monitoring on fractured zone and water inrush mechanism analysis of deep mining above aquifer in Xingdong coalmine. *Rock Soil Mech.* **2017**, *38*, 2335–2342.
15. Qian, M.G.; Xu, J.L.; Wang, J.C.; Wu, Y.P. *Rock Pressure and Strata Control in Mines*, 3rd ed.; China University of Mining and Technology Press: Xuzhou, China, 2021.
16. Wang, F. Study on the Theory and Application of 'Key Stratum-Loose Stratum Arch' of Mining Overburden Structure. Ph.D. Thesis, China University of Mining & Technology, Xuzhou, China, 2016.
17. Liang, Y.P.; Li, B.; Yuan, Y.; Zhou, Q.L.; Jia, L.X. Moving type of key strata and its influence on ground pressure in fully-mechanized mining face with large mining height. *J. China Coal Soc.* **2017**, *42*, 1380–1391.
18. Xu, J.L.; Ju, J.F. Structural morphology of key stratum and its influence on strata behaviors in fully-mechanized face with super-large mining height. *Chin. J. Rock Mech. Eng.* **2011**, *30*, 1547–1556.
19. Gong, P.L.; Jin, Z.M. Study on the structure characteristics and movement laws of overlying strata with large mining height. *J. China Coal Soc.* **2004**, *29*, 7–11.
20. Yu, B.; Gao, R.; Meng, X.B.; Kuang, T.J. Near-far strata structure instability and associate strata behaviors in large space and corresponding control technology. *Chin. J. Rock Mech. Eng.* **2018**, *37*, 1134–1145.
21. Yi, Y.J. Study on the development characteristics of overburden water-flowing fracture height in No.8 coal seam of Baode Coal Mine. *China Min. Mag.* **2023**, *32*, 121–126.
22. Chen, Z.H.; Xie, H.P.; Li, Q.S. Study on plate group mechanical model for main roof of longwall face. *J. China Coal Soc.* **2005**, *30*, 172–176.
23. Zhong, Y.; Yin, J.H. Theoretical Solution of Elastic Rectangular Thin Plate with Opposite Boundary Completely Clamped Support and Others Free. *J. Chongqing Jianzhu Univ.* **2005**, *27*, 29–32.
24. Song, Y.Q.; Yang, M.J.; Ma, H.F.; Zheng, J.J.; Shen, F.X. Study on deformation law and breaking span of main roof of isolated working face based on thin plate theory. *J. Saf. Coal Mines* **2022**, *53*, 201–207.
25. Yu, H. Numerical investigation on the basic roof weighting step of the first mining face based on thin plate theory. *Sci. Technol. Eng.* **2018**, *18*, 195–199.
26. Huang, C.; Hua, X. Using elastic thin plate theory to predict main roof weighting pace at mining face. *Saf. Coal Mines* **2013**, *44*, 206.
27. Liu, Z.C.; Li, W.L. Sheet model analysis of roof rupture in island working face. *Min. Saf. Environ. Prot.* **2014**, *41*, 104–106.
28. Liu, X.B.; An, L.; Zhang, F.P. Analysis on roof stability of gob area based on thin plate theory. *J. Northeast. Univ. (Nat. Sci.)* **2012**, *33*, 1628–1632.
29. Zhang, Y.D.; Zhang, C.; Fan, Z.Q.; Wang, X.X.; Jin, Z.Y. Analysis of fracture mechanism of main key strata in isolated working face based on plate model. *Coal Eng.* **2010**, *9*, 59–62.
30. Lin, H.F.; Li, S.G.; Cheng, L.H.; Zhang, W. Key layer distinguishing method of overlying strata based on the thin slab theory. *China Coal Soc.* **2008**, 1081–1085. [CrossRef]
31. Wu, F.F.; Yang, J.X.; Yu, B.; Chen, X.H. Determination of the roof caving heights of thick and extra thick coal seams. *China Univ. Min. Technol.* **2014**, *43*, 765–772.
32. Wang, X.; Lu, M.; Wei, R.; Wang, Z.; Li, S. Mechanical model of fracture mechanism of stope roof for working face with variable length. *J. China Univ. Min. Technol.* **2015**, *44*, 36–45.
33. Tu, H.; Tu, S.; Chen, F.; Wang, C. Study on the deformation and fracture feature of steep inclined coal seam roof based on the theory of thin plates. *J. Min. Saf. Eng.* **2014**, *31*, 49–54.
34. Yang, S.L.; Wang, J.C.; Li, L.H. Analysis of deformation and fracture characteristics of key strata based on the medium thick plate theory. *J. China Coal Soc.* **2020**, *45*, 2718–2727.
35. Chang, X.; Wang, M.; Zhu, W. Study on Height Development Characteristics of Water-Conducting Fracture Zone in Fully Mechanized Mining of Shallow Thick Coal Seam under Water. *Sustainability* **2023**, *15*, 7370. [CrossRef]
36. He, X.; Zhao, Y.; Zhang, C.; Han, P. A Model to Estimate the Height of the Water-conducting Fracture Zone for Longwall Panels in Western China. *Mine Water Environ.* **2020**, *39*, 823–838. [CrossRef]
37. Guo, W.; Zhao, G.; Lou, G. A New Method of Predicting the Height of the Fractured Water-Conducting Zone Due to High-Intensity Longwall Coal Mining in China. *Rock Mech Rock Eng.* **2019**, *52*, 2789–2802. [CrossRef]
38. Xu, D.; Peng, S.; Xiang, S.; He, Y. A novel caving model of overburden strata movement induced by coal mining. *Energies* **2017**, *10*, 476. [CrossRef]
39. Xu, Z.L. *Elastic Mechanics*; People's Education Press: Beijing, China, 1982; Volume II.
40. Chen, L.; Wu, B.; Xu, X.K.; Shang, R.Y. Determination of overburden failure height in alternate strata of mudstone and sandstone with fully mechanized caving method. *J. Min. Saf. Eng.* **2017**, *34*, 431–436+443.

41. Zhao, G.; Guo, W.; Lou, G.; Ma, Z. Study on the development height of water-conducting fracture zones based on overburden damage transfer. *Coalf. Geol. Explor.* **2019**, *47*, 144–150.
42. Guo, W.; Lou, G. Definition of the degree of sufficient mining for overburden destruction and discrimination method. *J. Coal* **2019**, *44*, 755–766.

applied sciences

MDPI

Article

Study on Shear Failure Process and Zonal Disintegration Mechanism of Roadway under High Ground Stress: A Numerical Simulation via a Strain-Softening Plastic Model and the Discrete Element Method

Peiju Yang, Shurong Zhang * and Changyou Liu

School of Mines, China University of Mining and Technology, Xuzhou 221000, China
* Correspondence: srzhang@cumt.edu.cn

Abstract: Fracture expansion in rock masses can be observed by monitoring the break of contacts between the bounding particles via the discrete element method. The latter's realization in this study via the PFC^{2D} program tracked the evolution process of the zonal disintegration in an exemplary roadway-surrounding rock affected by mining. Besides, the damage evolution pattern in a high-stress soft rock roadway was simulated by the FLAC^{2D} program using a strain-softening plastic model, revealing the effects of rock mass strength, stress state, and anchor support on the zonal disintegration of the roadway. Numerical simulation results show that in a roadway with high-level stress, the obvious fractures spread from the roadway surface to the depth of the surrounding rock along a series of geometric planes and cut the surrounding rock into rock mass blocks. Under high crustal stress, conjugate shear fractures occur near the roadway surfaces and form a closed-loop fractured zone after intersecting the conjugate fracture faces. The closed fractured zone becomes a free face, from which conjugate shear fractures develop, forming new closed fractured zones in the deep surrounding rock. By repeatedly generating the closed fracture zones, a fracture network appears in the roadway-surrounding rock. The development of zonal disintegration of roadway-surrounding rock mainly depends on the rock mass strength and its stress state. Zonal disintegration only occurs when the crustal stress of the roadway-surrounding rock exceeds its strength. When the horizontal stress is low and the vertical stress exceeds the rock mass strength, zonal disintegration only occurs on two sides of the roadway. When the vertical stress is low and the horizontal stress exceeds the rock's mass strength, it only appears on the roof and floor. When the values of cohesion, internal friction angle, and tensile strength are reduced in the same proportion, cohesion has the greatest impact on the expansion of the zonal disintegration zone, followed by the internal friction angle, while the tensile strength effect is the least. In anchor-supported roadways undergoing zonal disintegration processes, the intact zone blocks slide relatively along the fracture surface during the process of loosening and deformation of the surrounding rock, making the anchor rods susceptible to tensile, shear, and bending actions.

Keywords: 1roadway surrounding rock; zonal disintegration; strain-softening model; compression–shear failure; slip line

1. Introduction

The surrounding rocks in the roadway may show zonal disintegration characteristics at a great burial depth or under high stress, i.e., obviously fractured zones and complete zones appear alternately in shallow and deep surrounding rocks [1]. This phenomenon was revealed as early as 1972 by Borisovets [2]. Scholars from South Africa and China also observed zonal disintegration phenomena later in deep mining [3–6]. Zonal disintegration is currently recognized as a common phenomenon in deep roadways [7] and has become a hotspot in rock mechanics and rock engineering. In general, zonal integration is the key feature of roadway damage in deep mining. Gaining in-depth knowledge on its formation mechanisms and patterns can provide insightful references for addressing the control problems of the surrounding rocks in deep roadways [7–9].

Scholars have first investigated zonal integration based on field borehole measurements. By arranging the drilling holes along the periphery of the roadway in a cross-section, the alternate appearance of disintegration and complete zones was confirmed in the drilling holes via borehole TV survey and ultrasonic perspective examination. It was thus derived that the disintegration zone and the complete zones appeared alternately around the roadway, approximately in a circular-ring pattern [4–6]. The related studies were mainly focused on the reproduction of alternate distribution characteristics in the laboratory. Some scholars conducted physical experiments but failed to obtain the obvious zonal disintegration phenomenon [10–12]; others did not obtain annularly-distributed disintegration in the roadway-surrounding rocks from physical analog experiments but found the distribution of the fractures in the section parallel to the roadway [13]. Accordingly, they deduced that the spatial distribution of conjugated shear zones can explain zonal disintegration. Afterward, scholars placed their research emphasis on annular zonal disintegration. They concluded that the approximately annularly zonal disintegration effect was due to the nonuniform settings of the mechanical parameters of the surrounding rocks in the roadway [14]. Gao et al. simulated the development characteristics of the fractures in homogeneous surrounding rocks based on a bilinear strain-softening ubiquitous-joint model in FLAC^{2D}. Nevertheless, they did not observe annular disintegration zones in alternate patterns [15].

Investigating the zonal disintegration patterns of the surrounding rocks in the roadway via physical experiments or numerical simulation was crucial for validating field measurement results and provided the foundation for further analyzing the mechanisms and related influencing factors. Since fractures in the surrounding rocks involved the rock fracturing process, the latter was successfully simulated by the discrete element method. Based on the research results in roadway support engineering, this study first conducted a numerical simulation of the development of zonal disintegration in the surrounding rocks during deformation and fracture processes in high-stress and deep high-stress roadways via PFC^{2D} 5.00 and FLAC^{2D} 7.00 software packages. The performed numerical simulation clarified the effects of the surrounding rock's mechanical parameters, the distribution of stress fields, and the roadway's sectional shape/pattern on zonal disintegration.

2. Zonal Rupture Modeling of Roadway Perimeter Rock under High Stress

2.1. Base Range Modeling of Roadway Perimeter Rock Fragmentation Based on Discrete Element Approach

This study is focused on a belt conveyance haulage way in the working face of the Chahasu mine in Erdos, China, with a burial depth of approximately 650 m and a coal seam depth of 6.0 m. The immediate roof and floor were made up of mudstone with depths of 3.2 and 5.4 m, respectively, while the upper roof and floor were made up of medium-grained sandstone. The rectangular roadway cross-section was 5.8 m in width and 4.0 m in height. The surrounding rocks' deformation and failure process after the roadway's excavation was simulated using the PFC^{2D} (particle flow code in two dimensions) software produced by Itasca Consulting Group, Inc. (Minneapolis, MN, USA), via the established discrete element model shown in Figure 1. The model dimensions were 36.0 m $\times$ 24.0 m. In the numerical simulation, the radii of particles in the coal seam, immediate roof/floor, and upper roof/floor were set at 0.05, 0.1, and 0.2 m, respectively.

A parallel bonding model is used between particles, and bonding bonds are set between adjacent particles. The parallel bonding model is one of the bonding models built into the software, the core implementation of which is to generate a bonding material of a certain size between neighboring particles. When the bonding material is broken, the broken surface of the particles will become a linear stiffness model. In this model, the particles can be similar to springs with constant normal and tangential stiffness, uniformly distributed in a rectangular cross-section on the contact plane centered at the point of contact. These springs act in parallel with the springs of the linear assembly to produce a parallel bond, and then contact occurs where the bond breaks if the maximum stress exceeds the bond strength. It is possible to model the contact forces $\overline{F}_p$ and moments $\overline{M}_p$ expressed as follows:

$$\overline{F}_p = F_n\overline{n}_p + F_s\overline{t}_p \tag{1}$$

$$\overline{M}_p = M_n\overline{n}_p + M_s\overline{t}_p \tag{2}$$

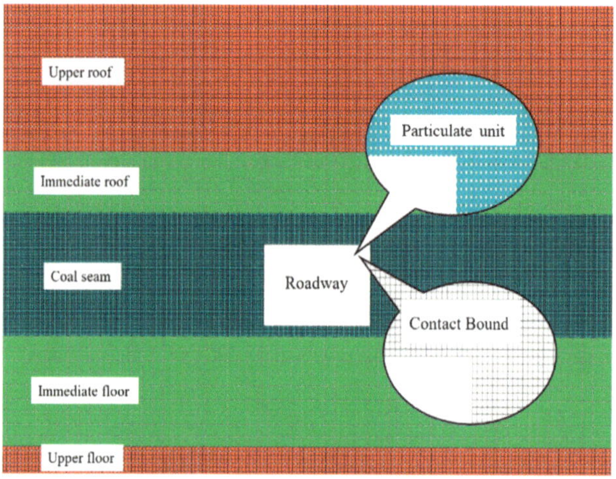

Figure 1. The discrete element model.

In models made using the parallel bonding ontological relationship, a large number of particles are usually bonded to form constructs of multiple shapes, while the mechanical properties of the constructs are described using the holistic approach. The fact that the constructs are composed of multiple particles also allows the constructs to be modeled with tensile, shear, and torsional mechanical characteristics. When the stress on the construct exceeds its corresponding bond strength, the bond relationship between particles will be broken, the bond relationship at the corresponding particle contact will be degraded to a linear relationship, and the construct as a whole becomes multiple smaller constructs that continue to move, as shown in Figure 2 [16].

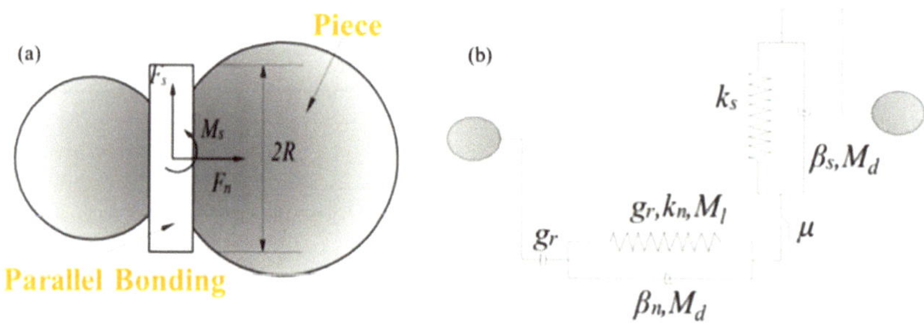

Figure 2. Schematic diagram of the principle of parallel bonding modeling. (**a**) Particle–parallel bonding relationship; (**b**) intrinsic relationship.

The model parameters were verified with reference to the simulation data from the literature [17], which is based on the experimental results of the mechanical parameters of the surrounding rock in the roadway of the Chahasu mine. The specific parameters are listed in Table 1.

Table 1. Statistical table of mechanical experiment results for the coal rock body.

Stratum	Normal Stiffness, GPa	Tensile Strength, MPa	Tangential Stiffness, GPa	Shear Strength, MPa	Frictional Coefficient
Upper roof/floor	2.4	1.0	2.4	5.0	0.35
Immediate roof/floor	1.2	0.3	1.2	2.5	0.25
Coal seam	0.8	0.15	0.8	1.0	0.2

Both sides and the floor obeyed the following displacement boundary conditions: the horizontal displacements were fixed at both sides, while the vertical displacement was fixed at the floor.

At this stage, the load on the roof plate will increase from the original rock stress to the situation of over-head support pressure affected by mining. Combined with the actual data measured in the field, we set the original rock stress of the roof slab to 18 MPa and the over-support pressure to 30 MPa. The original rock stress load was applied when the roadway in the model had not yet been excavated and the model reached an initial equilibrium, and the over-support pressure was applied when the roadway was further excavated.

We monitored the fracture of the bonding bonds between particles in the model, as shown in Figure 3, and used it to represent the fracture process of the roadway perimeter rock.

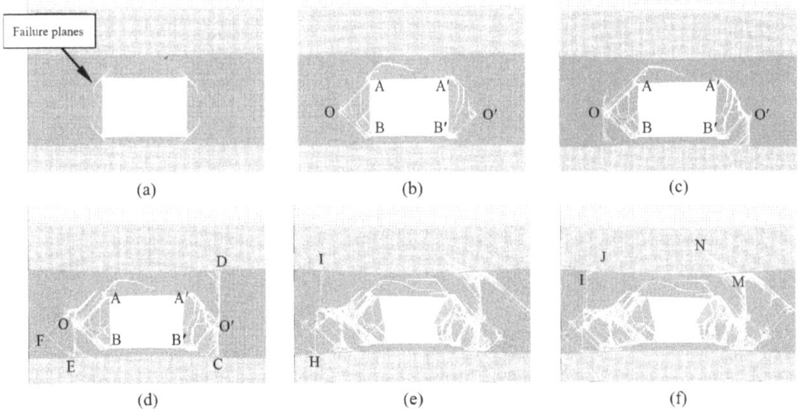

Figure 3. Simulation effect of rupture of roadway perimeter rock zones. (**a–f**) denote the rupture of each stage of the roadway model. Capital letters A, B, C, D, E, F, O, H, I, J, M, and N denote the intersections of the rupture lines during the zonal rupture.

After the rectangular roadway was excavated, the stress concentration was observed at four angular points [18], where fractures also began to appear, as shown in Figure 3a. Next, the fractures extended along two conjugate shear planes to the depth between the two sides and the roof, forming the conjugate fracture planes within two sides, depicted by OA, OB, O′A′, and O′B′ in Figure 3b. On both sides of the roadway, when the conjugate failure surfaces intersected at points O and O′, these surfaces stopped extending forward and started to develop vertically toward the floor (see Figure 3c). After reaching the floor, the fracture planes underwent two different development processes on two sides: the first one implied the development from the intersection point O′ to the roof until the coal-rock interface was reached, forming a vertical failure plane penetrating through the coal seam (see the plane CD in Figure 3d). The second process involved a continued development to the depth via a conjugate failure pattern (see the OF and EF fracture planes in Figure 3d), generating a vertical failure plane at the intersection point F that penetrated through the coal seam (see HI in Figure 3e). After forming the vertical failure plane from the coal seam to the roof/floor, the fractures continued developing toward both sides via a conjugate failure pattern towards the depth of the two sides (see Figure 3f). Besides, upward conjugate failures appeared on the roof (see the IJ and MN fracture planes in Figure 3f).

2.2. Perimeter Rock Damage Model of a Deep, High-Stress Soft Rock Tunnel Based on the Finite Difference Method

2.2.1. Modeling

The destruction process of the surrounding rock in the deep, high-stress soft rock roadway was adopted from the basic geological conditions of the belt transportation roadway in the mining area of the Quandian coal mine in Xuchang City, China. The section size of this roadway is 5.2 m × 4.0 m, the burial depth is about 600 m, and the top and bottom plates of the rock formation consist of low-strength mudstone and sandy mudstone.

Our model parameters were adopted from the data obtained from physical tests. As shown in Figure 4, the researcher collected core samples of the surrounding rock from the belt transportation roadway of the mining area at the Quandian coal mine. The uniaxial compressive test, the Brazilian splitting method tensile test, and the shear test were carried out on these samples using the MTS servo tester and the SYNS pressure shear machine, respectively. The mechanical parameters of the perimeter rock of this roadway were obtained as shown in Table 2.

Figure 4. Testing of mechanical parameters of surrounding rock: (**a**) test specimen; (**b**) servo testing machine; (**c**) shear tester; (**d**) uniaxial compressive test; (**e**) tensile test; and (**f**) shear test.

Table 2. Test results of the mechanical parameters of the surrounding rock.

Stratum	Densities, kg/m^3	Bulk Modulus, GPa	Shear Modulus, GPa	Cohesion, MPa	Internal Friction Angle, °
Mudstone	2300	1.2	0.3	1.0	16

The deformation and fracture processes after roadway excavation were numerically simulated in this study by a model with a size of 80.0 m × 64.0 m, as shown in Figure 5, which was realized via the commercial FLAC^{2D} software package. The roadway vicinity was represented by a refined mesh with a size of 0.2 m × 0.2 m.

Figure 5. Finite difference model.

In the simulations, we used a strain softening model for the reduction. The rock density was set at 2300 kg/m^3, while the bulk and shear moduli were set at 1.2 and 0.3 GPa, respectively. The strain softening model is a commonly used simulation method based on the Moore–Coulomb model that is not associated with the shear flow law but with the tensile flow law, the difference being that the cohesion, friction angle, shear expansion and dilatation, and tensile strength may change after the onset of plastic yielding. In the Moore–Coulomb model, such properties are assumed to remain constant. Users can define their own softening parameters, such as cohesion and friction angle, and set them as segmented linear functions (Figure 6). They can then use these softening parameters to measure plastic shear strains. In this model, we use data obtained from physical tests. This model calculates the total plastic shear and tensile strains by increasing the softening parameter at each time step and, in this way, drives the material properties to be consistent with the user-defined functions. For yield and potential functions, the plastic flow law and stress corrections are identical to the Moore–Coulomb model.

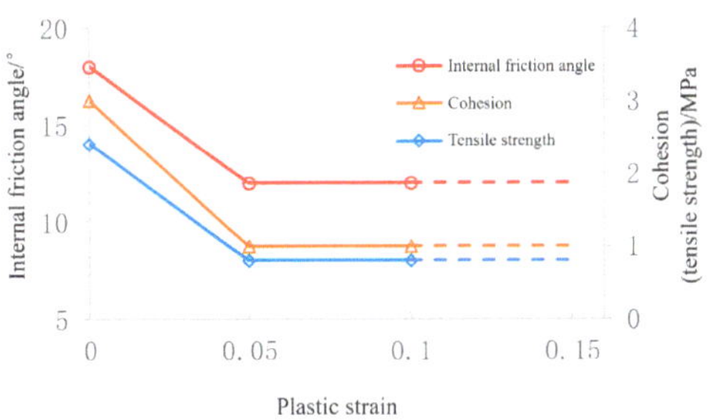

Figure 6. Variations in the tensile strength, cohesion, and internal friction of plastic strain.

Both vertical and horizontal in-situ stresses in the surrounding rock were set at 15.0 MPa. The horizontal displacement of both sides and the vertical displacement of the floor were fixed. The upper boundary was under a pressure of 15.0 MPa.

2.2.2. Simulation Results

Firstly, the calculation was made to bring the model to the initial equilibrium, then the roadway was excavated, and the calculation was continued for the new equilibrium state.

The deformation and damage degrees of the surrounding rock were evaluated via the shear strain increment ($\Delta\gamma$). A larger shear strain increment indicates more serious deformation and damage to the rock [14]. Figure 7 shows the evolution of the shear strain increment of the surrounding rock in the roadway, which was calculated via the proposed model.

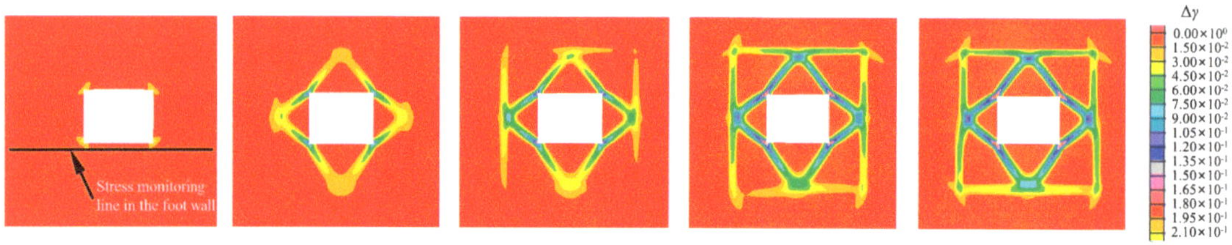

(a) (b) (c) (d) (e)

Figure 7. Evolution of shear strain increment $\Delta\gamma$ in the roadway-surrounding rock. (**a**–**e**) denote the rupture of each stage of the roadway model.

According to the discrete element analysis predictions, the fracturing of the surrounding rocks in the roadway first occurred at four angular points (see Figure 7a) and developed in the form of compression–shear conjugate damage deep into the two sides, the roof, and the floor, until the fractures became intersected (see Figure 7b), thereby forming a tilted rectangular fracture loop. After the conjugate fracture planes intersected, the fractures developed from the intersection point toward the deep surrounding rocks along vertical and horizontal directions (see Figure 7c) until they intersected (see Figure 7d). Accordingly, a new rectangular fracture loop was formed outside the first layer of the fracture loop. Next, the fracture planes developed from the four angular points of the second layer of rectangular fracture loop deep towards the sides, the roof, and the floor in a shear fracture mode (see Figure 7e).

3. Analysis of Results

3.1. Zonal Rupture Pattern of the Roadway Enclosure

According to numerical simulation results, the development of the fracture planes of the surrounding rocks in the roadway followed the compression–shear conjugate fracture patterns of rock mechanics [18]. In the rectangular roadway after excavation, as shown in Figure 8, the shallow surrounding rocks became loose toward the roadway, and the stresses perpendicular to the roadway surface were gradually released. Accordingly, the shallow surrounding rocks underwent bidirectional compression. Therefore, the stress condition of the roadway could be reduced to a plane strain state. The stress conditions on the roof and the floor are shown in Figure 8 by ABCD, while AEFG corresponds to the stress conditions on both sides. When the horizontal σ_H and vertical σ_V stresses in the surrounding rocks exceeded the compressive strength value, conjugate failure occurred along the AM, GM, AN, and DN lines. After the conjugate fracture planes intersected, they formed an enclosed fracture loop MNOP in the surrounding rocks. The shallow surrounding rocks in the roadway were fractured and cut into wedge-shaped blocks by the fracture loop MNOP, as shown by the triangle AMG in Figure 8. Under high-stress conditions, the surrounding rocks could easily slip toward the roadway. After this loosening, the MNOP acted as a new roadway boundary: the stressed body under plane strain conditions (such as MUVN) continued to be isolated along each side of the MNOP. When the pressure σ_s exceeded the strength of the surrounding rocks, the conjugate shear fracture occurred, and a new fracture loop was formed outside the MNOP.

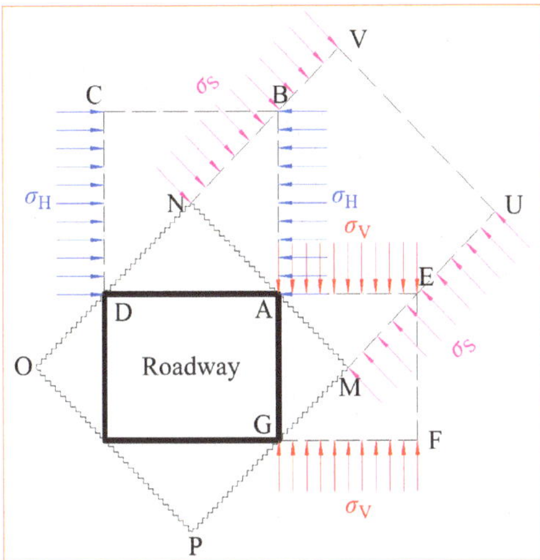

Figure 8. Fracture expansion pattern in a high-stress soft rock roadway. Capital letters A,B,C,D,E,G,O,M,U,V,P denote geometric intersections.

According to the simulation results depicted in Figure 4, the nonuniform strength of the surrounding rocks can affect the development of the fracturing process. As shown in Figure 9, the roadway was located in soft rock strata, while its roof and floor were in hard rock strata.

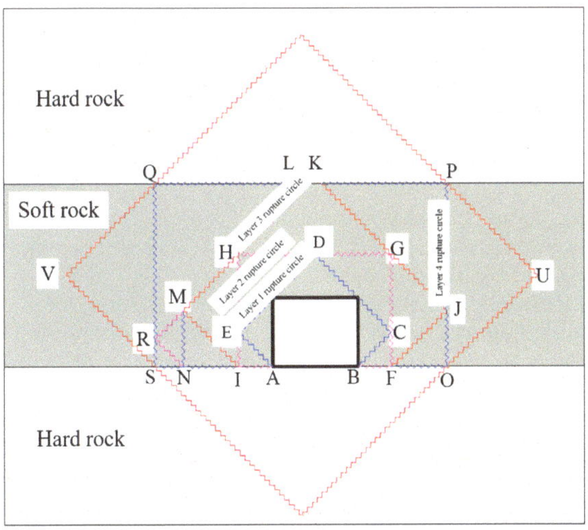

Figure 9. The expansion process of a fractured zone in the roadway with a strong-soft-strong composite surrounding rock. Capital letters A,B,C,D,E,F,G,H,I,J,K,L,M,N,O,P,Q,R,S,U,V denote geometric intersections.

A comprehensive analysis of the simulation results presented in Figure 4 and the above fracture patterns revealed that the first and fourth layers of the fracture loops developed only in soft strata without penetrating the hard strata. With the expanding boundary of the loosening loop, the vertical pressures acting on the roof and floor gradually dropped. For example, during the process shown in Figure 6, the vertical stress variation along the preset stress-monitoring line (Figure 7a) set up in the roadway floor occurred, as shown in Figure 10. After forming the second layer of the fracture loop, the mean vertical stress on the floor between the left and right boundaries dropped by approximately 40% compared with the value reached after the formation of the first layer of the fracture loop. After forming the first layer of the fracture loop, as shown in Figure 8, the stress on the floor within the range AB was close to zero, while both sides AI and BF were located in the stress-reduction zone. After forming the second layer of the fracture loop, the stresses within the AI and BF ranges further decreased. Accompanied by the decline in vertical pressure, the floor became prone to compression–shear failure. It was assumed that the roof and the floor reached their ultimate condition when the loosening fractured zones developed into OPQS. Afterward, the fractures developed towards the deep roof, the floor, and both sides in a compression–shear conjugate failure pattern.

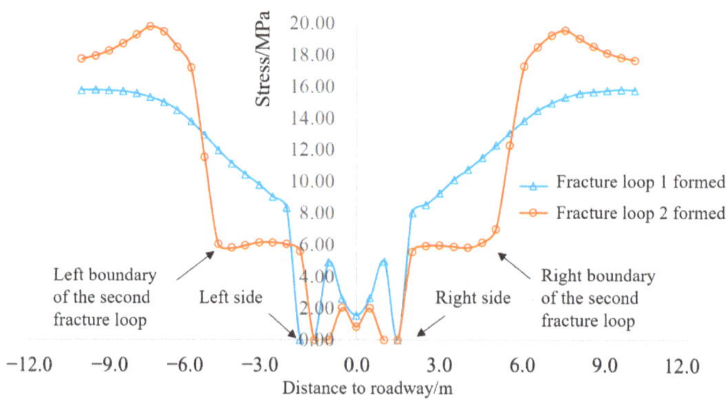

Figure 10. Vertical stress variation in the roadway floor versus distance to the roadway.

It can be inferred from the failure processes depicted in Figure 4d,e that the development patterns of the fracture planes on the two sides were not completely symmetrical, even in the strata with identical lithological properties. For example, after forming the third layer of the fracture loop, as shown in Figure 9, the JO, FO, JP, and KP fracture planes could develop on the right side. Meanwhile, after forming the NI and MN fracture planes on the left side, the NR and MR fracture planes were next to appear, with a further gradual development of the RS, NS, RQ, and LQ fracture planes.

Overall, on account of the difference in lithology and stress states, the development of the fracture planes in the surrounding rocks shows a certain complexity; however, after repeated compression–shear conjugate damage, the intact and fractured zones with varying fracture degrees are systematically obtained.

3.2. Mechanical Principle of Zonal Disintegration

Besides the physical analog modeling and numerical analysis of zonal disintegration, its mechanical mechanisms were also widely explored. Zhu et al. revealed the existence of unloading waves in deep strata after the excavation of high-stress strata, thereby inducing discontinuous fracturing of the surrounding rocks in the roadway [19]. Qi et al. analyzed the stress field evolution of the surrounding rock based on the pressure fluctuation at the inner wall of the roadway to explore the alternate manifestation patterns of fractured and intact zones [20,21]. Based on the energy conversion characteristics of deep rocks, some scholars have analyzed the mechanical principle of zonal disintegration by modifying the conventional elastoplastic model [22,23]. Chanyshev et al. performed a theoretical exploration of fractured and intact zones in the roadway based on damage mechanics and non-Euclidean geometry [24]. Other studies mostly focused on the heterogeneity and discontinuity of rock masses. Thus, using the rock strength-related parameters that obeyed the Weibull distribution law, Wang et al. designed a nonuniform strain-softening model of rocks and obtained annular zonal disintegration results via numerical simulation [14]. The factors of interest in each mechanical mechanism or model can lead to non-homogeneous rock mass damage. Accordingly, scholars hold different opinions regarding zonal disintegration mechanisms.

The disputes on the zonal disintegration mechanism essentially lie in the fact that the alternate appearance of disintegration and intact zones in the annular pattern was speculated based on limited measurement results. Despite the alternate occurrence of fractured and intact zones that could be observed in multiple drilling holes of the same roadway section, the stressed state of these zones between boreholes was unknown. The existing empirical model was obtained by collocating them in the order of their occurrence [8]. At the same time, the scientific validity of this data-processing method has not been effectively confirmed in the laboratory.

Wang et al. constructed the characteristic lines of the roadway perimeter based on the elastoplastic flow theory and pointed out that the zonal disintegration can be obtained by reproducing the cave shape using a family of characteristic lines [24], as shown in Figure 11. Based on the actual measurement results of Meng [25], Shemyakin et al. reported that the zonal disintegration occurred in a circular damage zone, with a certain distance (spacing) between damaged zones and a circular intact zone between adjacent damaged zones. Moreover, the neighboring disintegration regions were separated by annular intact ones, while slip or failure lines intersected to a certain degree, which was not consistent with the actual condition of zonal disintegration. According to field measurements, not all disintegration zones were annular or completely separated from each other. Figure 12a shows the measured results in [25], where different numbers of disintegration zones can be observed in the drilling holes 2, 3, 4, and 5, suggesting fairly complex spatial relationships between fractured zones. The data in [5] can also validate the measured results (see Figure 12b).

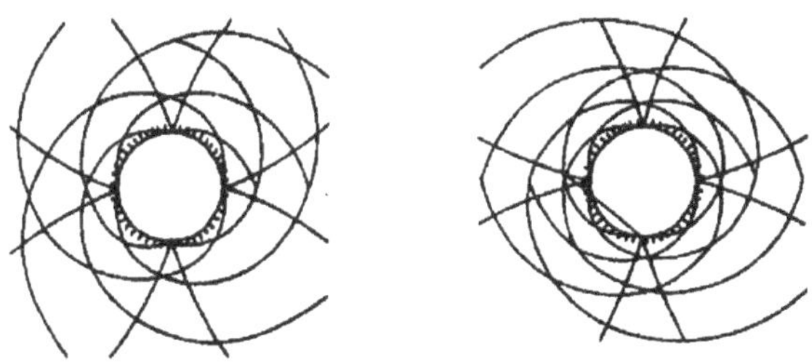

(a) (b)

Figure 11. Characteristic lines constructed via the elastoplastic flow theory. (**a**,**b**) denote different types, respectively.

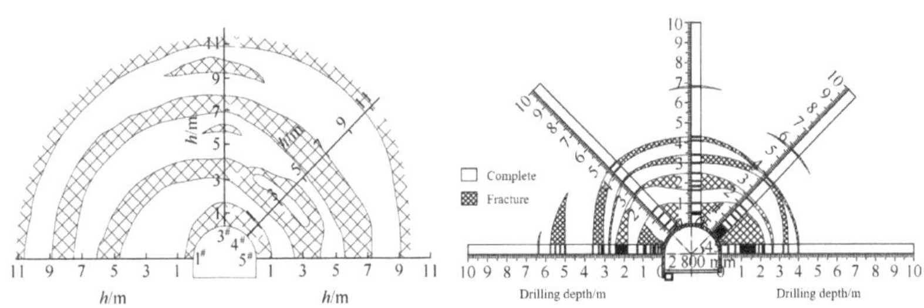

(a) (b)

Figure 12. Measured results on zonal disintegration: (**a**) measured results in [25]; and (**b**) measured results in [5].

The following conclusions can be drawn from the above numerical simulation results: Firstly, the fractures in the surrounding rocks gradually developed from the roadway surface to deeper parts. Secondly, the fracture planes developed towards the deep surrounding rocks from the free plane (the roadway surface), forming an enclosed fracture loop around the roadway. Thirdly, at sufficiently high stress, the fractures developed toward the depth of the surrounding rock from the boundary (i.e., the outermost failure loop) in a compression–shear conjugate failure pattern. According to the numerical simulation results, various fracture loops were connected, while the finally formed fractured zone was no longer a closed ring but an interconnected network structure. It can be concluded that the compression–shear conjugate damage surface was the result of the damage development along the slip line based on elastoplastic theory, which was consistent with the theoretical predictions depicted in Figure 11.

Figure 13 shows the failure mode of the surrounding rocks in the roadway obtained via physical analog modeling in the study [26]. Specifically, Figure 13a is the physical image, while Figure 13b is a digital photograph depicting shear deformation in the surrounding rocks. The conjugate failure in the physical analog test and the fracture planes in the numerical simulation results reflect similar development processes, confirming the rationality of interpreting zonal disintegration via the compression–shear conjugate failure pattern.

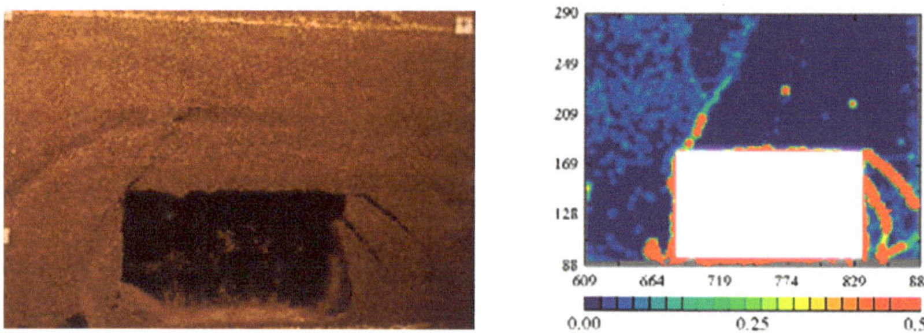

(a) (b)

Figure 13. Fracture state of roadway-surrounding rock in [26]: (**a**) physical image; and (**b**) shear stress.

3.3. Factors Influencing the Roadway Zonal Disintegration

3.3.1. Effect of the Surrounding Rock Strength

Based on the parameter settings depicted in Figure 6, new simulation schemes were designed by varying the surrounding rocks' internal friction angle, the cohesive force, and the uniaxial tensile strength, respectively, as shown in Table 3. In schemes 1–3, both the initial and residual values of the internal friction angle, cohesive force, and uniaxial tensile strength were doubled, respectively. In schemes 4~6, the above parameters were reduced by 1/3, respectively. The remaining physical and mechanical parameters in all schemes, such as density and bulk modulus, remained unchanged.

Table 3. Strength parameters of rock mass in the designed schemes.

Scheme No.	Initial Strength			Residual Strength		
	Internal Friction angle, °	Cohesive Force, MPa	Tensile Strength, MPa	Internal Friction Angle, °	Cohesive Force, MPa	Tensile Strength, MPa
1	36	3	2.4	24	1	0.8
2	18	6	2.4	12	2	0.8
3	18	3	4.8	12	1	1.6
4	18	2	2.4	12	0.8	0.8
5	12	3	2.4	9	1	0.8
6	18	3	1.6	12	1	0.5

Figure 14 shows the distribution pattern of the significantly fractured zones in the roadway-surrounding rocks for the strength parameters in schemes 1–6. As seen in Figure 14, the increased rock strength parameters hindered the development of the disintegration zones and vice versa. After the internal friction angle was doubled, significant fractured zones developed only on both sides (see Figure 14a), and the maximum shear strain increment of 0.06 appeared at two footings, reaching about 29% of the value in the original scheme (see Figure 7e). After the cohesive force was doubled, significantly fractured zones also developed on two sides (Figure 14b). The maximum shear strain increment appeared on two shoulders and footings and approximately equaled 0.105, which was 50% of the original scheme's value. After the tensile strength was doubled, the distribution pattern and range of the fractured zones (see Figure 14c) were similar to those in the original scheme, while the maximum shear strain increment remained unchanged (0.21). Therefore, the effect of tensile strength on the development of fractured zones was weaker than that of the internal friction angle and cohesive force.

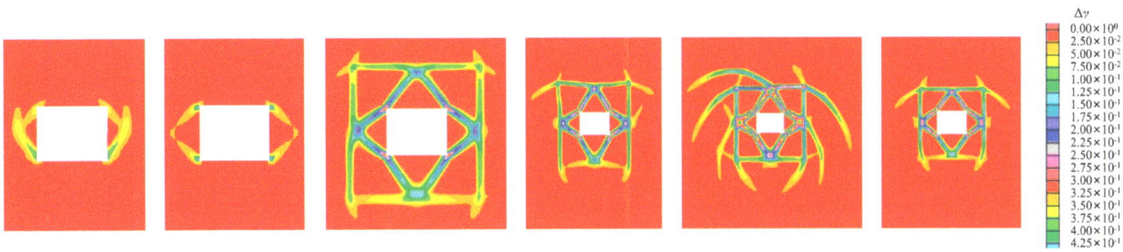

(**a**) (**b**) (**c**) (**d**) (**e**) (**f**)

Figure 14. Fractured zone state in roadway-surrounding rock with different strength levels. (**a**–**f**) denote the rupture of each scenario of the roadway model, respectively.

When the original scheme's values of the internal frictional angle and tensile strength were reduced by 2/3, the distribution pattern of significant fractured zones showed a slight difference (see Figure 14e,f). Nevertheless, the maximum fracturing depths in the surrounding rocks increased by approximately 52 and 39%, respectively. The maximum shear strain increments increased by approximately 43 and 64% compared to the original scheme's values. After the cohesive force was reduced by 2/3, the fracture planes pointed toward deeper parts of the surrounding rocks appeared at the angular points of the second layer of the fracture loop and the first/second layer intersection points of the fracture loop in the conjugate fracture plane direction (see Figure 14d). The maximum depth for the fractured zone was doubled, and the maximum shear strain increment increased by 107% compared to the original scheme's value (see Figure 6e). Conclusively, the cohesive force imposed the most significant effect on the degree of zonal disintegration.

3.3.2. Effect of the Stress Environment

Based on the stress settings in the primary rock, as shown in Figure 7, the simulation was performed under new stress conditions by varying the vertical and horizontal stress settings. Table 4 shows the designed simulation schemes under different stress conditions.

Table 4. Stress levels in designed schemes.

Design scheme	Scheme 7	Scheme 8	Scheme 9	Scheme 10
Vertical stress, MPa	10.0	10.0	17.0	25.0
Horizontal stress, MPa	10	15.0	12.0	25.0

Figures 15 and 16 show the simulation results under different stress conditions. The zonal disintegration morphology was closely related to the stress state. When both vertical and horizontal stresses were 10.0 MPa, the significant fractures only occurred at four corners of the roadway (at two shoulders and two footings), despite plastic failure on two sides, the roof, and the floor of the roadway (see Figure 17). At a vertical stress of 10.0 MPa and a horizontal stress of 15.0 MPa, the fractured zones were observed only within the roof and the floor (see Figure 15b). The conjugate failure first started from the free surfaces (i.e., the roof and the floor) such as the AE, BE, CF, and DF fracture planes; next, conjugate failures occurred with the AE, BE, CF, and DF fracture planes as the boundaries; finally, the MN and UV fracture planes developed to form conjugate failures in the deep roof and floor. At a vertical stress of 17.0 MPa and a horizontal stress of 12.0 MPa (i.e., at high vertical but low horizontal stress), fractured zones only occurred on two sides (Figure 15c). The development of the fracture planes on two sides was similar to that in the roof and the floor (see Figure 15b).

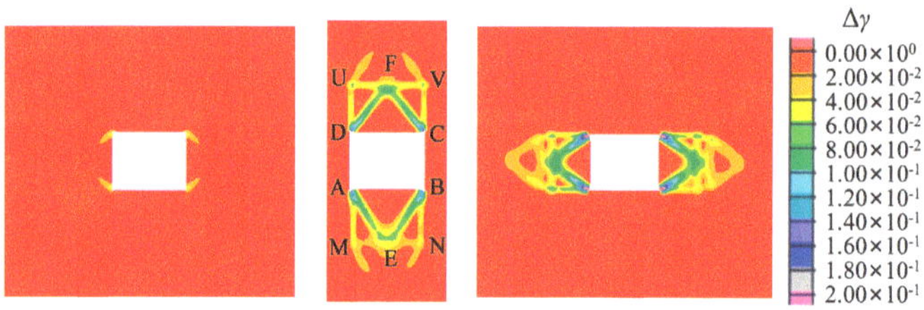

(a) (b) (c)

Figure 15. Fractured zone state in roadway surrounding rock under different stress levels: (**a**) scheme 7; (**b**) scheme 8; and (**c**) scheme 9.

(a) (b) (c) (d) (e)

Figure 16. The expansion process of a fractured zone in scheme 10. (**a–e**) represent the rupture at each stage of the roadway model, respectively.

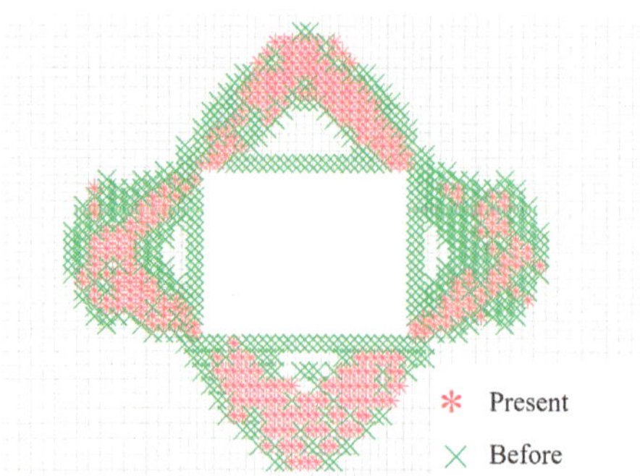

Figure 17. Distribution of the plastic zone in scheme 7: current plastic state and previous plastic state.

Figure 16 shows the development of the fractured zones at identical vertical and horizontal stresses of 25.0 MPa. The first layer of the fracture loop was first formed in the surrounding rocks (see Figure 16a), which then developed to form the second layer of the fracture loop (see Figure 16b,c). After forming the AEBFCGDH fracture loop (see Figure 16c), the fracture planes developed from the AE, EB, BF, FC, CG, GD, DH, and DA boundaries toward the deep surrounding rocks via the compression–shear conjugate failure pattern (see Figure 16d). Finally, the AI, IE, EJ, JB, BK, KF, FL, LC, CM, MG, GN, ND, DO, OH, HP, and PA fracture planes were formed in the third layer of the fracture loop. The fractures outside the third layer of the fracture loop continued their outward development along LCM, NGM, and NDO boundaries in a conjugate failure pattern.

3.4. Effect of the Roadway Cross-Section

Previous theoretical studies on zonal disintegration of the roadway-surrounding rocks were mainly focused on circular roadways and derived the annular distribution pattern of the zonal fracturing zones around the roadway [20–30]. Based on elastoplastic theory and numerical simulation results, it can be observed that failures were not always centrosymmetric and mostly presented a large damage depth near the two shoulders and two footings, while the damage depth at the roof, floor, and two sides was small [31,32].

In this study, the roadway with a circular cross-section was simulated using the PFC^{2D} and $FLAC^{2D}$ software packages, with the results shown in Figure 18. The wedge-shaped shear deformation failures first appeared on the roadway wall at the beginning of the failure in the circular roadway. The discrete finite simulation results showed that the initial failures were concentrated around two shoulders and two footings. As shown in Figure 18a, the wedge-shaped failures appeared at points A, B, C, and D, while the boundary of the ABCD failure zone can be approximated by a rectangle. According to finite difference simulation results, the significant compression–shear deformation also showed wedge-shaped failure at the beginning, with uniform distribution around the roadway perimeter (see Figure 18c). As the deformation increased, the shear strain increment near the two shoulders and two footings began to increase. Therefore, the significant failure zone boundary can be approximated by a rectangle (see Figure 18d).

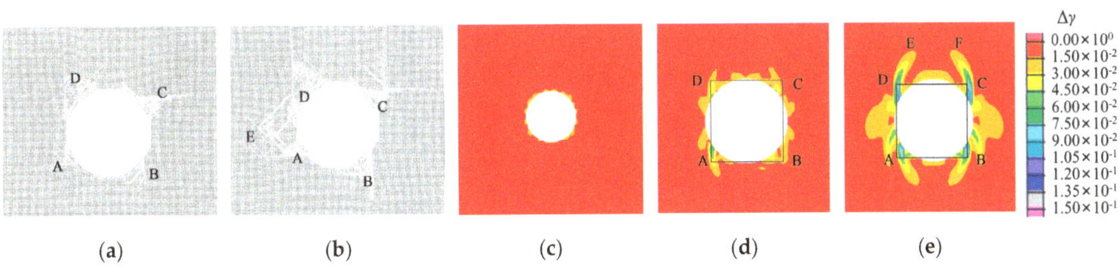

Figure 18. Fracture form in a circular roadway-surrounding rock. (**a**,**b**) denote the simulation of the fracture of the surrounding rock of the roadway in the PFC model, respectively. (**c**–**e**) denote the fracture of the roadway surrounding rock at each stage in the FLAC model, respectively.

After the formation of the initial failure boundary, which was approximately rectangular in shape, the initially formed boundaries comprised the free surface for the subsequent damage development of the roadway, and a significant failure developed towards the deep surrounding rocks in the compression–shear failure pattern (see the AE and DE fracture planes in Figure 18b and the DE and CF fracture planes in Figure 18d).

The initial position of the zonal disintegration development in the surrounding rocks of the circular roadway was subjected to the effect of the initial failure boundary. In contrast, the shape of the initial failure boundary was related to the physical and mechanical properties of the surrounding rocks and the stress state [33,34]. This inevitably increased the complexity of zonal disintegration in the circular roadway.

3.5. Compression Mechanisms of Anchor Rods

Fang et al. measured the force condition on the anchor rods and found that many anchor rods were under extension; additionally, tension and compression segments co-existed on some anchor rods [28]. Wang et al. concluded that tension and compression alternately appeared in the surrounding rocks based on field-measured results [29]. Li et al. also conducted field measurements and found that both loosening and compression zones existed in the roadway-surrounding rocks; moreover, they derived that the co-existence of loosening and compression regions was tightly related to the appearance of zonal disintegration in the surrounding rocks [8]. Explaining the alternate appearance of tension and compression in the roadway-surrounding rocks has become a basic problem that should be addressed in investigating zonal disintegration mechanisms.

Figure 19 shows the numerical simulation results obtained via the discrete element method for the failure of the surrounding rocks in the roadway reinforced by anchor rods and the forces acting on the anchor rods of 2.2 m in length and a 100.0 kN bearing capacity.

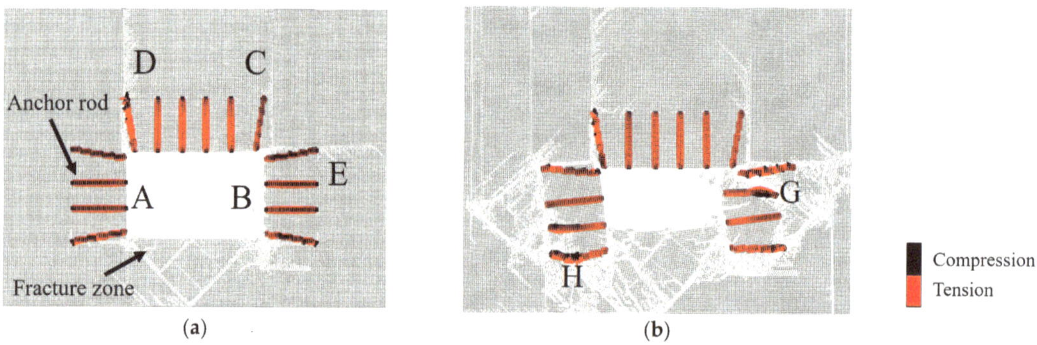

Figure 19. Stress state of anchor rods. (**a**,**b**) indicate the different stages of the lane, respectively.

After the reinforcement of the roof and two sides by the anchor rods, the initial failure of the surrounding rocks occurred on the floor. The conjugate failure plane was formed and extended from two shoulders of the roadway in the vertical or horizontal directions, as illustrated by the AD, BC, and BE fracture planes in Figure 19a. As the stress in the roof increased, the compression–shear failure began to appear on two sides and developed deeper toward the surrounding rocks in a conjugate-failure pattern (see Figure 19b). By monitoring the forces acting on the anchor rods, it can be concluded that the anchor rods were mainly subjected to tension. This occurred because the radial displacement of the shallow surrounding rocks was larger than that of the deep rocks when the surrounding rocks at different depths underwent deformation and movement toward the roadway, thus putting each anchor rod as a whole in a tensile state. When the surrounding rocks were fractured, the integrated rocks moved along the shear-fractured zones, and the anchor rods were subjected to bending moments. In the bending deformation of the anchor rods, the bulk of the bent segment was under compression, as shown by stress conditions near points H and G in Figure 19b. Therefore, the compression of the anchor rods in the roadway-surrounding rocks in the zonal disintegration state could be measured via strain gauges attached to the anchor rod surface.

The compression phenomenon in the roadway-surrounding rocks was closely related to zonal disintegration. However, the compression state was not the result of radial compression of the surrounding rocks but was induced by the bending of the anchor rods or the testing devices when the integrated rocks after compression–shear cutting were loosened toward the roadway.

4. Discussion

There are the following elements of this study that need to be discussed in order to provide hints for further research:

1. The results discussed in this paper are based on the simulation effect of numerical simulation software, so the conclusions may have idealized results. Including the homogenization and isotropy of the surrounding rock of the roadway, these are conditions that are difficult to exist in the real environment.
2. The geological conditions of the mine are certain, and there is a risk of distortion by reducing a certain rock mechanical parameter one by one in the simulation test.
3. In the process of excavation, the surrounding rock of the tunnel may be damaged, and this damage is usually shown on a three-dimensional scale, which is ignored in the two-dimensional direction of the simulation conducted in this paper.

5. Conclusions

The development of fractures in the rock can be judged by intermittent zonal fracturing via the discrete element method. The damage to roadway-surrounding rocks can be subdivided into distinct fractured and intact zones. The zonal disintegration characteristics were obtained in this study by numerical simulation using the FLAC^{2D} software. The development of fractured zones and the appearance of intact ones exhibited a certain regularity. The main conclusions were drawn as follows:

(1) Under high-stress conditions, the damage to a rectangular roadway started at its four corners. The fractured zones extended towards the deeper parts of surrounding rocks in a compression–shear conjugate fracture mode, forming a nearly rectangular fracture loop, whereas the four edges (the roof, the floor, and two sides) served as free surfaces. The rock masses within the fracture loop were extruded and loosened into the roadway, and the boundary stress of the fracture loop dropped, thereby forming a new free surface. Under high stress, the fractures developed toward the deep parts of the surrounding rocks in a compression–shear conjugate fracture mode along various sides of the fracture loop to form a new fracture loop. For a circular roadway, the loosening failure first appeared near the two shoulders and footings, accompanied by the formation of a nearly rectangular failure boundary. Next, zonal disintegration evolved and developed in the surrounding rocks. Thus, the rectangular and circular roadways showed similar development patterns of zonal disintegration.

(2) Rock strength is a crucial factor in influencing zonal disintegration. Under fixed stress conditions, significantly fractured zones appeared only in shallow surrounding rocks and could not develop deep when rock strength was high. According to the Mohr–Coulomb model, the main factors controlling the development of zonal disintegration could be ranked in decreasing order as follows: cohesive force, internal friction angle, and tensile strength.

(3) The stressed state is a key factor controlling the development of zonal disintegration. At low vertical and horizontal stresses, significantly fractured zones appeared only near the roadway perimeter surfaces. At high vertical stresses and low horizontal ones, zonal disintegration appeared on the roadway's two sides. At high horizontal stresses and low vertical ones, zonal disintegration developed only in the roof and the floor. When both vertical and horizontal stresses increased, the extension range of the fractured zone grew steadily, while its structure became increasingly complex.

(4) The fractured zones in the surrounding rocks formed under the compression–shear conjugate fracture mode originated from the roadway surface. Various layers of fracture loops were interconnected, forming a network, which was consistent with the elastoplastic slip line theory prediction.

(5) In anchor-supported roadways undergoing zonal disintegration processes, the intact zone blocks slid relatively along the fracture surface during the process of loosening and deformation of the surrounding rock, making the anchor rods susceptible to tensile, shear, and bending actions.

Author Contributions: Conceptualization, P.Y. and C.L.; methodology, P.Y.; software, P.Y. and S.Z.; validation, P.Y. and C.L.; formal analysis, C.L.; investigation, P.Y. and S.Z.; resources, P.Y.; data curation, P.Y.; writing—original draft preparation, P.Y.; writing—review and editing, P.Y. and S.Z.; visualization, S.Z.; supervision, C.L.; project administration, C.L.; funding acquisition, C.L. All authors have read and agreed to the published version of the manuscript.

Funding: This research was funded by the National Natural Science Foundation of China, grant number 52074267.

Institutional Review Board Statement: Not applicable.

Informed Consent Statement: Not applicable.

Data Availability Statement: Data are contained within the article.

Conflicts of Interest: The authors declare no conflicts of interest.

References

1. Qian, Q.H.; Li, S.C. A review of research on zonal disintegration phenomenon in deep rock mass engineering. *Chin. J. Rock Mech. Eng.* **2008**, *27*, 278–1284.
2. Borisovets, V.A. Wave-type inhomogeneities in rocks near drilling-and-blasting zones. *Min. Constr.* **1972**, *9*, 7–11. (In Russian)
3. Cloete, D.R.; Jager, A.J. The nature of the fracture zone in gold mines as revealed by diamond core drilling. *S. Afr. Assoc. Mine Manag.* **1974**, 103.
4. Adams, G.R.; Jager, A.J. Peteroscopic observations of rock fracturing ahead of slope face in deep-level goldmines. *J. S. Afr. Inst. Min. Metall.* **1980**, *80*, 204–209.
5. Li, S.C.; Wang, H.P.; Qian, Q.H.; Li, S.C.; Fan, Q.Z.; Yuan, L.; Xue, J.H.; Zhang, Q.S. In-situ monitoring research on zonal disintegration of surrounding rock mass in deep mine roadways. *Chin. J. Rock Mech. Eng.* **2008**, *27*, 1545–1553.
6. Li, S.C. *Field Observation Analysis and Numerical Simulation. 21st Anthology of Academic Salon on New Ideas and Theories: Zonal Disintegration Effect of Surrounding Rock in Deep Rock Engineering*; China Science and Technology Press: Beijing, China, 2008; pp. 82–88.
7. Qi, C.Z.; Qian, Q.H.; Wang, M.Y.; Luo, J. Advance in investigation of zonal disintegration phenomenon. *J. PLA Univ. Sci. Technol. Nat. Sci. Ed.* **2011**, *12*, 472–479.
8. He, Y.N.; Zhang, H.Q. Discussion on theory and practice of zonal disintegration in surrounding rocks of deep roadways. *Chin. J. Rock Mech. Eng.* **2008**, *27*, 2369–2375.
9. Zhu, D.K.; Zuo, Y.J.; Chen, C.C. Analysis of deep surrounding rock supporting with zonal disintegration tendency. *Nonferrous Met. (Min. Sect.)* **2012**, *64*, 34–38.
10. Pan, Y.S.; Li, Y.J.; Tang, X.; Zhang, Z.H. Study on zonal disintegration of rock. *Chin. J. Rock Mech. Eng.* **2007**, *26* (Supp. 1), 3335–3341.
11. Gu, J.C.; Gu, L.Y.; Chen, A.M.; Xu, J.; Chen, W. Model test study on mechanism of layered fracture within surrounding rock of tunnels in deep stratum. *Chin. J. Rock Mech. Eng.* **2008**, *27*, 433–438.
12. He, Y.N.; Jiang, B.S.; Han, L.; Shao, P.; Zhang, H. Study on intermittent zonal fracturing of surrounding rock in deep roadways. *J. China Univ. Min. Technol.* **2008**, *37*, 300–304.
13. Zhang, Z.H.; Wang, X.B.; Pan, Y.S.; Su, A.G. Experiment on layered observation of zonal disintegration of split similar material. *J. Water Resour. Water Eng.* **2010**, *21*, 1–5.
14. Wang, X.B.; Bai, X.Y.; Ma, B.; Zhang, Z.H.; Lv, J.G. Effects of the heterogeneity of surrounding rock on the zonal rupture of roadway. *J. China Univ. Min. Technol.* **2019**, *48*, 78–86.
15. Gao, F.Q.; Kang, H.P.; Lin, J. Numerical simulation of zonal disintegration of surrounding rock mass in deep mine roadways. *J. China Coal Soc.* **2010**, *35*, 21–25.
16. Yang, P.; Zhang, S.; Wang, X. Numerical Study on the Characteristics and Control Method of Coal Leakage between Supports in Integrated Mining of Extremely Loose and Soft Coal Seams. *Energies* **2024**, *17*, 1013. [CrossRef]
17. Li, L.H. Mechanism and Control of Lateral Strong Ore Pressure in Large and Thick Roof. Master's Thesis, China University of Mining and Technology, Xuzhou, China, 2019.
18. Zhang, S.K.; Lu, Q.K.; Zhang, X.D. A study on the stability of rectangular cross-section tunnel. *J. Guangxi Univ. Nat. Sci. Ed.* **2012**, *37*, 210–214.
19. Zhu, Z.M.; Liu, K.; Kang, J.M.; Zhang, X.Y. Study of slowly unloading P-waves causing discrete fracture mechanism. *Chin. J. Rock Mech. Eng.* **2014**, *33* (Suppl. S2), 3948–3955.
20. Zhou, X.P.; Qian, Q.H. Zonal fracturing mechanism in deep tunnel. *Chin. J. Rock Mech. Eng.* **2007**, *26*, 877–885.
21. Qi, C.Z.; Qian, Q.H.; Wang, M.Y.; Chen, J.J. Internal variable gradient plasticity model for zonal disintegration of surrounding rocks in deep tunnels. *Chin. J. Rock Mech. Eng.* **2012**, *31* (Suppl. S1), 2722–2728.
22. Wang, M.; Chen, H.; Li, J.; Li, X. Theoretical research on zonal disintegration of rock masses around deep tunnels and comparisons with in-situ observations. *Chin. J. Rock Mech. Eng.* **2018**, *37*, 2209–2218.

23. Zhou, X.; Zhou, M.; Qian, Q. Influence of rock damage on zonal disintegration around deep circle tunnels. *Chin. J. Solid Mech.* **2012**, *33*, 242–250.
24. Chanyshev, A.I. *Plastic Deformation and Zonal Disintegration of Deep Surrounding Rock. 21st Anthology of Academic Salon on New Ideas and Theories: Zonal Disintegration Effect of Surrounding Rock in Deep Rock Engineering*; China Science and Technology Press: Beijing, China, 2008; pp. 95–108.
25. Shemyakin, E.I.; Fisenko, G.L.; Kurlenya, M.V.; Oparin, V.N.; Reva, V.N.; Glushikhin, F.P.; Rozenbaum, M.A.; Tropp, E.A.; Kuznetsov, Y.S. Zonal disintegration of rocks around underground workings. Part I: Data of in-situ observation. *J. Min. Sci.* **1986**, *22*, 157–168. [CrossRef]
26. Su, Y.J. Study on Deformation Law and Control of High Stress Roadway in Lower Part of Multicoal Seam Mining in Close Distance in Jinhuagong Mine. Master's Thesis, China University of Mining and Technology, Xuzhou, China, 2023.
27. Wang, X.B.; Zhang, Z.H.; Pan, Y.S.; Zhang, C.Y. Numerical simulation of the influence of post-peak brittleness on the failure and energy liberation in the surrounding rock of a circular tunnel—Discussion on the difference between slip lines and zonal disintegration. *J. Disaster Prev. Mitig. Eng.* **2013**, *33*, 11–17.
28. Li, S.P. Observation report of anchor test in roadways of Quantai coal mine and discussion on new viewpoint of anchor characteristics and parameter selection. *J. China Inst. Min. Technol.* **1979**, *3*, 19–57.
29. Fang, Z.L. *Maintenance Mechanism and Control Measures in Soft Roadways. Support Theory and Practice of Soft Mine Roadways in China*; University of Mining and Technology Press: Beijing, China, 1996; pp. 64–69.
30. Wang, M.Y.; Zhou, Z.P.; Qian, Q.H. Tectonic deformation and failure problems of deep rock mass. *Chin. J. Rock Mech. Eng.* **2006**, *25*, 448–455.
31. Ma, N.N.; Li, J.; Zhao, Z.Q. Distribution of the deviatoric stress field and plastic zone in circular roadway surrounding rock. *J. China Univ. Min. Technol.* **2015**, *44*, 206–213.
32. Wang, H.Y.; Zhang, Q.; Zhang, Y.; Jiang, K. Numerical simulation on disintegration of surrounding rock mass in deep mine roadways. *J. China Coal Soc.* **2010**, *35*, 535–540.
33. Fang, Q.; Zhang, D.; Wong, L.N.Y.; Li, P.; Li, Q. Model test study of failure modes of surrounding rock for circular caverns. *Chin. J. Rock Mech. Eng.* **2011**, *30*, 564–571.
34. Li, Y.; Zhou, Y.; Wu, S.; Wang, C.; Wang, Y. Deformation and failure of surrounding rock of circular tunnel using coupled continuous-discrete method. *Chin. J. Rock Mech. Eng.* **2015**, *34*, 1849–1858.

applied sciences

Article

Characteristics of Perimeter Rock Damage in a Bottom-Pumping Roadway under the Influence of Mining Activities and Rational Location Studies: A Case Study

Libin Bai [1,2], Fengfeng Wu [1,*], Peiju Yang [1], Shurong Zhang [1] and Bin Li [1]

1 School of Mines, China University of Mining and Technology, Xuzhou 221000, China
2 Liuta Coal Mine, China National Energy Shendong Coal Group, Ordos 221000, China
* Correspondence: wufengfeng@cumt.edu.cn

Abstract: With the aim of determining the damage characteristics and a reasonable positional arrangement of the surrounding rock in a bottom-pumping roadway influenced by mining in a high-gas mine, the boundary equation for the plastic zone of the surrounding rock in a circular roadway under an unequal compressive stress field was adopted to analyze the relationship between the distribution characteristics of the plastic zone of the bottom-pumping roadway and the stability of the rock surrounding the bottom-pumping roadway under different bidirectional stress ratios. This was carried out in the bottom-pumping roadway of the working face of Licun coal mine 3301 as the engineering background, where the nature of the coal seams mined is bituminous coal, and the absolute gas outflow is 0.5 m^3/min^{-1}. A numerical simulation was used to analyze the distribution characteristics of the surrounding rock stress and the bidirectional stress ratio, as well as the deformation and damage characteristics of the surrounding rock at different positions in the bottom-pumping roadway. A numerical simulation was applied to analyze the distribution characteristics of the surrounding rock stress and the two-way stress ratio, as well as the deformation and damage characteristics of the rock surrounding the bottom-pumping roadway when the bottom-pumping roadway was arranged in different locations. The results show that, with an increase in the bidirectional stress ratio, the plastic zone of the perimeter rock in the bottom-pumping roadway shows nonuniform "butterfly" distribution characteristics, which seriously affects the stability of the rock on the perimeter of the roadway; the stress on the bottom plate of the working face after excavation can be divided into four areas according to the size of the bidirectional stress ratio and the stress loading and unloading states. In addition, the size of the perimeter rock deformation can be sorted into four areas according to the damage range of the perimeter of the rock plastic zone in the bottom-pumping roadway. The size of the deformation in the surrounding rock can be sorted as follows: unpressurized high-stress ratio > unpressurized stress ratio stable area > pressurized low-stress ratio area > original rock stress ratio area. Accordingly, we found that the reasonable location of the bottom-pumping roadway is arranged at the 15 m position outside the hollow area below the coal pillar, along the limestone upper medium-grained sandstone layer along the bottom. The study's results were applied to the field. The industrial experiments on the site show that the deformation of the surrounding rock is reasonable when the bottom-pumping roadway is dug along the limestone roof and arranged 15 m outside the fault of the mining hollow area below the coal pillar.

Keywords: bidirectional stress ratio; bottom-pumping alley; reasonable location; surrounding rock deformation damage characteristics

1. Introduction

With the gradual expansion of deep mining for coal resources in China, coal and gas outbursts in mines have become more and more serious. To ensure safe production in mines, it is extremely important to arrange the bottom pumping roadway on the floor of the working face for gas drainage from the upper working face [1–5]. However, when the working face is being mined back, the bottom plate is affected by the mining of the working face, and the stress is transferred to the deep part of the bottom plate of the working face of the roadway, which redistributes the stress in the rock surrounding the bottom plate. Then, there is a nonuniform stress field, i.e., a two-way unequal compressive stress field, in the rock surrounding the bottom plate, which results in different degrees of damage to the rock surrounding the bottom plate in different locations [6–8]. The damage to the bottom plate rock at different locations is caused by the nonuniform stress field. Therefore, obtaining the distribution law and damage characteristics of the two-way unequal compressive stress field in the base plate rock under the influence of mining in the working face is the key factor in determining the reasonable location of the bottom-pumping tunnel.

Experts and scholars at home and abroad have conducted much research concerning the distribution law and damage characteristics of two-way unequal compressive stress fields in the rock surrounding the base plate under the influence of mining in the working face and have achieved remarkable results. Among them are Li Yong'en et al. [9–12]. Given the contradiction between the bottom-pumping lane and the pressurized water, they established a boundary equation for the plastic zone of the circular lane under a two-way unequal compressive stress field, analyzed the damage characteristics of the plastic zone of the surrounding rock in the bottom-pumping lane with the different values of the two-way stress ratio, and determined the reasonable location of the bottom-pumping lane according to the damage characteristics of the plastic zone. Li Ang et al. [13–15] applied a research method combining theoretical analysis and numerical simulation to analyze the stress distribution characteristics of the bottom plate of the working face under the influence of mining and the damage evolution characteristics of the surrounding rock in the plastic zone. They also obtained the distribution law of the supporting pressure of the rock layer of the bottom plate and the damage range of the rock layer in the bottom plate under the influence of mining. Wang et al. [16–19] created a tangential stress equation for the surrounding rock in a two-way unequal pressure stress field of a circular roadway, which was derived using the method of angle-preserving transformation in the complex function, and the formula for calculating the plastic zone of the circular roadway under the condition of two-way unequal pressure was further obtained based on the Moore–Coulomb strength criterion. Zhu Ye et al. [20–22] investigated the stability of the base plate under different stress field environments through theoretical analysis and numerical calculation and pointed out that the stresses in the base plate peripheral rock under the influence of mining were significantly different from each other, and the variability of the stress environments was the main reason for the variability in the stresses in the base plate's peripheral rock. By establishing a physical similarity model, Gao Shigang et al. explored the reasonable arrangement of mining roadways in middle and lower coal seams in close-distance coal seam mining [23]. Through numerical calculations, Chi Xiaolou et al. [24,25] analyzed the distribution law of stress fields and the evolution characteristics of the plastic zone of a rock mining face's bottom plate with a change in the advancing degree. They concluded that, in the process of the advancing face, the rock at the periphery of the bottom plate appeared to be the supporting stress zone, and with the forward advancement of the working face, there were differences in the stress zone supporting the bottom, and the plastic zone of the bottom plate peripheral rock is different from the different supporting stress zone. In addition, many scholars [26,27] have researched the stress environment of the peripheral rock of the floor roadway and the damage characteristics of the peripheral rock. They proposed using the theory of elastic mechanics to deduce the stress distribution law of the peripheral rock of the floor, and they combined it with the relevant damage criterion to calculate the maximum depth of destruction of the floor. Some scholars have also analyzed the layer relationship between the floor roadway of the working face and the coal pillar above it, as well as the quarry, by collecting on-site data [28,29]. The damage mechanism of the bottom roadway under the influence of mining and the deformation of the roadway's peripheral rock are also analyzed in depth. However, the peripheral rock of the bottom roadway under the influence of mining is in the nonuniform stress area, and the peripheral rock of the bottom roadway is affected by the two-way unequal compressive stress [30]. In the past, most scholars have focused on the change in support pressure when they studied the damage characteristics of the bottom plate and neglected the influence of a two-way unequal compressive stress field on damage to the bottom plate roadway. In recent years, a few scholars have deduced the boundary equation of the surrounding rock plastic zone under a two-way unequal compressive stress field and more comprehensively analyzed the damage characteristics of the surrounding rock plastic zone of a subgrade roadway under a nonuniform stress field, which put forward a novel idea for analyzing the damage characteristics of the surrounding rock plastic zone of a subgrade roadway under nonuniform stress fields.

To summarize, different two-way unequal compressive stress fields in the base plate under the influence of mining are the main reason for the differences in the damage characteristics of the plastic zone of the surrounding rock in the base plate roadway. There is currently less research on the distribution law of two-way unequal compressive stress fields in the base plate and the damage characteristics of the plastic zone of the surrounding rock in the base plate under two-way unequal compressive stress fields. The authors take the bottom-pumping roadway of the working face of Licun coal mine 3301 as the engineering background, use theoretical analysis to analyze the damage characteristics of the plastic zone of the surrounding rock under different bidirectional stress ratios in the bottom plate, and apply FLAC^{3D} 5.0 numerical simulation software. This is carried out to analyze the distribution law of the stress in the surrounding rock of the bottom plate and the characteristics of the distribution of bidirectional stress ratios. Furthermore, on the basis of the above, research is performed on the damage characteristics of the plastic zone of the bottom-pumping roadway when it is in different positions in the bottom plate, and then the reasonable position of the bottom-pumping roadway in the Licun coal mine is determined. The reasonable location of the bottom-pumping tunnel in the coal mine is determined, and the research results are used in the field to provide a reference basis for mines with the same geological conditions.

2. Research Background

2.1. Engineering Background

In the past, the Licun coal mine has had difficulty with gas extraction and drilling operations, along with low extraction efficiency. To ensure that the mine can meet the production requirements for the regionalization management of coal seam gas, there is an urgent need for the efficient extraction of coal seam gas in the bottom layout of the gas extraction roadway of the No. 3 coal. The working face of 3301 is mainly for mining No. 3 coal, the depth of the coal seam is about 540 m, and the average thickness of the coal is about 6 m. The No. 3 seam is stable and simple in structure, with a thickness variation coefficient of 0.24 and a recoverability coefficient of 100%, making it a stable and regionally recoverable coal seam. The average gas content is 14.52 m^3/t. The nature of the coal seam is bituminous coal, and the absolute gas outflow is 0.5 m^3/min^{-1}, which makes it a high-gas mine. The length of the working face is designed to be 160 m, and between the two working faces, there is a 30 m protective coal pillar. The roof is mainly composed of dark gray mudstone and sandy mudstone. The design length of the working face is 160 m, and there is a 30 m protection pillar between the two working faces. The whole caving method is used to deal with the goaf in the working face. According to the survey results, it can be seen that the No. 3 coal seam is a stable and simple structure within the whole mining coal seam, and the average dip angle of the coal seam is 10°. Its top plate is mainly dominated by dark gray mudstone, sandy mudstone, and siltstone, and the bottom plate is mainly black mudstone, sandy mudstone, and dark gray siltstone. Its working face schematic and the drilling column diagram of the top and bottom rock layers are shown in Figure 1. The plane diagram of the working face layout of 3301 is shown in Figure 2.

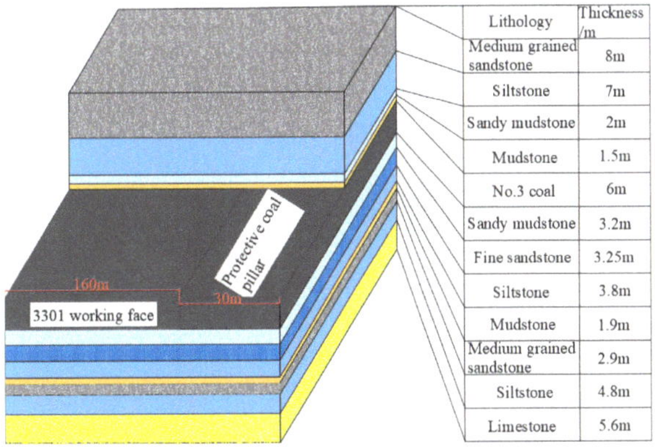

Figure 1. Schematic of 3301's working face and a column diagram of drill holes in the top and bottom rock strata.

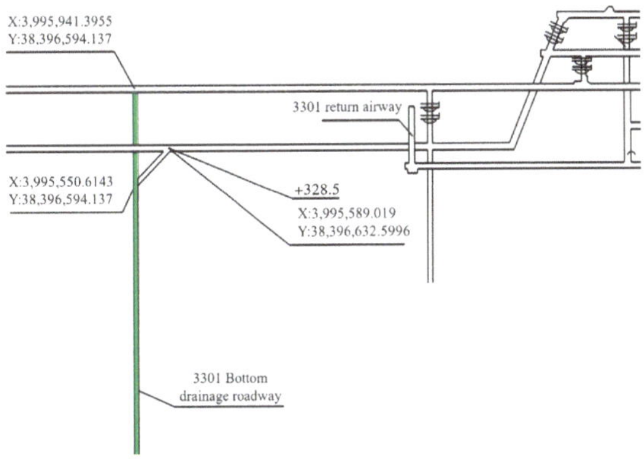

Figure 2. Plane diagram of the working face layout of 3301.

The process of roadway excavation may be affected by the SX anticline and its secondary structure. Within the scope of this influence, the lithology of the roadway side may change. The structure in the area is relatively simple, consisting of mainly wide and gentle folds. The direct water-filled aquifer of No. 3 coal is the roof sandstone fissure aquifer, and the unit water inflow of the borehole is the weak water content of the aquifer.

2.2. Determination of Mechanical Parameters of Surrounding Rock

To analyze the distribution of the bidirectionally unequal compressive stress field in the base plate of the working face and the damage characteristics of the plastic zone of the base plate's surrounding rock, samples of a coal rock body were collected from the top and bottom plates of the working face of 3301 where there was no obvious tectonic zone. Furthermore, the rock samples were processed into standard specimens, as shown in Figure 3 below, and the specimens were tested to determine the surrounding rock's physical and mechanical properties. The specimen size for the uniaxial compression test was h × φ = 100 mm × 50 mm, the specimen size for the Brazilian splitting test was h × φ = 25 mm × 50 mm, and the test specimen for the shear strength test was a 50 mm × 50 mm cube. Three specimens were tested in each group, and the test results were averaged. The displacement loading mode was used in the test, and the loading rate was 0.2 mm/min.

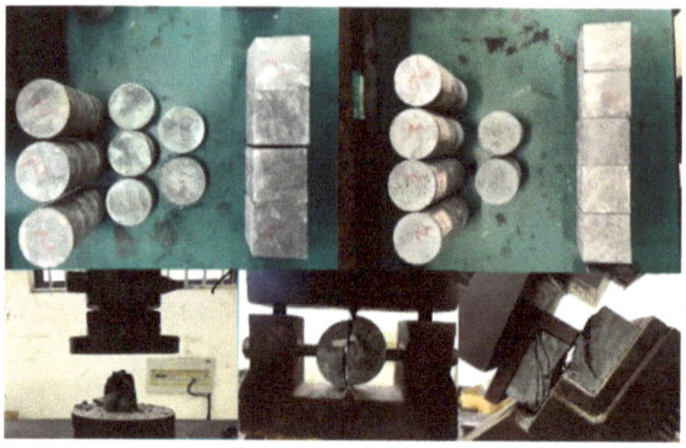

Figure 3. Coal rock body mechanical properties are part of the standard rock sample test damage map.

The test results were obtained by testing the physical and mechanical properties of the surrounding rock at the top and bottom of 3301's working face, as shown in Table 1 below.

Table 1. Statistical table of the results of the mechanical experiment on the coal rock body.

Lithology	Density ($kg \cdot m^{-3}$)	Elastic Modulus (GPa)	Normal Strength (GPa/m)	Tangential Strength (GPa/m)	Cohesive Force (MPa)	Internal Friction Angle (°)	Tensile Strength (MPa)
No. 3 Coal	1400	2.6	112	44.8	1.25	20	0.64
Mudstone	2510	5.4	228	91	9.34	27	1.99
No. 4 Coal	1400	2.7	113	45.1	1.2	19	0.59
Siltstone	2550	9.7	351.2	140	7.1	31	2.15
Calcareous Limestone	2800	10.4	464.4	185.7	16.08	36	8.05

3. Characteristics of Damage in the Plastic Zone of the Rock Surrounding the Base Plate under the Influence of Mining

When the working face is mined, the stress on the surrounding rock of the rock layer around the working face is transferred to the bottom plate due to mining. As a result, the stress field of the surrounding rock of the bottom plate is changed, and a nonuniform stress field is generated at different locations of the bottom plate. This results in different degrees of damage to the bottom plate of the working face at different locations. Furthermore, because of different positional relationships between the bottom-pumping roadway and the working face above it, the peripheral rock of the bottom-pumping roadway will inevitably be damaged to a different extent. Therefore, analyzing the damage characteristics of the peripheral rock of the bottom-pumping roadway under the influence of mining in the working face is crucial to determine the reasonable positional relationship between the bottom-pumping roadway and the working face.

3.1. Mechanical Mechanism of the Formation of the Plastic Zone in the Rock Surrounding the Floor Gas Drainage Roadway

From the theory of elastic mechanics, when the roadway is buried more than or equal to 20 times the radius of the roadway, the self-weight of the rock within the influence of the roadway can be ignored. Then, the original horizontal and vertical rock stress around the roadway can be simplified as a homogeneous load. Therefore, the model of the roadway strain on the circular hole, shown in Figure 4 below, is established, and the model's boundary conditions are as follows: the internal geometric boundary is a circular hole with a radius of r_0; P_X and P_Z are the horizontal and lead stresses, respectively, in the regional stress field under the state of the original rock stress, which are both the dominant stresses; and the stress boundary of the circular hole's internal boundary is the branching resistance, but since the roadway support resistance is much smaller than the original rock stress, it can be neglected.

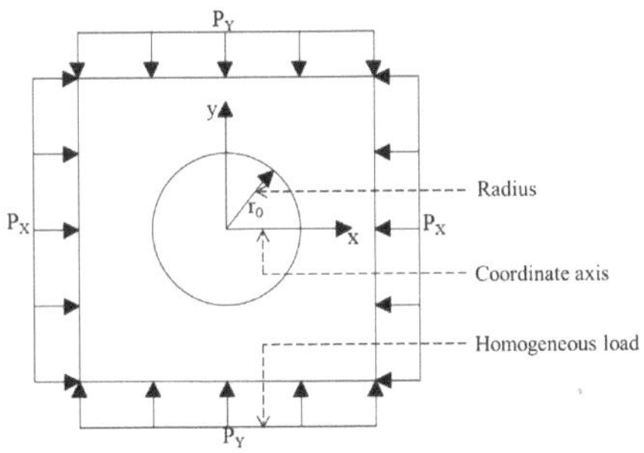

Figure 4. Mechanical model of the perimeter rock of the circular tunnel.

According to the theory of elastic mechanics, the polar coordinate expression for the stress solution at any point of the circular hole enclosure is

$$\begin{cases} \sigma_r = \frac{P_Y+P_X}{2}(1-\frac{r_0^2}{r^2}) + \frac{P_Y-P_X}{2}(1-4\frac{r_0^2}{r^2}+3\frac{r_0^4}{r^4})\cos 2\theta \\ \sigma_\theta = \frac{P_Y+P_X}{2}(1+\frac{r_0^2}{r^2}) - \frac{P_Y-P_X}{2}(1+3\frac{r_0^4}{r^4})\cos 2\theta \\ \tau_{\tau\theta} = \frac{P_Y-P_X}{2}(1+2\frac{r_0^2}{r^2}-3\frac{r_0^4}{r^4})\sin 2\theta \end{cases}, \tag{1}$$

where σ_r is the radial stress at any point outside the circular borehole, σ_θ is the circumferential stress, $\tau_{r\theta}$ is the shear stress, (r, θ) is the polar coordinates of any point, r_0 is the radius of the roadway, P_Y is the vertical stress, and P_X is the horizontal stress.

There is also the conversion relation between the stress expression in elastic mechanics in the right-angle coordinate system and the stress expression in the polar coordinate system there:

$$\begin{cases} \sigma_x = \frac{\sigma_r+\sigma_\theta}{2} + \frac{\sigma_r-\sigma_\theta}{2}\cos 2\theta - \tau_{r\theta}\sin 2\theta \\ \sigma_x = \frac{\sigma_r+\sigma_\theta}{2} - \frac{\sigma_r-\sigma_\theta}{2}\cos 2\theta + \tau_{r\theta}\sin 2\theta \\ \tau_{xy} = \frac{\sigma_r-\sigma_\theta}{2}\sin 2\theta + \tau_{r\theta}\cos 2\theta \end{cases}, \tag{2}$$

The expressions for the maximum and minimum principal stresses at any point in elastic mechanics are

$$\begin{cases} \sigma_1 = \frac{\sigma_x+\sigma_y}{2} + \frac{1}{2}\sqrt{(\sigma_x-\sigma_y)^2+4\tau_{xy}} \\ \sigma_3 = \frac{\sigma_x+\sigma_y}{2} - \frac{1}{2}\sqrt{(\sigma_x-\sigma_y)^2+4\tau_{xy}} \end{cases}, \tag{3}$$

where σ_1 is the maximum principal stress and σ_3 is the minimum principal stress.

The polar coordinate expressions for the maximum and minimum principal stresses at any point within the surrounding rock can be obtained by combining Equations (2) and (3):

$$\begin{cases} \sigma_1 = \frac{\sigma_r+\sigma_\theta}{2} + \sqrt{(\frac{\sigma_r-\sigma_\theta}{2})^2+(\tau_{r\theta})^2} \\ \sigma_3 = \frac{\sigma_r+\sigma_\theta}{2} - \sqrt{(\frac{\sigma_r-\sigma_\theta}{2})^2+(\tau_{r\theta})^2} \end{cases}, \tag{4}$$

It is also known from the Moorcullen damage criterion that the limiting damage condition of the tunnel envelope is

$$\sigma_1 = 2C\frac{\cos\varphi}{1-\sin\varphi} + \frac{1+\sin\varphi}{1-\sin\varphi}\sigma_3, \tag{5}$$

Therefore, substituting Equation (4) into Equation (3) yields an expression for the surrounding rock stress at a point in the roadway:

$$\frac{2+\sin\varphi}{2-2\sin\varphi}\sqrt{(\sigma_r-\sigma_\theta)^2+4\tau_{r\theta}^2} - 2C\frac{\cos\varphi}{1-\sin\varphi} - \frac{\sigma_r+\sigma_\theta}{2}\frac{3\sin\varphi}{1-\sin\varphi} = 0, \tag{6}$$

Substituting Equation (2) into Equation (5) and associating it with Equation (3) yields the implicit equation for the boundary of the plastic zone of the circular roadway enclosure:

$$\left\{\begin{array}{l}\frac{2+\sin\varphi}{2-2\sin\varphi}\sqrt{\left[(1+k)P_Y\frac{r_0^2}{r^2}+2(k-1)P_Y\cos 2\theta\frac{r_0^2}{r^2}\right]^2+4\left[\frac{(1-k)P}{2}\left(1+2\frac{r_0^2}{r^2}-3\frac{r_0^2}{r^4}\right)\sin 2\theta\right]^2}, \\ -2C\frac{\cos\varphi}{1-\sin\varphi}+\frac{2(1+k)P_Y\frac{r_0^2}{r^2}}{2}\times\frac{3\sin\varphi}{1-\sin\varphi}=0\end{array}\right. \quad (7)$$

3.2. Characterization of the Plastic Zone of the Base Plate Surrounding the Rock

The role of the bottom-pumping roadway is mainly to extract gas from the working face above, so it needs to be considered when determining the reasonable longitudinal layer of the bottom-pumping roadway. ① Article 21 of the Measures for Preventing Coal and Gas Outbursts stipulates that the roadway less than 10 meters away from the prominent coal seam must be excavated while exploring and the distance between the roadway and the coal seam must be ensured to be not less than 5 m. Therefore, in determining the longitudinal layer of the bottom-pumping roadway, it must be ensured that the bottom-pumping roadway roof from the coal seam above the distance of the cloth is less than 10 m [31]. ② To satisfy the use of gas-extraction equipment in the bottom roadway, the section size of the bottom roadway of the Licun coal mine is designed to be 5000 mm × 4000 mm in width × height. Combined with these two relevant provisions, and according to the rock stratum column diagram of the Licun coal mine, the main rock layers of the coal seam floor are found to be sandy mudstone, fine-grained sandstone, siltstone, mudstone, medium-grained sandstone, and limestone. In addition, the distance from the coal seam floor to the limestone is found to be 15 m. It can be seen that with the less optimal bottom-pumping roadway at the bottom of the longitudinal layer of the floor, along with the limestone lithology of the harder endowment of the stability of the bottom of the pumping roadway arranged in the limestone formation, the roadway excavation is bound to increase labor, and the bottom-pumping roadway arranged at the bottom of the roadway will also increase labor. If the bottom-pumping roadway is arranged in the limestone rock layer, the roadway excavation will certainly increase the amount of labor, and if it is arranged below the bottom-pumping roadway, the amount of drilling work will also increase when the bottom-pumping roadway is subsequently drilled for extraction. Therefore, the comprehensive consideration is to arrange the bottom-pumping tunnel above the limestone, i.e., along the medium-grained sandstone, which can not only maintain the stability of the bottom plate of the bottom-pumping tunnel but also ensure the safety distance between the bottom-pumping tunnel and the coal seam.

From the geological data of Licun coal mine 3301's working face, the buried depth of medium-grained sandstone is 540 m, i.e., the vertical stress above the bottom-pumping lane is 1.35 MPa. The designed section size of the bottom-pumping lane is width × height = 5000 mm× 4000 mm, i.e., its radius is 2.5 m. Finally, from the above physical and mechanical experiments of the surrounding rock, it can be seen that the cohesion of the medium-grained sandstone is 1.25 MPa, and the angle of internal friction is 20°. According to related scholars [32], $\frac{104}{h}+0.75<k<\frac{265}{h}+0.98$. According to the expression of the distribution law of ground stress with burial depth, it can be known that the value range of the pressure coefficient of the bottom-pumping tunnel is 0.95–1.47, and k = 1.4 for the convenience of the study, so it can be known that the horizontal stress of the surrounding rock under the original rock stress condition of the bottom-pumping tunnel is Px = 1.89 MPa.

Due to the mining in the working face, the bottom plate is affected by the mining of the stress redistribution in the surrounding rock, the formation of a nonuniform stress field in the bottom plate of the working face, and the two-way stress ratio of the surrounding rock; these are obvious differences [33]. The two-way stress ratio of the surrounding rock shows obvious differences, and the bottom-pumping lane is arranged in the rock layer of the bottom plate. Its two-way stress ratio is bound to be different under the influence of the mining of the working face. Therefore, to analyze the bottom-pumping lane with different bidirectional stress ratios under the surrounding rock's plastic-zone damage characteristics, with the original rock stress's bidirectional stress ratio $k = 1.4$ as the dividing line, we take $k = 1.0, 1.5, 2.0, 2.5, 3.0$, which can be obtained from the bottom-pumping lane surrounded by the plastic zone damage characteristics of the plastic zone, as shown in Figure 5.

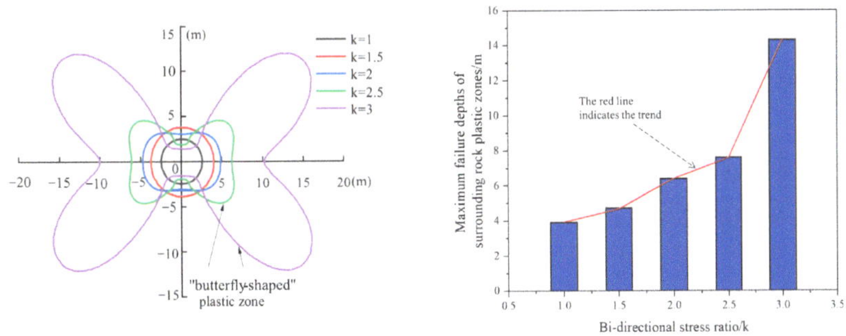

(**a**) (**b**)

Figure 5. Distribution characteristics of the plastic zone of the roadway's surrounding rock under different bidirectional stress ratios: (**a**) distribution curve of the plastic zone of surrounding rock; (**b**) distribution characteristics of the plastic zone of surrounding rock.

From Figure 5a, different two-way stress ratios under the roadway surrounding the rock plastic zone boundary curve can be seen. With the gradual increase in the two-way stress ratio k, the shape of the roadway surrounding the rock plastic zone presents obvious characteristics of a circular–elliptical–butterfly-shaped morphology change, and the surrounding rock plastic zone has "butterfly" plastic zone morphology characteristics. The expansion of the surrounding plastic zone with the two-way stress ratio has a strong sensitivity to the expansion of the plastic zone, i.e., with the increase in the two-way stress ratio, the expansion of the plastic zone is more obvious. When the plastic zone of the surrounding rock reaches the "butterfly-shaped" plastic zone, the expansion of the plastic zone of the surrounding rock has a strong sensitivity to the bidirectional stress ratio, i.e., the expansion of the plastic zone of the surrounding rock becomes more and more obvious with the increase in the bidirectional stress ratio. Combined with Figure 5b, different bidirectional stress ratios of the surrounding rock's plastic zone damage area can be seen, with a bottom-pumping lane in the bidirectional stress ratio of the larger nonuniform stress field. Furthermore, the plastic zone damage area is significantly larger than the bidirectional stress ratio of the smaller nonuniform stress field of the plastic zone's damaged area. It can be seen that the bidirectional stress ratio of the larger nonuniform stress field is a serious threat to the bottom of the pumping lane surrounding the stability of the rock. Among the conditions, $k = 1$ is the most ideal, and the degree of damage to the roadway is the smallest. As the bottom-pumping lane is arranged below the bottom plate of the working face, when the working face is mined and affected by the mining of the working face, it is bound to experience a variety of nonuniform stress fields, so choosing a reasonable location for the bottom-pumping lane and preventing it from experiencing a large nonuniform stress field are the key to maintaining the stability of the perimeter rock of the bottom-pumping lane.

4. Distribution Law of Floor Stress and Bidirectional Stress Ratio in the Working Face

From the above analysis, it can be seen that the damage characteristics of the plastic zone of the surrounding rock are mainly determined by the stress environment in which it is located, and the damage characteristics of the surrounding rock under different bidirectional stress ratios have obvious differences. In addition, under the influence of mining, the bidirectional stress ratios are bound to be different in different locations of the bottom plate of the working face. Therefore, the distribution of bidirectional stress ratios is the key to determining the reasonable location of the bottom-pumping roadway. The numerical simulation of the bottom plate after mining is analyzed using FLAC^{3D} to analyze the distribution characteristics of the bidirectional stress ratios. In this section, FLAC^{3D} numerical simulation is used to analyze the distribution characteristics of the two-way stress ratio in the surrounding rock of the bottom plate after mining to determine the reasonable location of the bottom-pumping roadway.

4.1. Modeling

According to the geological engineering conditions of the Licun coal mine, the length of the 3301 working face is 160 m, and the width of the coal pillar protected by the working face is 30 m. To avoid the influence of the boundary effect, the numerical model was established as shown in Figure 6 below, and the size of the model was 360 m × 420 m × 200 m. The intrinsic Mohr–Coulomb model was adopted for the model. In addition, horizontal displacement constraints were applied on both sides and the front and rear boundaries, vertical displacement constraints were applied on the bottom boundary, and uniformly distributed loads were applied on the upper boundary of the model. The load set degree was determined to be 13 MPa based on the depth of the burial of the coal seam, and the physical and mechanical parameters of each rock stratum from the rock mechanical parameters shown in Table 1 were adopted.

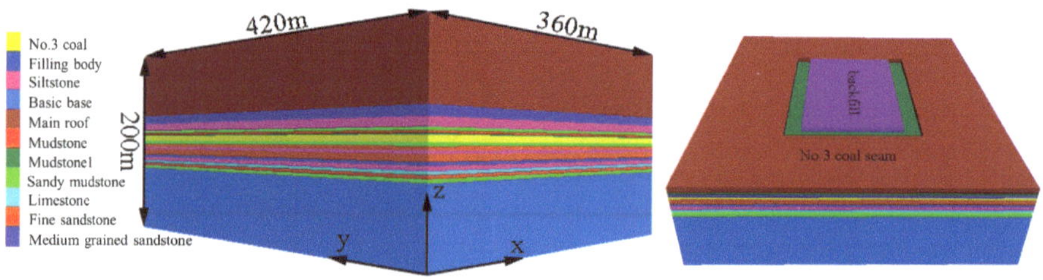

Figure 6. Numerical model diagram.

Through on-site research, we learned that the step interval of 3301's working face with cycle pressure was 20 m, and the step interval of the initial pressure was 38 m. The simulation corresponded to the real situation in the field, and to make the stress of the working face's top plate transfer to the bottom plate in the hollow area, we simulated the recovery of the bottom plate stress in the hollow area after the initial pressure of the working face and the cycle pressure. Therefore, we determined that the working face must be excavated as follows. The model was excavated along the x-axis for 160 m to simulate the length of the working face, and from the position of the y-axis at 100 m, the excavation was carried out in 12 steps, with an excavation interval of 20 m, and a total of 240 m was excavated. Then, after the third step of the excavation of the working face, the filling body of length × width × height = 120 m × 20 m × 6 m was filled at the lagging position of the working face, which was simulated using the double yield model, and its physical and mechanical parameters are listed in Table 2 below, including the simulation and its physical and mechanical parameters.

Table 2. Physical and mechanical parameters of the filling body.

Name of Rock Formation	Density/kg·m^{-3}	Bulk/GPa	Shear/GPa	Friction Angle/(°)	Tensile Strength/MPa	Resistance Max/MPa
filling body	2000	4.3556	0.068	32	0.1	0

4.2. Validation of the Reasonableness of the Numerical Simulation

From the above distribution, it can be seen that the bottom-pumping roadway is arranged in the medium-grained sandstone of the bottom plate of the working face, and the distance from the working face is 15 m. Therefore, the vertical stress cloud map at the position of 15 m of the bottom plate of the working face is intercepted as shown in Figure 7 below. The vertical stress of monitoring line 1 is arranged in the middle of the position of 15 m at the bottom plate of the working face, and the stress monitoring curve of the vertical stress monitoring line is extracted as shown in Figure 8 below.

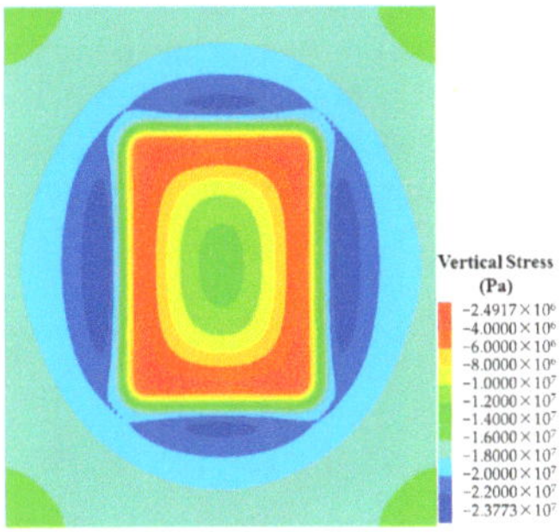

Figure 7. Vertical stress cloud at 15 m on the bottom plate of the working face.

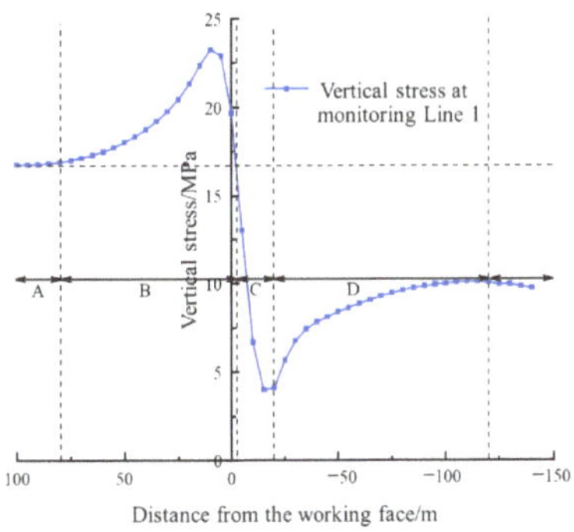

Figure 8. Vertical stress distribution curve of monitoring line 1.

According to the vertical stress cloud diagram at 15 m on the bottom plate of the working face in Figure 7, it can be seen that within the projection range of the coal pillar outside the hollow area, the vertical stress appears to be a stress-concentration phenomenon, the vertical stress of the surrounding rock outside the filling area at the edge of the projection of the hollow area decreases greatly, the vertical stress of the surrounding rock in the range of the filling area is gradually restored, and the closer to the middle of the hollow area, the better the effect of the vertical stress. Combined with the vertical stress distribution curve of monitoring line 1 in Figure 8, it can be seen that the vertical stress of the surrounding rock in the range of 80 m in front of the working face increases gradually, the vertical stress reaches a maximum of 23.5 MPa at the position of 10 m in front of the working face, the vertical stress decreases sharply in the range of 23 m behind the working face, the vertical stress reaches a minimum of 3.6 MPa at this position, the vertical stress gradually increases in the range of 24–100 m behind the working face, and the vertical stress decreases gradually in the range of 24–100 m behind the working face. In the range of 24–100 m behind the working face, the vertical stress of the surrounding rock increases gradually, and the vertical stress of the surrounding rock stays above and below 11 MPa. Therefore, the vertical stress from the edge of the model to the middle of the mining area can be divided into the original rock stress area, the stress increase area, the stress decrease area, and the stress stabilization area, and its change is in line with the distribution law of the supporting pressure of the working face's bottom plate after the coal seam is mined [34]. The change occurs due to the distribution of support pressure in the working face after mining, which verifies the accuracy of the numerical simulation.

4.3. Characteristics of the Bidirectional Stress Ratio Distribution of the Bottom Plate of the Working Face

Vertical and horizontal stress monitoring lines 2, 3, and 4 were arranged in the middle of the working face (Figure 9), and the distance from the working face is 15 m, the distance to the right of the middle of the working face is 40 m, and the distance outside the projection of the coal pillar is 15 m. The monitoring lines all went along the normal direction of the working face, and the distribution of the peripheral rock stress and the bidirectional stress ratio at different locations of the bottom plate of the working face were obtained as shown in Figure 10.

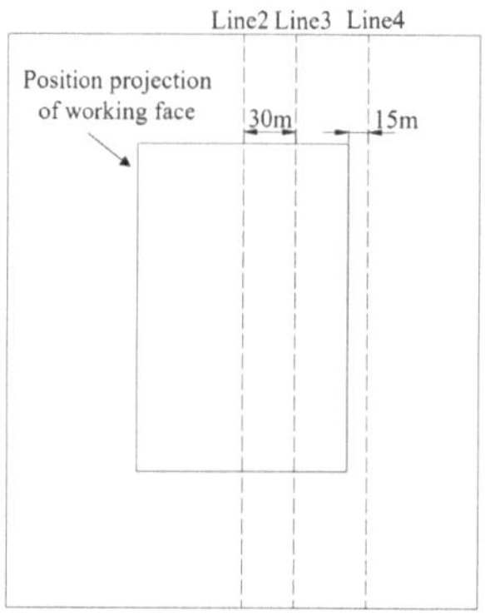

Figure 9. Line layout diagram.

According to the monitoring curve of the stress and the bidirectional stress ratio at different locations in Figure 10, it can be seen that when the vertical stress increases sharply, the horizontal stress increases slightly, and when the vertical stress decreases sharply, the horizontal stress decreases less, which means that that there is a weakly sensitive feature of the vertical stress and the horizontal stress of the surrounding rock at the bottom plate of the working face. Therefore, due to the existence of this weak sensitivity, the two-way stress ratio of the surrounding rock changes drastically at different locations of the base plate.

The monitoring line 2 stress and bidirectional stress ratio monitoring curve can be seen in Figure 10a. In the area far away from the mining range (0–60 m), the surrounding rock is in the original rock stress state, and the bidirectional stress ratio is almost unchanged, so this area is defined as the "original rock stress ratio area". In the area close to the mining range (60–100 m), the vertical stress of the surrounding rock is significantly increased by the mining. In the area near the mining range (60–100 m), the vertical stress of the surrounding rock increases significantly, the horizontal stress increases less, the bidirectional stress ratio decreases from 1 to 0.7, and the surrounding rock is in a pressurized state in this range, so this area is defined as a "pressurized low-stress ratio area". In the mining range (100–140 m), due to the mining of the working face, the surrounding rock is in a decompression state, the vertical stress decreases significantly, and the vertical stress decreases significantly, so this area is defined as an "original rock stress ratio area". In the mining range (100–140 m), due to the mining of the working face, the surrounding rock is in an unpressurized state, and the vertical stress decreases greatly, but the horizontal stress decreases less. At this time, the bidirectional stress ratio increases from 1 to 3.5, and the bidirectional stress ratio is larger. Therefore, this area is defined as the "unpressurized high-stress ratio zone". In the mining range (140–260 m), due to the top plate of the working face collapsing and compacting, the stress in the top plate is transferred downward, and the vertical stress in the bottom plate is transferred to the bottom plate. In the mining range (140–260 m), due to the collapse and compaction of the working face roof, the stress on the top plate is transferred downward, and the vertical stress on the bottom plate gradually recovers and tends to stabilize. At this time, the change in bidirectional stress is relatively small, changing from 1.4 to 1.6. Therefore, this tendency is defined as the "unpressurized stress ratio stabilization zone". The distribution of the bidirectional stress ratio is the same in the range of 260–420 m and the range of 0–260 m and will not be described again here.

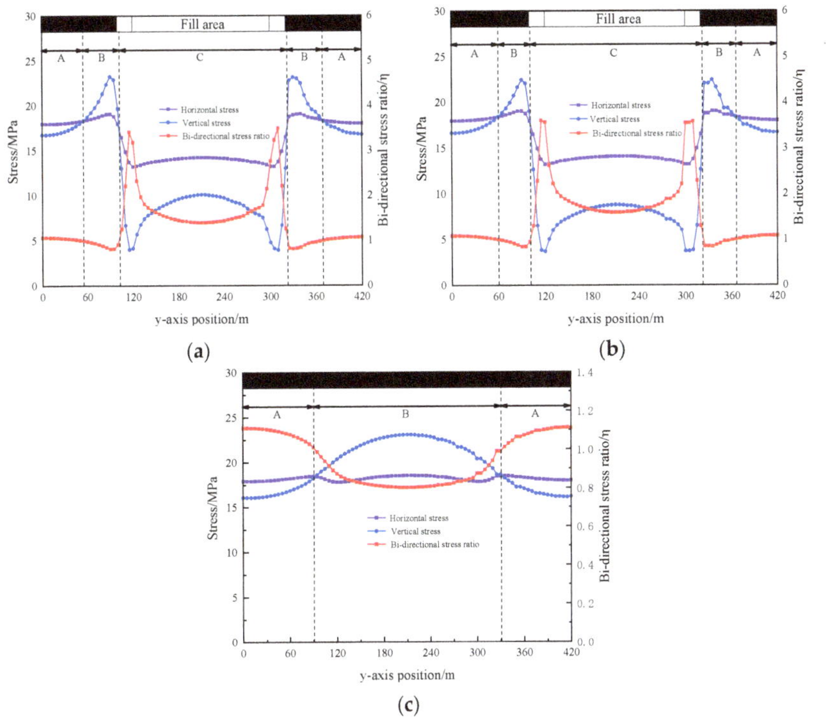

Figure 10. Monitoring curves of stress and bidirectional stress ratios at different locations: (**a**) distribution curve of stress and bidirectional stress ratio for monitoring line 2; (**b**) distribution curve of stress and bidirectional stress ratio for monitoring line 3; (**c**) distribution curve of 4-stress versus bidirectional stress ratios in the monitoring line.

From the monitoring curve of stress and the bidirectional stress ratio in monitoring line 3 in Figure 10b, it can be seen that the distribution law of the bidirectional stress ratio in the surrounding rock is similar to that in monitoring line 2, and there are the "original rock stress ratio zone", "pressurized low-stress ratio zone", "unpressurized high-stress ratio zone" and "unpressurized stress ratio stable zone" from the edge of the model to the mining area. The only difference between monitoring line 2 and monitoring line 3 is that the distribution range of the "unloading stress ratio stabilization zone" is smaller. The distribution range of monitoring line 3's "unpressurized stress ratio stabilization zone" is smaller, and the bidirectional stress ratio is larger.

As shown in Figure 10c, showing the monitoring line 4 stress and bidirectional stress ratio distribution curve, we can see that in the range of 0–90 m, the surrounding rock is in the original rock stress state, and the bidirectional stress ratio of the surrounding rock varies between 1.05–1.1, so this area is called the "original rock stress ratio area". In the range of 90–330 m, the vertical stress is significantly increased by the mining influence, but the horizontal stress increases less, and the bidirectional stress ratio increases from 0.8 to 1.0, so the bidirectional stress ratio is smaller and this area is the "pressurized low-stress ratio area". In the range of 90–330 m, the vertical stress is increased by mining, but the increase in horizontal stress is small, and the bidirectional stress ratio increases from 0.8 to 1.0, which is a small bidirectional stress ratio, so this area is called the "pressurized low-stress ratio area". In the range of 330–420 m, the surrounding rock is in the state of original rock stress, and the bidirectional stress ratio of the surrounding rock varies between 1.05–1.1, so this area is also called the "original rock stress ratio area". Therefore, this area is also called the "original rock stress ratio area". In the range of 330–420 m, the surrounding rock is in the original rock stress state, and the two-way stress ratio of the surrounding rock varies from 1.05 to 1.1, so this area is also the "original rock stress ratio area".

4.4. Characteristics of the Distribution of the Bidirectional Stress Ratio of the Working Face Floor under Distributed Excavation

The above analysis shows that when the bottom-pumping tunnel is arranged below the mining area, the bottom-pumping tunnel will experience the influence of the "original rock stress ratio area", the "pressurized low-stress ratio area", the "unpressurized high-stress ratio area", and the "unpressurized stress ratio stabilization area". If the bottom-pumping roadway is arranged under the coal pillar, it will only experience the influence of the "original rock stress ratio zone" and the "pressurized low-stress ratio zone". To analyze the influence of the bidirectional stress ratio on the bottom-pumping tunnel during the whole excavation process, the vertical and horizontal stresses after excavation are extracted from monitoring line 2, monitoring line 3, and monitoring line 4, and the distribution of the bidirectional stress ratio under excavation is obtained from different locations of the bottom plate of the working face, which is shown in Figure 11 below.

In Figure 11, it can be seen that the bidirectional stress ratios of the surrounding rock under a distributed excavation are completely different at different locations below the floor of the working face.

At monitoring line 2, before the initial pressure, the rock surrounding the bottom plate experiences the environment with the largest bidirectional stress ratio in the whole process of advancing the working face, and after the initial pressure, the bidirectional stress ratio of the surrounding rock will gradually change from a "single-peak-like" to a "double-peak-like" state, and the rock body at a depth of 50 m from the projection of the cutting eye will experience the influence of an "unpressurized high-stress ratio" during the advancing process of the working face, and it will finally be under the influence of the "unpressurized stress ratio stabilization zone".

At monitoring line 3, like at monitoring line 2, the rock surrounding the bottom plate will experience the stress environment with the largest bidirectional stress ratio in the whole process of the advancement of the working face before the initial pressure, and after the initial pressure, the bidirectional stress ratio of the surrounding rock will gradually change from a "single-peak-shape" distribution to a "double-peak-shape" distribution, and the surrounding rock will experience the influence of the "unpressurized stress ratio stabilization zone" in the process of the advancement of the working face. After the initial pressure, the bidirectional stress ratio of the surrounding rock will gradually change from a "single-peaked" distribution to a "bimodal" distribution, and the surrounding rock will experience the influence of the "unpressurized stress ratio stabilization zone" during the process of the advancement of the working face. However, what is different from the bidirectional stress ratio at monitoring line 2 is that during the advancement of the working face, the surrounding rock is located in a stress environment where the bidirectional stress ratio ranges from 2 to 9, with a larger value of the bidirectional stress ratio.

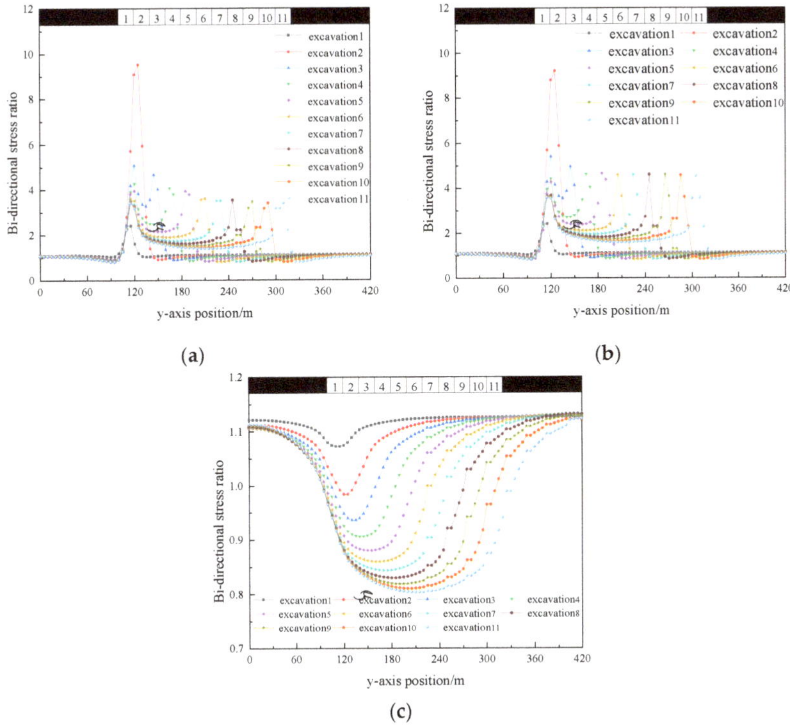

Figure 11. Bidirectional stress ratios at different locations under a distributed excavation: (**a**) bidirectional stress ratio curve for monitoring line 2; (**b**) bidirectional stress ratio curve for monitoring line 3; (**c**) distribution of bidirectional stress ratios for monitoring line 4.

At monitoring line 4, as the working face advances, the stress in the surrounding rock of the bottom plate gradually changes from the horizontal stress-led original rock stress to the vertical stress-led pressurized stress, and as the working face advances, the pressurized state of the surrounding rock gradually increases and finally remains stable, and finally, the bidirectional stress ratio of the surrounding rock is maintained at about 0.8. Therefore, in the whole process of working face advancement, the surrounding rock experiences the influence of the "original rock stress ratio area" and "pressurized low-stress ratio area".

As the bottom-pumping roadway is arranged in the rock layer of the bottom plate of the working face, the above analysis shows that when the bottom-pumping roadway is arranged in the middle of the projection of the hollow area, it is affected by the mining movement of the working face and experiences the influence of the "unpressurized high-stress ratio zone" and the "unpressurized stress ratio stabilization zone". When the bottom-pumping tunnel is arranged at the position 40 m to the right of the center of the projection of the hollow mining area, it experiences the influence of a "pressure relief high-stress ratio area" and a "pressure relief stress ratio stabilization area", with a larger bidirectional stress ratio. When the bottom-pumping tunnel is arranged at the position 15 m outside the projection of the mining hollow area below the coal pillar, it only experiences the influence of the "original rock stress ratio area" and the "pressure relief stress ratio stabilization area". When the bottom drawer is arranged below the coal pillar, i.e., 15 m outside the projection of the mining area, it only experiences the influence of the "original rock stress ratio area" and the "pressurized low-stress ratio area". According to the theory of the "butterfly plastic zone" in Section 3, the larger the bidirectional stress ratio is, the larger the surrounding rock plastic zone is. Therefore, it can be seen that when the bottom-pumping roadway is arranged below the coal pillar, i.e., 15 m outside the wrong position of the mining area, the surrounding rock plastic zone is the smallest, and its position is the most reasonable, which is verified by numerical simulation below.

5. Verification of the Reasonable Location of the Bottom-pumping Alley

The three schemes designed are shown in Figure 12. In order to verify the reasonableness of the above location of the bottom-pumping roadway arrangement, the perimeter's rock damage characteristics when the bottom-pumping roadway is arranged at different locations are simulated.

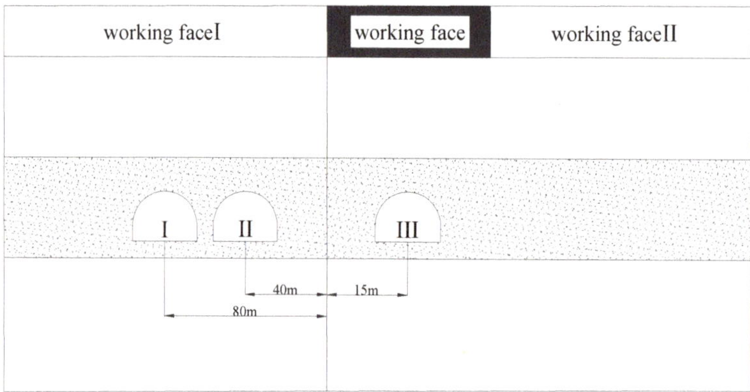

Figure 12. Simulation scheme of the bottom-pumping roadway.

Scheme 1: the bottom-pumping roadway is arranged at the center line of the working face. Scheme 2: the bottom-pumping roadway is arranged at the position 40 m outside the middle of the mining hollow area. Scheme 3: the bottom-pumping roadway is arranged at the position 15 m outside the mining hollow area. Considering the influence of the adjacent back-to-mining of the working face, the 30 m protective coal pillar will be retained after the back-to-mining of working face 1, followed by the back-to-mining of working face 2. The numerical model developed is shown in Figure 13.

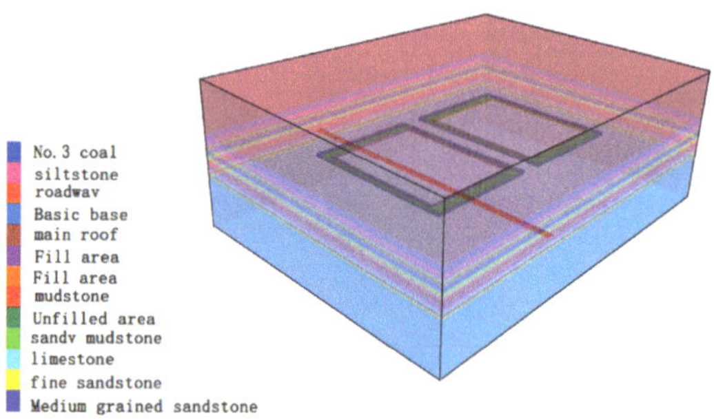

Figure 13. Bottom pumping roadway layout model.

Figure 14 below shows, for each scheme, the interception of the cut at the maximum damage area of the plastic zone of the surrounding rock, which is in the bottom draw roadway of each scheme, along with the statistics of the damaged area of the plastic zone of the surrounding rock in the bottom draw roadway.

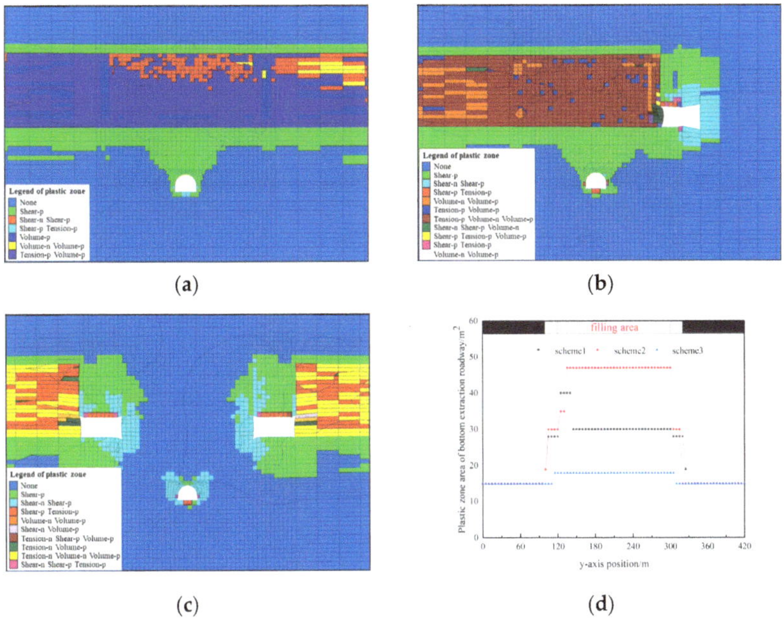

Figure 14. Distribution characteristics of the plastic zone in the bottom-pumping roadway of different schemes and statistics for the area of the plastic zone in the surrounding rock. (**a**) Scheme 1: y = 125 m; (**b**) Scheme 2: y = 220 m; (**c**) Scheme 3: y = 220 m. (**d**) Statistics for the area of the plastic zone of the surrounding rock in the bottom-pumping tunnel.

Figure 14d shows the statistics for the bottom-pumping roadway surrounding the rock plastic area. In the distribution of excavation calculations after the balance, the destruction area of the bottom-pumping roadway surrounding the rock plastic area of each program shows obvious differences. Outside the mining range of the working face, due to the small impact of mining on the working face, the damage area within each program's bottom-pumping row perimeter rock plastic area is 15 m^2. In the mining orientation of the working face, option 2 in the damage area of the bottom-pumping roadway's peripheral rock plastic area appears to increase significantly. In the working face excavation from the opening eye in the range of 20 m–45 m, the damage area within the bottom-pumping roadway peripheral rock plastic area reaches a maximum of 40 m^2. In the range of 20 m from the opening eye and the end mining penetration, the area of the plastic zone of the surrounding rock in the bottom-pumping tunnel is 28 m^2, and the area of the plastic zone of the surrounding rock in the rest of the location of the bottom-pumping tunnel is 30 m^2. In Scheme 1, the damaged area of the plastic zone of the surrounding rock of the bottom-pumping tunnel increases drastically, and the maximum damage area of the surrounding rock of the plastic zone of the bottom-pumping tunnel in the range of 25 m–200 m from the opening eye reaches 47 m^2. In Scheme 3, due to the arrangement of the bottom-pumping tunnel below the coal pillar, the bottom plate rock is affected by the pressurization and by the environment having a low stress ratio, which is the most important factor in the development of the bottom-pumping tunnel. In Scenario 3, due to the layout of the bottom roadway below the coal pillar, the surrounding rock of the bottom slab is affected by the environment being pressurized and having a low-stress ratio, and the plastic zone of the surrounding rock of the bottom roadway is less affected by the mining of the working face. In addition, the damaged area of the surrounding rock plastic zone is the smallest, and the damaged area of the surrounding rock plastic zone of the bottom drawway is only 18 m within the mining range of the working face^2. Finally, the damaged area of the surrounding rock plastic zone of the bottom roadway is the same as those of Scenarios 1 and 2, which are both 15 m^2. Combined with the different locations of the graphs in Figure 14a–c, it can be seen that the destructive characteristics of the plastic zone of the surrounding rock in the bottom-pumping tunnel, the destructive range of the plastic zone of the surrounding rock in the bottom-pumping tunnel of Scheme 1 and Scheme 2 are connected to the destructive range of the plastic zone of the bottom plate of the open area. It can also be seen that the destructive range of the plastic zone of the surrounding rock in the bottom-pumping tunnel of Scheme 3 is not connected with the plastic zone of the bottom plate of the open area of the two sides. In Scheme 1 and Scheme 2, two arrangement schemes of the bottom-pumping roadway peripheral rock are always affected by the unloading pressure at a high-stress ratio range. In addition, the bottom-pumping roadway peripheral rock plastic area under the mining air zone damaged area is large, and the roadway's top plate plastic area and the bottom plate of the mining air zone are connected to the plastic area. Scheme 3 in the bottom-pumping roadway is affected by the pressurized stress environment, but the bottom-pumping roadway is far from the peripheral rock stress concentration area, the bottom-pumping roadway peripheral rock is always in the pressurization of the low-stress ratio environment, the damaged area within the peripheral plastic area is small, and the bottom-pumping roadway is not connected with both sides of the mining air zone plastic area. At the same time, there is a large intact rock layer between the plastic zone of the top plate of the bottom-pumping roadway and the plastic zone of the bottom plate of the hollow area, which ensures the stability of the surrounding rock of the bottom-pumping roadway. Therefore, according to the degree of influence on the stability of the peripheral rock in the bottom-pumping tunnel, the order of the four stress states is depressurized high-stress ratio area > depressurized stress ratio stable area > pressurized low-stress ratio area > original rock stress ratio area. Therefore, the reasonable location of the bottom-pumping tunnel is 15 m below the coal pillar, which is the most reasonable location outside the mining area, and the correctness of the numerical analysis in Section 4 is also verified.

6. Field Application and Effect Analysis

The results of the above analysis show that the reasonable location of the bottom-pumping roadway in the working face of Licun coal mine 3301 is below the coal pillar and dug along the limestone roof plate. The roadway is supported by anchor rods + anchor cable joint support, the row spacing between anchor rods is 900 mm × 1400 mm, and the row spacing between anchor cables is 2400 mm × 1400 mm; the supporting schematic diagram is shown in Figure 15 below.

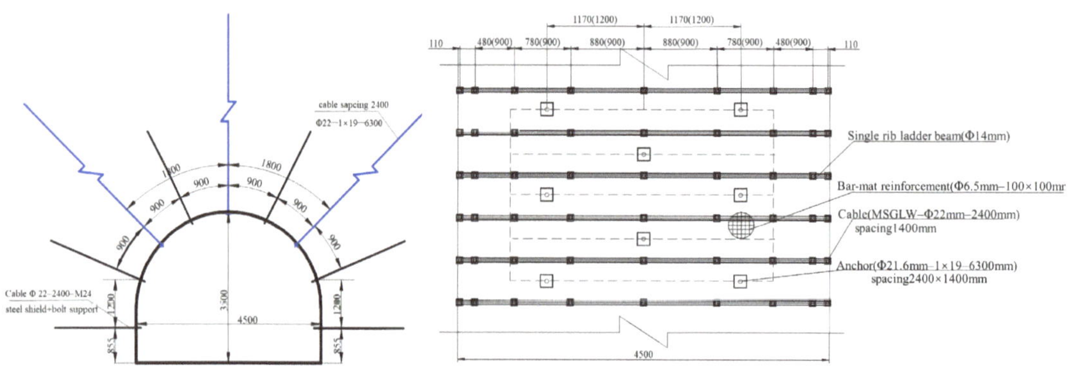

(a) (b)

Figure 15. Schematic diagram of the bottom draw alley support: (**a**) cross-section of bottom drawer support; (**b**) plan view of bottom drawer support.

The real-time monitoring of the deformation of the surrounding rock and the force of the anchor rods and cables in the bottom-pumping roadway was carried out in a range of 200 m, and the monitoring results are shown in Figure 16 below.

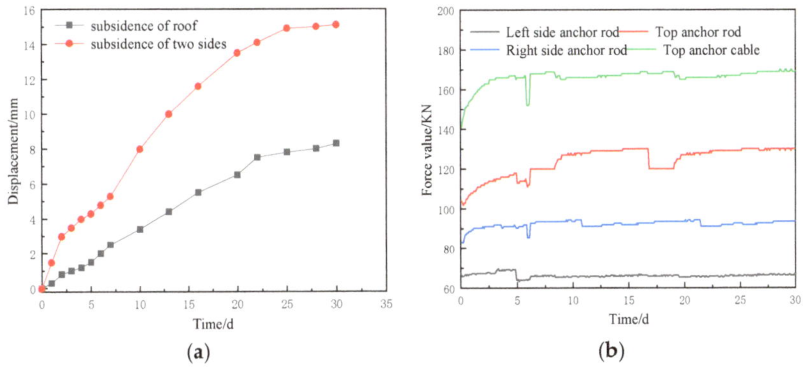

Figure 16. Deformation of the surrounding rock in the bottom-pumping roadway and the force on the anchor rods and anchors: (**a**) deformation of the roadway perimeter rock; (**b**) strength of the anchor rods and cables.

According to the monitoring results, the maximum value of the roadway roof subsidence is 9 mm. When the deformation of the two sides and the roof and floor of the roadway does not exceed 10% of the designed section size of the roadway, the roadway can serve normally. The monitoring results show that the actual roadway deformation is less than 1%, the maximum value of the two gangs approaching is 16 mm, the deformation of the roadway is larger within 0~20 d, and the deformation of the roadway is smaller. The deformation rate is low after 20 d, and the peripheral rock of the roadway remains stable. The roadway gradually slows down the deformation rate at 120 m from the headland (the slope rate becomes smaller). The roadway is stabilized 170 m away from the headland, and the anchor rods and anchor cables are subjected to synchronous force during the deformation of the peripheral rock. Furthermore, anchor rods and an anchor cable in the peripheral rock experience a deformation synchronization force, and the overall force presents the first increase after the trend of stability. When the force of the anchor cable is less than 50% of the initial force, the support effect is considered to be lost. At this time, the anchor rods and anchor cable have a role in the roadway above the peripheral rock, limiting its overall displacement and interlayer misalignment. The formation of the peripheral rock and the support community give full play to the role of the anchor rods and the anchor cable support. They also support the deformation of the roadway peripheral rock, which does not increase significantly after the whole is in a stable state. This proves that the choice of the location of the bottom draw and the support program design is reasonable.

7. Discussion

At present, the layout method of bottom-drainage roadways mainly includes theoretical analysis and empirical methods. In terms of theoretical calculation, due to the simplification of relevant specific mining and geological conditions in the calculation process, there is a certain deviation between the final results and the actual situation. The empirical method is strongly constrained by the testing conditions, the actual deviation is very large, and it is even easy for engineering accidents to occur. Therefore, the method of combining theory with numerical simulation is used to study the layout of the bottom-drainage roadway. This method is not limited by specific mining conditions and has strong applicability. The collection of rock data involved in this method is also relatively easy. However, it is necessary to point out that for numerical simulation, because it is difficult to select accurate parameters when establishing numerical models, there is a certain deviation in reflecting the actual situation on site. Therefore, the feasibility of the method should be verified through subsequent engineering practices and further improved and optimized.

8. Conclusions

(1) In the nonuniform stress field under the influence of mining in the upper working face, and with the increase in the bidirectional stress ratio, the damage characteristics of the roadway surrounding the rock show the characteristics of a "butterfly" nonuniform distribution, and the "butterfly" nonuniform damage to the rock surrounding the roadway significantly reduces the stability of the rock surrounding the roadway.

(2) The working face of 3301 is mainly for the mining of No. 3 coal, and its roof is dominated by dark gray mudstone and sandy mudstone, while the floor is mainly black mudstone and sandy mudstone. After the excavation of the working face, according to the size of the two-way stress ratio and the state of the loading and unloading of the surrounding rock stress, the floor of the coal bed can be divided into four kinds of stress states, namely "original rock stress ratio area", "unloading high-stress ratio area", "pressurized low-stress ratio stable area", and "pressurized high-stress ratio area". When the bottom drawer is arranged below the working face, it will experience an "unloading high-stress ratio area" and a "pressurized low-stress ratio stable area". When the bottom-pumping road is arranged below the working face, it will experience an "unpressurized high-stress ratio zone" and an "unpressurized stress ratio stability zone". When the bottom-pumping tunnel is arranged below the working face, it will experience the influence of a "decompression high-stress ratio area" and a "decompression stress ratio stabilization area". When the bottom-pumping tunnel is arranged below the coal pillar, it will experience the influence of a "pressurized low-stress ratio area".

(3) The theory of the "butterfly shape" of the plastic zone indicates that the reasonable location of the bottom-pumping roadway is 15 m below the coal pillar. The damage range and deformation characteristics of the plastic zone at different locations of the bottom-pumping tunnel in the coal seam bottom plate are in the following order: "unpressurized high-stress ratio zone" > "unpressurized stress ratio stable zone" > "pressurized low-stress ratio zone" > "original rock stress ratio zone", which verifies the reasonableness of the bottom-pumping roadway arranged at the 15 m position outside the faulty hollow area below the coal pillar.

(4) Industrial experiments show that the bottom-pumping roadway is arranged in the coal pillar below the hollow area outside the wrong 15 m. Along where the digging of the limestone roof plate is carried out, the maximum value of the roof sinks by 9 mm, the maximum value of the two gangs approaches 16 mm, the deformation of the roadway's perimeter rock is in a reasonable range, and the location of the bottom-pumping roadway has a reasonable layout.

Author Contributions: Conceptualization, L.B. and F.W.; methodology, B.L.; software, P.Y. and S.Z.; validation, P.Y.; formal analysis, L.B.; investigation, L.B.; data curation, B.L.; writing—original draft preparation, L.B.; writing—review and editing, L.B., P.Y. and F.W.; visualization, L.B., P.Y., S.Z. and F.W.; funding acquisition, F.W. All authors have read and agreed to the published version of the manuscript.

Funding: This research received no external funding.

Institutional Review Board Statement: Not applicable.

Informed Consent Statement: Not applicable.

Data Availability Statement: The data presented in this study are available in article.

Conflicts of Interest: Author Libin Bai was employed by the company Liuta Coal Mine, China National Energy Shendong Coal Group.The remaining authors declare that the research was conducted in the absence of any commercial or financial relationships that could be construed as a potential conflict of interest.

References

1. Sotskov, V.; Dereviahina, N.; Malanchuk, L. Analysis of operation parameters of partial backfilling in the context of selective coal mining. *Min. Miner. Depos.* **2019**, *13*, 129–138. [CrossRef]
2. Wu, F.F.; Zhang, S.R.; Yue, X.; Liu, C.; Liu, Q.; Zhang, J.; Lv, B.; Gao, Z. Study on mechanism and improvement technology of Top coal loss in horizontal sublevel fully mechanized caving mining. *Energy Explor. Exploit.* **2022**, *40*, 1394–1408. [CrossRef]
3. Moomivand, H.; Moosazadeh, S.; Gilani, S.O. A new empirical approach to estimate the ratio of horizontal to vertical in-situ stress and evaluation of its effect on the stability analysis of underground spaces. *Rud.-Geol.-Naft. Zb.* **2022**, *38*, 97–107. [CrossRef]
4. Dychkovskyi, R.; Shavarskyi, I.; Saik, P.; Lozynskyi, V.; Falshtynskyi, V.; Cabana, E. Research into stress-strain state of the rock mass condition in the process of the operation of double-unit longwalls. *Min. Miner. Depos.* **2020**, *14*, 85–94. [CrossRef]
5. Yang, P.J.; Zhang, S.R.; Wang, X.F. Numerical Study on the Characteristics and Control Method of Coal Leakage between Supports in Integrated Mining of Extremely Loose and Soft Coal Seams. *Energies* **2024**, *17*, 1013. [CrossRef]
6. Zholmagambetov, N.; Khalikova, E.; Demin, V.; Balabas, A.; Abdrashev, R.; Suiintayeva, S. Ensuring a safe geomechanical state of the rock mass surrounding the mine workings in the Karaganda coal basin. *Min. Miner. Depos.* **2023**, *17*, 74–83. [CrossRef]
7. Xie, J.L.; Xu, J.L.; Wang, F. Mining-induced stress distribution of the working face in a kilometer-deep coal mine-a case study in Tangshan coal mine. *J. Geophys. Eng.* **2018**, *15*, 2060–2070. [CrossRef]
8. Zhang, Y.J. Distribution law of floor stress during mining of the upper protective coal seam. *Sci. Prog.* **2020**, *103*, 0036850420930982. [CrossRef] [PubMed]
9. Xiong, T.G.; Chang, J.C.; He, K.; Su, Y.F.; Qi, C. Study on deformation and failure characteristics of surrounding rock of overlying roadway under upward mining in the deep mine. *Geofluids* **2021**, *2021*, 5667183. [CrossRef]
10. Shan, R.L.; Wei, Y.H.; Wang, C.H.; Li, Z.; Zhao, Y.; Tong, X.; Xiao, S. Research on the evolution of the deviatoric stress and control of a gas extraction roadway under mining influence. *Environ. Earth Sci.* **2023**, *82*, 425. [CrossRef]
11. Guo, X.F.; Li, Y.G.; Zhou, G.D.; He, Z.; Yu, H.; Xu, Y. Stability analysis and reasonable layout of floor drainage roadway above confined water and under mining influence. *Geofluids* **2021**, *2021*, 5578717. [CrossRef]
12. Li, Y.E.; Ma, N.J.; Ma, J.; Zhang, H.; Gao, Z. Characteristics of perimeter rock damage and reasonable location of deep pressurized water upper bottom pumping tunnel. *J. Coal* **2018**, *43*, 2491–2500.
13. Li, A.; Ma, Q.; Lian, Y.Q.; Ma, L.; Mu, Q.; Chen, J. Numerical simulation and experimental study on floor failure mechanism of typical working face in thick coal seam in Chenghe mining area of Weibei, China. *Environ. Earth Sci.* **2020**, *79*, 118. [CrossRef]
14. Lu, C.J.; Xu, J.P.; Zhao, H.; Zhang, H.; Zhang, Y.; Chen, M. Floor disturbance and failure characteristics of super-large mining height working face. *Geofluids* **2022**, *2022*, 1279642. [CrossRef]
15. Zhang, D.M.; Sun, G.Y.; Li, G. Stability analysis and maintenance program design of roadway in remining coal seams under the influence of double mining. *J. Eng. Geol.* **2021**, *29*, 1028–1036.
16. Xiao, T.Q.; Bai, J.B.; Xu, L.; Zhang, X. Characteristics of stress distribution in floor strata and control of roadway stability under coal pillars. *Min. Sci. Technol.* **2011**, *21*, 243–247. [CrossRef]
17. Tang, Y.S.; Sun, W.C.; Zhang, X.; Liu, P. Effect of Advancing Direction of Working Face on Mining Stress Distribution in Deep Coal Mine. *Adv. Civ. Eng.* **2021**, *2021*, 7402164. [CrossRef]
18. Yu, Y.; Zhang, L.Y.; Lu, J.F.; Chen, D.; Zhao, X.; Liu, L. Research on the evolution characteristics of floor stress and reasonable layout of roadways in deep coal mining. *Geofluids* **2021**, *2021*, 6631444. [CrossRef]
19. Wang, Z.Q.; Wu, C.; Shi, L.; Su, Z.H.; Wang, P.; Huang, X. Analysis of surrounding rock stress and plastic zone of two-way unequal pressure circular roadway based on complex variable theory. *J. Coal* **2019**, *44*, 419–429.
20. Peng, Y.J.; Li, Q.F. Floor failure and roadway deformation induced by contiguous coal seams mining at Huopu Mine. *Arab. J. Geosci.* **2020**, *13*, 601. [CrossRef]
21. Cui, J.F.; Wang, W.J.; Jia, Q.; Peng, G.; Wu, H. Measurement and analysis of roadway deformation and stress under mining-induced stress. *Shock. Vib.* **2021**, *2021*, 5561093. [CrossRef]
22. Zhu, Y.; Yang, K.; Li, Y.L.; Li, W. Stability analysis and control countermeasures of laminated bottom plate in mining roadway under different coal pillar sizes. *J. Min. Saf. Eng.* **2023**, *40*, 467–479.
23. Gao, S.; Yang, J.; Gao, D.; Ouyang, Y.B.; Chai, J. Lay-out pattern and deformation law of mining roadway incontiguous coal seam mining. *J. Xi'an Univ. Sci. Technol.* **2022**, *42*, 191–199.
24. Chi, X.L.; Yang, K.; Fu, Q.; Dou, L. The mechanism of mining-induced stress evolution and ground pressure control at irregular working faces in inclined seams. *Geotech. Geol. Eng.* **2020**, *38*, 91–107. [CrossRef]
25. Dinmohammadpour, M.; Nikkhah, M.; Goshtasbi, K.; Ahangari, K. Application of the Kaiser Effect in In-situ Stress Measurement in Rocks-an overview. *Rud.-Geol.-Naft. Zb.* **2022**, *37*, 1–6. [CrossRef]
26. Sun, Y.X. Research on the deformation mechanism of mining roadway stratiform surrounding rock with nonuniform stress field. *Shock. Vib.* **2021**, *2021*, 6624045. [CrossRef]
27. Nehrii, S.; Nehrii, T.; Zolotarova, O.; Volkov, S. Investigation of the geomechanical state of soft adjoining rocks under protective constructions. *Rud.-Geol.-Naft. Zb.* **2021**, *36*, 14–22. [CrossRef]
28. Wang, P.; Jiang, L.S.; Ma, C.Q.; Yuan, A. Evolution Laws of Floor Stress and Stability of Floor Roadway Affected by Overhead Mining. *Earth Sci. Res. J.* **2020**, *24*, 45–54. [CrossRef]
29. Zhang, D.; Bai, J.B.; Yan, S.; Wang, R.; Meng, N.; Wang, G. Investigation on the failure mechanism of weak floors in deep and high-stress roadway and the corresponding control technology. *Minerals* **2021**, *11*, 1408. [CrossRef]

30. Tian, M.L.; Han, L.J.; Yang, X.X.; Feng, Q.; Meng, Q.; Xie, Y.; Li, W.; Ma, C. Asymmetric deformation failure mechanism and support technology of roadways under non-uniform pressure from a mining disturbance. *Bull. Eng. Geol. Environ.* **2022**, *81*, 211. [CrossRef]
31. State Administration of Work Safety; State Administration of Coal Mine Safety. *Regulation on Prevention and Control of Coal and Gas Outburst*; Coal Industry Press: Beijing, China, 2009.
32. Wang, Z.Q.; Yan, E.; Lu, G.D.; Gao, L.T.; Zhang, Q.M.; Tang, R.X. Statistical analysis of the distribution pattern of ground stress field in the reservoir site area of groundwater sealed hole reservoirs in mainland China. *Geotechnics* **2014**, *35*, 251–256.
33. Cao, W.X.; Liu, H.L.; Hang, Y.G.; Wang, H.; Li, G. Similarity simulation on the movement characteristics of surrounding rock and floor stress distribution for large-dip coal seam. *Sensors* **2022**, *22*, 2761. [CrossRef] [PubMed]
34. Song, Z.Q.; Wen, Z.J.; Jiang, Y.J.; Jiang, J.Q.; Shi, Y.K. Key theories and engineering applications of mining dynamics and rock control. *J. Coal* **2023**, 1–21. [CrossRef]

applied sciences

MDPI

Article

Numerical and Analytical Determination of Rockburst Characteristics: Case Study from Polish Deep Copper Mine

Witold Pytel [1], Krzysztof Fuławka [1,*], Bogumiła Pałac-Walko [2] and Piotr Mertuszka [1]

1 KGHM CUPRUM Ltd., Research & Development Centre, 53-659 Wrocław, Poland; witold.pytel@kghmcuprum.com (W.P.); piotr.mertuszka@kghmcuprum.com (P.M.)
2 Faculty of Geoengineering, Mining and Geology, Wrocław University of Science and Technology, 50-421 Wrocław, Poland; bogumila.palac-walko@pwr.edu.pl
* Correspondence: krzysztof.fulawka@kghmcuprum.com

Abstract: A simplified analytical method useful for ductile ground support design in underground mine workings is presented. This approach allows for maintaining the stability of sidewalls in rectangular openings extracted in competent and homogeneous rocks, especially in high-pressure conditions, favoring rockburst event occurrence. The proposed design procedure involves the typical assumptions governing the limit equilibrium method (LEM) with respect to a triangular rock block expelled from a sidewall of a long mine excavation subjected to normal stresses of the values determined based on the Maugis's analytical solution concerned with stress distribution around the elliptical opening extracted within the homogeneous infinite elastic space. This stage of the local assessment of rock susceptibility to ejection from the walls of the excavation allowed for determining the geometry of the block whose ejection is most likely in a given geological and mining situation. Having extensive information about the geometry of the excavations and the properties of the surrounding rocks, it was possible to make an exemplary map of the risk from rockburst hazard, developed as the 2D contours of safety indexes' values, for special-purpose excavations such as heavy machinery chambers, main excavations, etc. in conditions of selected mining panel of the deep copper mine at Legnica-Głogów Copper Basin, Poland. Another important element of the obtained results is the calculated values of the horizontal forces potentially pushing out the predetermined rock blocks. These forces are the surplus over the potential of frictional resistance and cohesion on the surfaces of previously identified discontinuities or on new cracks appearing as a result of overloading of the sidewalls. Finally, the presented algorithm allows us to perform quantitative tracking of rockburst phenomena as a function of time by determination of acceleration, velocity, and displacement of expelled rocks. Such information may be useful at the stage of designing the support for underground workings.

Keywords: rockburst hazard; numerical modeling; limit equilibrium method; finite element method; underground mining; ground control

1. Introduction

Ensuring the stability of mine workings located at a great depth is one of the most challenging issues facing the mining industry [1]. High stresses within the rock mass have a negative effect on the productivity and safety of mining operations, and in some cases may jeopardize the entire mining project [2,3]. One of the most serious hazards associated with mining at great depths is a rockburst phenomenon, which may be described as an uncontrolled violent event associated with a sudden release of strain energy, resulting frequently in deformation or even destruction of mine workings [4–7]. After such an event, the excavation walls usually take the form of a curvilinear cavity, which substitutes the original flat surfaces [8,9]. As it was pointed out by Wang et al. [10] as well as Małkowski and Niedbalski [11], the problem of rockburst occurrence is present on a global scale and has been recently noticed in deep mines over the world. However, the knowledge of a rock mass's potential for dynamic and destructive behavior, such as rock and pillar bursts in underground excavations, still has not yet been sufficiently explained, formulated, or consistently applied in practice [12,13]. Therefore, rockburst hazard has been increasingly common in the mining sector worldwide, often leading to huge social and economic losses [14,15]. In order to minimize this risk, it is necessary to develop and implement new methods of monitoring and estimating the rockburst risk, especially at great depths, where the stress level is higher and higher. Locating places with an increased probability of instability may be the basis for taking appropriate preventive measures in advance.

Within this paper, the complex evaluation of rockburst risk in the conditions of Polish underground copper mines has been proposed. Based on the analytical as well as numerical formulations, the characteristics of rockbursts occurring in underground galleries of a selected copper mine have been described and verified. The developed approach is particularly useful for structurally controlled instability events where the rock mass surrounding the opening fails by sliding on the discontinuities and crushing of rock pieces. For estimation of the rockburst potential, a "rockburst factor of safety" has been formulated. Generally, it is a ratio of so-called "critical depth" to the depth at which the mining operation is currently carried out. The value of the critical depth depends on geological and mining conditions at which a rockburst may develop if the driving forces are equal to the rock mass resistance. For the purposes of local rockburst characteristic determination, a parameter called "safety index" has also been introduced. In general, this parameter describes the ratio of resisting forces to driving forces which affects the movement of the wedge. The results of the analysis have been compared with historical cases of rockbursts observed in one of the Polish underground copper mines.

2. Recent Advances in the Evaluation of Rockburst Characteristics

One of the significant parameters that allows for the sufficient description of the rockburst mechanism as a function of time is the motion's velocity (v_e), which may be assessed from the following relationship based on the empirical back-calculation approach [16]:

$$v_e = d\sqrt{\frac{g}{2hcos^2\theta + dsin^2\theta}} \quad (1)$$

where θ—initial ejection angle measured upwards from the horizontal; h—1/2 height of excavation; d—observed horizontal range of the motion; g—gravity. This equation will be later used for validation of the usefulness of the below-proposed approach for the rockburst hazard estimation.

Appropriate calculations proved that the value of the above velocity reaches about 5 m/s in the case of serious events. For the cases in which the ejected rock mass is smaller, this velocity is greater and reaches a value of about 10 m/s [17,18]. As pointed out by Barton [19], in the case of good-quality rocks, one may assume that under high-stress conditions, multiple joints may develop, and their spatial orientations should fulfill the condition of the minimum value of a safety index. The issue mentioned herein is a typical problem occurring during underground excavation and should be taken into consideration during the planning of mining works. Thus, it is recommended to develop small-scale models at the first stage for a better understanding of the considered processes rather than large models that are theoretically able to give an absolutely accurate answer, but at the same time, they use a lot of parameters whose values are characterized by unknown reliability levels [20]. Therefore, the implementation of an analytical approach that permits the evaluation of rockburst safety indexes around the excavation is highly recommended at the preliminary risk evaluation stage. Such an analytical approach may be based on the modified limit equilibrium method (LEM), which has been used for years, particularly for obtaining approximate solutions for different stability problems in rock mechanics and geotechnical engineering [21–23].

It should be kept in mind that LEM also requires some assumptions in terms of the stress distribution along the failure/wedge surface, which allows for providing a general equation of mechanical equilibrium addressed to different structures. If more complicated mechanisms of failure need to be considered, such as translation integrated with rotation, the literature proposes different more complex algorithms, e.g., Discontinuous Deformation Analysis (DDA) [24–26] or Block Stability 3D (BS3D) [27,28], which also permits considering the dynamic events. However, these numerical methods require reliable knowledge of joint systems, which unfortunately is not always available, especially near the mining fronts. Therefore, in most cases, these methods are simplified and limited just to translation mechanisms [29] without any severe losses in accuracy [30,31] and may be utilized in typical geologic conditions in underground mines.

In turn, as highlighted by Wang et al. [10], from numerical approaches and codes in rock mechanics applied for rockburst prediction in the last 20 years, continuous methods such as the finite element method (FEM) and finite difference method (FDM) are still the most common. These methods involved over 77% of the published works, whereas the discontinuous methods (e.g., DEM—discrete element method) were utilized in more than 20% of publications. The first successful attempts at numerical rockburst predictions were presented in 1972 when Blake determined possible locations of rockburst occurrence [32]. Then rapid development of rockburst analyses with dynamic FEM was observed which resulted in the improvement of reliable constitutive models for underground mining conditions [33–35]. Today, FEM, FDM, and DEM seem to be the most suitable methods for determining the shape and characteristics of rockburst formation under high-stress levels. However, in a situation where a general view of the risk of rock outbursts over a very large area of mine workings is needed, as well as data on the geometry of excavations and current fracture systems are very limited, the advantages of the abovementioned methods may be offset by inconveniences such as the necessary huge amount of labor-intensive and time-consuming research work carried out underground and at the computer. A similar situation is related to the quantity and quality of geotechnical data, which are routinely limited to four parameters describing the deformation and strength properties of the rocks surrounding the excavations. Taking into account the abovementioned limitations, the authors finally decided to implement a simplified method based on LEM as a main rockburst prediction approach and validate its reliability with FEM-based simulations. The calculations were performed for a 2-dimensional case relating to the tunnels and gallery point located far enough from the intersections and other underground workings, utilizing both an LEM and FEM approach and a 3-dimensional case representing conditions observed within the pillars at the intersections (FEM).

3. Material and Methods

The stability analyses were performed for specific mining-geologic conditions adequate for one of the Polish underground copper mines. These conditions comprise typical geometries of the openings (depth, height, width), typical surrounding rock types with their deformation/strength parameters, and the mining technology which is the most often applied, i.e., a single-level room-and-pillar-mining method with roof deflection at the depth of 1000–1200 m below the ground surface. The generalized scheme of the utilized mining system is presented in Figure 1, where one may also find the epicenters of strong seismic events as well as the location of massive rockbursts that occurred after them.

The geometry of the numerical model and geologic profile has been prepared based on 12 boreholes which were drilled from the surface to the level of excavation. In turn, the strength parameters of rocks have been obtained with laboratory tests.

In the direct roof of the analyzed mine, stiff layers of dolomite and limestone are located, whereas, above them, strong strata of dolomite and anhydrite are present. Uniaxial compressive strength in both cases reaches the value of 220 MPa. At the same time, floor layers consist of strata of red sandstone with medium strength and UCS up to 80 MPa. Strong roof layers with the simultaneous presence of medium-strength layers in the floor generate conditions that *favor* the softening and yielding of the floor and pillars and result in the accumulation of elastic energy above the mining area.

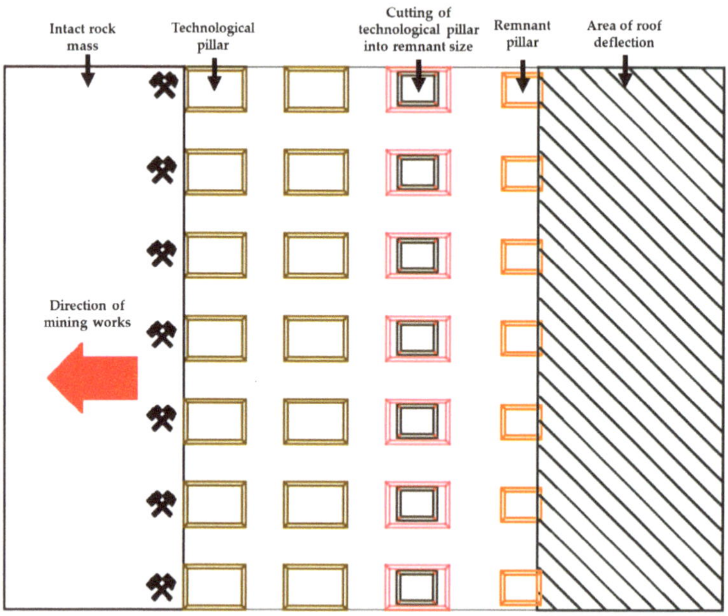

Figure 1. Scheme of utilized room and pillar-mining system.

Figure 2 shows the mining situation in the analyzed area and indicates the places where high-energy tremors and rockbursts occur.

Figure 2. Example of room-and-pillar mining with roof deflection in Polish copper mines—the site where the proposed approach was validated.

The general flowchart diagram presenting the used methodology for rockburst hazard estimation is shown in Figure 3.

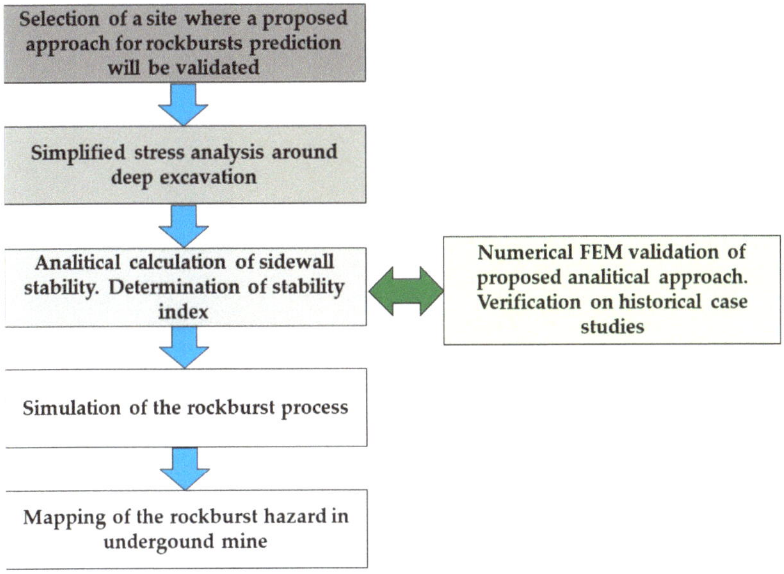

Figure 3. General flowchart diagram explaining the proposed methodology.

3.1. Analytical Solution

The general closed-form solution of Maugis [36] concerning stress distribution around the elliptical excavation within the homogeneous and infinite elastic space was the base for the analytical LEM formulation developed with respect to mine opening safety. This solution has several undisputable advantages such as conciseness, simplicity, and ease of coding in computer programs, which led it to be used willingly in many technical applications until today. Newer studies in this field (e.g., [37–39]), based mainly on complex potential functions and conformal mapping representation, are characterized by considerable complexity, which means that they so far are rarely used in practical applications.

Using the Maugis formulation, the relationships describing normal stress σ_z and σ_x distributed at the plane $z = b = 0.5 \times H$, which are the result of the presence of the uniformly distributed horizontal pressure p_x and the vertical pressure $p_z = \gamma_{av} H_0$, have been obtained. For the normal stresses acting in the $z = b = 0.5 \times H$, the following relationships were obtained:

$$\sigma_z = \frac{1}{2} \times \left(\overline{A} + \overline{B} + \overline{C} + \overline{D}\right) \tag{2}$$

$$\sigma_x = \frac{1}{2} \times \left(\overline{A} + \overline{B} - \overline{C} - \overline{D}\right) \tag{3}$$

where $\overline{A}$, $\overline{B}$, $\overline{C}$, and $\overline{D}$, the algebraic types of functions of the elliptical opening size (half-axis a and b), coordinate x and the external load, p_x and p_z.

A formulated stability condition for the sidewalls of rectangular excavations that are prone to rockburst occurrence introduces the assumptions that normal stress σ_x and σ_z distribution at height $z = 0.5 \times H$ in the sidewalls of the excavation of width B and height H are approximately equal to the corresponding values of the principal stresses obtained for the excavation of the elliptical cross-section with the half-axes equal to a and b ($a > b$). Also, the ejected rock wedge is of the shape of a triangle defined by the angles of α and β that identify the lines AB and CB, i.e., the envelope of the area where the stability index value is less than 1 (Figure 4):

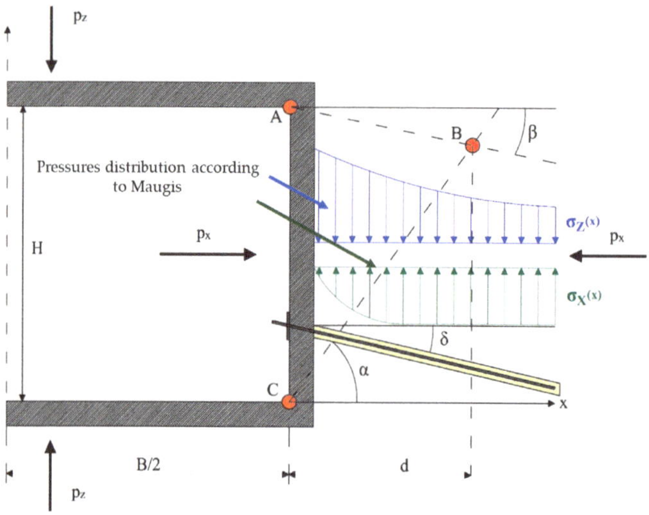

Figure 4. Physical model of the sidewall loading for the case of a deep opening excavated within a homogeneous or joined rock mass.

The stability index for the rock wedge ABC (Figure 3) and unbalanced ejecting force may be determined using the relationship presented below, based on the limit equilibrium principles:

$$\text{stability index}: \quad \overline{F} = \frac{F_p}{F_a} = A_f + A_c + A_b \tag{4a}$$

$$\text{or unbalanced force}: \; P_k = F_a - F_p \tag{4b}$$

where A_f—relative resistance of internal friction forces; A_c—relative resistance of cohesion forces; and A_b—relative resistance of the forces mobilized within n rods of rock bolts, each of the capacity N_i and inclination angle δ_i. The safety index requires maintaining the following limit conditions: $\overline{F} \geq 1.0$ or $P_k \geq 0.0\ (MN)$. The resistance and the driving potentials are described as follows:

$$tg(\phi + dil)[(cos\beta + cos\alpha)\int_1^{1+\frac{d}{a}} \sigma_z(t)dt + (sin\beta + sin\alpha)\int_1^{1+\frac{d}{a}} \sigma_x(t)dt] + cd\left(\frac{1}{cos\alpha} + \frac{1}{cos\beta}\right) + \sum\nolimits_{i=1}^{n}(N_i cos\delta_i) \tag{5}$$

$$F_a = (sin\beta + sin\alpha)\int_1^{1+\frac{d}{a}} \sigma_z(t)dt - (cos\beta + cos\alpha)\int_1^{1+\frac{d}{a}} \sigma_x(t)dt \tag{6}$$

where $t = x/a$, *dil*—dilation angle on the joint's surface; *c*—cohesion at the discontinuity surface, i.e., cohesion of the continuous (intact) rock mass; ϕ—average angle of friction at the discontinuity surface; n—number of rock bolts installed within 1 m of the excavation length; N_i—rock bolt capacity (MN); and δ_i—inclination angle from the horizontal plane.

The triangular-shaped wedge of the ejected volume of rock seems to be more adequate for brittle rocks, revealing higher values of the angle of internal friction. Weaker and heavily jointed rocks may behave in different ways (Figure 5—right), for which the expelled rock block's edge can be described by a curvilinear (elliptic, parabolic) envelope *z* = *f*(*x*).

Figure 5. Views of damaged mine workings with inclined sidewall surfaces: (**left**)—long pillar at an intersection in Rudna mine, (**right**)—sidewall of a rectangular gallery with inclusion of competent rocks—Polkowice-Sieroszowice mine after blasting works in the close neighborhood, February 2016.

In the case of intact rock, the critical value of the stability index should be determined directly based on Equation (4a,b), with the use of iterative computations for different values of angle α and the rock wedge's height d. Analyses conducted on the model presented in Figure 4 reveal that for the case of homogeneous rock mass, the critical rock wedge ABC has the shape of an isosceles triangle ($\alpha = \beta$). Moreover, the shape of the rock wedge expelled from the sidewall may be defined as regular wedge ABC, or a group of triangle elements, such as stiff wedge ADC and plastic wedge ABCD, within which the value of the stability index $\alpha_2 < \pi$ is always less than 1, independently of the slip planes' direction (for a case where the rocks reveal a finite cohesion value); see Figure 6.

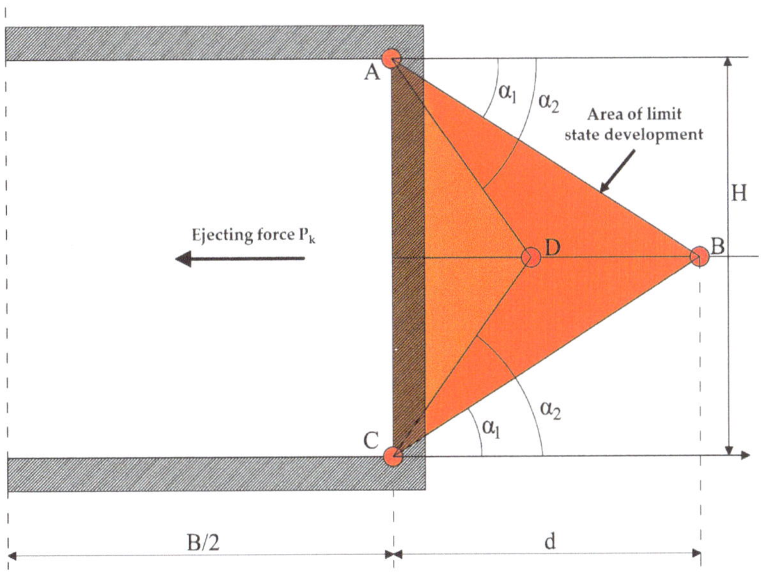

Figure 6. Rock mass ejection from the excavation sidewall.

In order to better match the considered rectangular excavation to the stress field around an elliptical model, equivalent dimensions of the first excavation are determined as follows:

$$H = \sqrt{3}b, B = \frac{\pi a}{\sqrt{3}} \tag{7}$$

Therefore, Equations (3)–(7) may be conditionally utilized for rating the stability of the excavation's sidewall after substituting the ellipse's half-axes by the following dimensions of the equivalent rectangular excavation:

$$a = 0.55B, t = \frac{x}{0.55B}, \left(\frac{b}{a}\right) = 1.05\left(\frac{H}{B}\right) \tag{8}$$

3.2. Numerical Calculations

Numerical calculations, in general, provide more detailed and reliable results in comparison to the analytical solution, but at the same time are more complex and time consuming. As a result, it is challenging to implement such solutions to the regular evaluation of rockburst risk and its characteristics in terms of the continuously progressing exploitation front. Still, knowing the strengths of FEM analyses, the authors proposed to use them for the validation of results obtained with simplified analytical solutions.

Calculations were performed for 2-dimensional cases relating to the points in tunnels and galleries located far enough from the intersections and other underground workings, with the use of RS2 software by Rocscience Inc., Toronto, Canada. The geometry of the numerical model and geologic profile were obtained based on 12 boreholes that were drilled from the surface to the level of the excavations. The strength parameters of modeled rocks have been determined in the laboratory. The dimensions of workings are consistent with the geometries of workings in the analyzed area. The models have been solved using the following input data:

- the galleries' rectangular cross-section size: $B \times H$ = 6 m × 3 m;
- overburden vertical pressure p_z = 20 MPa;
- strength parameters of the rock mass: cohesion c = 8 MPa, angle of internal friction ϕ = 35°;
- deformation parameters of the rock mass: modulus of deformation: E = 40 GPa, Poisson's ratio ν = 0.2 (p_x = 5 MPa).

For 2D FEM analysis, the areas prone to rockburst occurrence were identified on the basis of the maximum shear strain distribution.

4. Results

4.1. Calculation of Ejecting Force Pk and Angles α_1, α_2

The analysis was performed based on 625 unique input datasets composed of values of four basic parameters: p_x, p_z, B, and φ. Each parameter had been represented by a set of five different values. The results of calculations have been visualized as diagrams which allow immediately evaluating the values of angles α_1 and α_2 and values of unbalanced forces as well (Figure 7). In order to assess the sensitivity to the variations of base parameters, a sensitivity analysis has been carried out. As a basic model for such an analysis, the solution with the model characterized by the following parameters was adopted: B = 6 m, p_x = 5 MPa, p_z = 20 Mpa, Φ = 45 deg., H = 4 m, and c = 4 MPa. This solution is treated as a reference and its result values are P_k = 4.64 MN, α_1 = 64.1 deg., and α_2 = 86.1 deg. The obtained changes describing the basic model sensitivity are shown in Table 1.

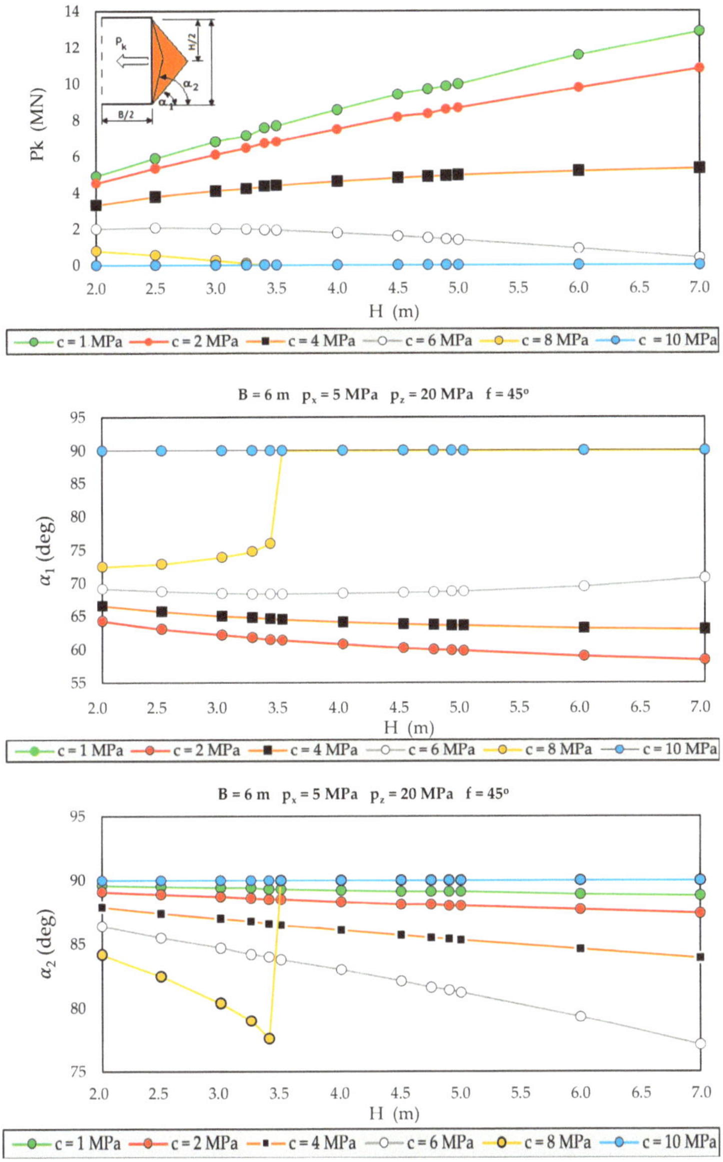

Figure 7. Values of unbalanced forces P_k which should be compensated by the appropriate ground support system (**top**), angle α_2 (**middle**), and angle α_1 (**bottom**), assessed for a specific case of excavation geometry, load, and strength characteristics of surrounding rocks.

As one may notice, the value of the force expelling the detached wedge is positively correlated with the value of the excavation height. For small values of cohesion, it is almost a linear relationship. For large values of cohesion, this correlation will become negative until the P_k value reaches zero. Moreover, in the LGCB region, the values of the base parameters remain approximately in the range $B = 6 \div 10$ m, $p_x = 5 \div 35$ MPa, $p_z = 15 \div 30$ MPa, and $\phi = 31.5 \div 62.7$ deg. Therefore, using the data presented in Table 1, the results are as follows: $0 \leq P_k \leq 16.7$ MN, $50.4 \leq \alpha_1 \leq 90$ deg. and $80.5 \leq \alpha_2 \leq 90$ deg.

Table 1. The unitary increment of values of the model's basic parameters.

Reference Result Values	Increment of the Basic Parameters			
	$\Delta B = 1$ m	$\Delta p_x = 1$ MPa	$\Delta p_z = 1$ MPA	$\Delta \varphi = 1$ deg.
$P_k = 4.64$ MN	0.375 MN/m	−0.167 MN/MPa	1.13 MN/MPa	−0.42 MN/deg.
$\alpha_1 = 64.1$ deg.	0.45 deg./m	0.260 deg./MPa	−0.52 deg./MPa	0.82 deg./deg.
$\alpha_2 = 86.1$ deg.	0.35 deg./m	−0.12/deg./MPa	0.32 deg./MPa	−0.02 deg./deg.

The presented procedure for assessing the stability of mining excavations turned out to be effective because the determined shape of an ejected wedge matches the real case studies observed in Polish copper mines. The information provided by the proposed approach in terms of ejecting forces as a function of geological and mining conditions is extremely valuable in terms of the possibility of assessing the rockburst risk. Moreover, knowing the dynamic forces of wedge ejection, an attempt at rockburst risk mitigation may be utilized by the rock mass reinforcement; for example, by using a ductile rock bolt support system.

In the top of Figure 6, a selected set of diagrams corresponding to the unique combination of parameters p_z, p_x, B, and φ, are presented. The first diagram shows values of the unbalanced by friction and cohesion, forces P_k which may be ejected from the rock wedges ABCD and ACD. The next two diagrams deal with the values of the angles describing the possible geometry of the ejected rock wedge (Figure 7—center and Figure 7—bottom). Analyzing Figure 7, it can be seen that the value of the stability index increases with increasing cohesion and decreases with increasing the excavation height. For a fixed cohesion value, the stability index reaches a value equal to 1.0 when the angles α_1 and α_2 reach the same value, which means that the area where a limit state could have developed disappears (see Figure 7, curve c = 8 MPa, H = 3.44 m, $\alpha_1 = \alpha_2 = 76.8$ deg)

4.2. FEM Validation Resuts and Identification of Rockburst-Prone Areas

Because the developed approach is very promising, the FEM-based validation of the LEM-based stability analyses was performed to ensure the reliability of the proposed solution. The 2D FEM-based calculation was performed with the use of RS2 software developed by RocScience [40]. Simulations were conducted for the four scenarios differing in the values of cohesion and angles of internal friction that characterize the rock mass surrounding the considered underground excavation. These models were presented in Table 2.

Table 2. Strength parameters for the considered models.

Model	Cohesion c (MPa)	Angle of Internal Friction ϕ (deg)
1	8	35
2	4	35
3	8	25
4	4	25

To increase the reliability of the numerical calculations modulus of intact rocks deformation, $E_0 = 40$ GPa was reduced to rock mass modulus acc. to as follows: $E_m = 36.0 \div 40.0$ GPa, depending on the nodes' distance from the opening walls envelope (Figure 8). The Poisson's ratio value was equal to $\upsilon = 0.2$ ($p_x = 5$ MPa). Then, based on [41], it was assumed that the zone of disturbance around the excavation ranges to the depth of $1.25 \times$ H from the sidewall and therefore the rock mass surrounding the opening was divided into several zones of different values of Geological Strength Index (GSI), as shown in Figure 7. At the same time, it was assumed that:

$$GSI = 100\left(1 - \frac{D}{2}\right) \tag{9}$$

where D—the disturbance factor [41], the value for which the analyzed case changes from zero (intact rocks) to 0.8 (rocks slightly damaged by the effect of the production blasting), 100—GSI for intact rocks.

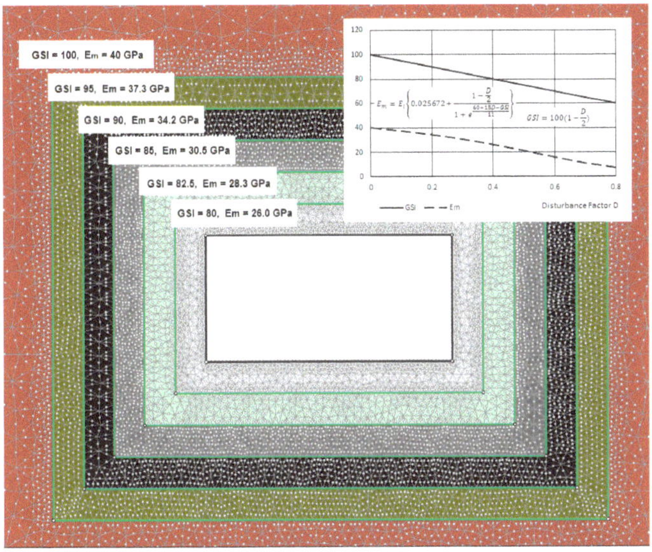

Figure 8. Modulus of deformation of surrounding rock mass, scaled down using the Hoek–Diederichs approach [41].

Finally, the rock mass modulus of deformation E_m, appropriate for the distinguished zones, was assessed based on the following relationship:

$$E_m = E_0 \left\{ 0.025672 + \frac{1 - \frac{D}{2}}{1 + exp^{\frac{6 \mp 15D - GSI}{11}}} \right\} \tag{10}$$

It was assumed that the most appropriate criteria for the comparison of the developed LEM-based solution and FEM numerical model are the degree of matching of the expected shape of the critical wedges identified by angles α_1 and α_2, and the bands of the maximum shear strain's values distributed within the excavation sidewalls. The shear strain is treated here as a fundamental factor that induces the rockburst phenomena. It is also assumed that the shape of the shear stress envelope does not change itself significantly with the stress increment. The qualitative representation of results obtained with the use of FEM-based numerical simulations for Model 1 and Model 4 are presented in Figure 9, whereas the compliance of the results obtained with the numerical and analytical solutions are presented in Table 3.

Table 3. Validation of the proposed limit equilibrium method with FEM-based numerical simulations.

No.	Proposed LEM Approach				FEM Solution		Correlation	
	$\overline{F}_{min}$	$\overline{F}_{av}$	α_{1LEM} (deg.)	α_{2LEM} (deg.)	α_{1FEM} (deg.)	α_{2FEM} (deg.)	α_{1LEM}–α_{1FEM}	α_{LEM}–α_{2FEM}
1	0.84	0.89	56.9	78.8	51	79	0.86	0.97
2	0.55	0.69	49.3	85.4	47	84		
3	0.73	0.81	47.8	79.8	50	79		
4	0.47	0.62	40.9	85.5	38	86		

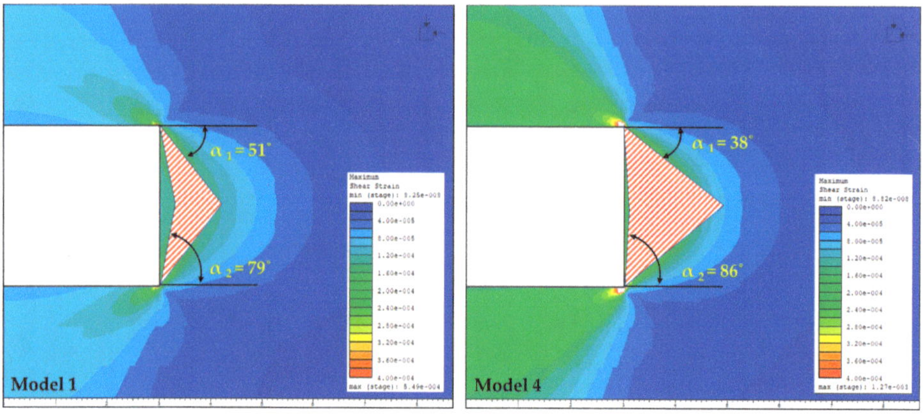

Figure 9. Maximum shear strain distribution obtained numerically (model 1, top; model 4, bottom).

The presented results, obtained using both independent approaches, indicate their high correlation, particularly in terms of the shapes of the ejected rock wedge and the envelope of maximum shear strain in a function of the angle of internal friction and cohesion of the rock mass. Also, the presence of "locally" stable triangular areas within the sidewall as the cohesion effect is visible. It should be highlighted that when analyzing the results presented in Table 3, one may notice that the correlation between the proposed approach and numerical solutions is relatively high and reach values of 0.86 in the case of α_1 and 0.97 in the case of α_2. The abovementioned results permit accepting the LEM as an appropriate and sufficiently accurate approach in preliminary modeling rockbursts in deep mines.

Areas of potential rockburst were identified [42] based on the calculated 3D distribution of the safety margin, formulated based on the Mohr–Coulomb theory of shearing failure, the mathematical notation of which has the following form:

$$S_m = -\sigma_3 \frac{1 + sin\phi}{1 - sin\phi} + \frac{2c\ cos\phi}{1 - sin\phi} + \sigma_1 \quad (11)$$

where σ_3 and σ_1—minor and major principal stresses.

The calculation results refer to two alternative cases:

- a 2D case relating to the gallery point, far enough from the intersections (cross-cuts, ramps, etc. of the cavities excavated from the galleries)—(Figure 10 left);
- a 3D case, relating exclusively to intersections (Figure 10 right).

The obtained distribution of the value of the safety margin for the 2D case allows us to conclude that the shape of rock volume ejected from the sidewalls ($S_m < 0$) is a longitudinal segment of an elliptic cylinder (Figure 10 left) supported at the corners of the excavation. The straight line tangent to the surface of this solid drawn from these corners is inclined in relation to the horizontal plane at the angle of $\alpha_1 \cong 57$.deg, so in a similar way to that calculated for Scenario 1 (Table 3), but with the use of the FEM numerical model formulated strictly for a plane deformation state. The geometry of the ejected three-dimensional solid at the intersection corner (Figure 10 right) is more complex, since it may be approximately described as a kind of hexahedron—a geometric solid with six faces (slip planes), which are formed by the interpenetration of two mutually perpendicular triangular prisms (2D LEM solutions). This statement is valid only if the local geomechanical conditions permit rockburst development in the galleries sufficiently far away from the intersection. Otherwise, only 3D FEM or similar computer codes can be utilized, since 3D analytical extension of Maugis's solution (Equation (4)) to three dimensions is not available so far. Despite the mentioned differences in the problem formulation, the similarities and compatibility proved between the two approaches, FEM and LEM, permit accepting the LEM as an appropriate and sufficiently accurate approach in modeling rockbursts in deep mines.

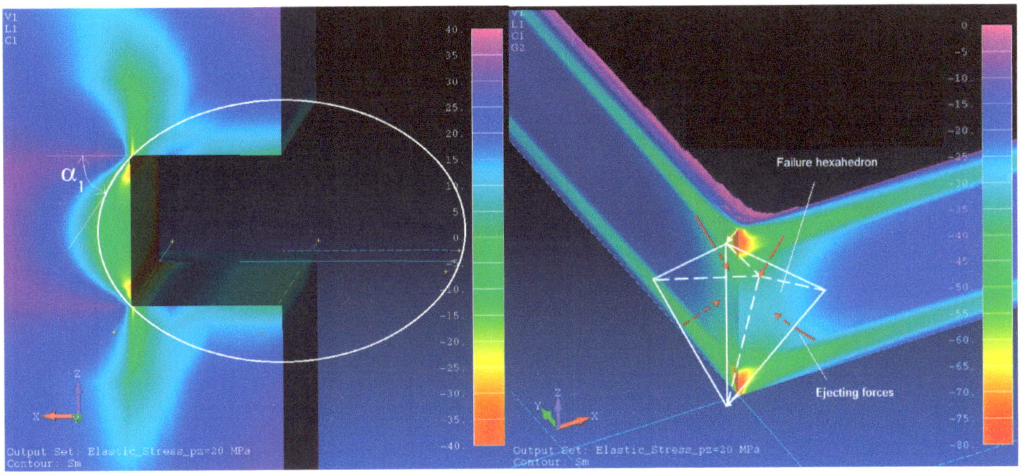

Figure 10. Quarter of the entire intersection; contour of the values of safety margin within sidewalls of the gallery (**left**); contours of slip planes of the failure hexahedron at the underground galleries' intersection (**right**).

4.3. Simulation of the Rockburst Process

It is assumed that during the detachment process, the rock wedge is subjected to acceleration with the initial value $a(0) = a_o$ which linearly attenuates (in time) up to zero after t = t_o. One of the simplest preliminary relations describing such a transformation process is the following equation:

$$\left.\begin{array}{c} a(t) = a_0\left(1 - \frac{t}{t_0}\right) \, for \, t \le t_0 \\ a(t) = 0 \, for \, t > t_0 \end{array}\right\} \quad (12)$$

Knowing the value of the ejection force P_k, one may also estimate the initial acceleration associated with the rockbursts as well as the velocity of the rock movement. Using the fundamental principle of dynamics, the acceleration that may potentially eject the rock wedge, unsecured by any support (free movement), may be calculated from the following relationship:

$$a_0 = \frac{P_k}{m_s} \quad (13)$$

where m_s—mass of the moving rock wedge and P_k—unbalanced force assessed from the limit equilibrium condition formulated for the detached rock block (see Figure 6, top). In turn, the rock wedge's velocity and its kinetic energy (without retardation) may be determined from the following expressions:

$$\left.\begin{array}{c} v(t) = a_0\left(t - \frac{t^2}{2t_0}\right) = \frac{P_k}{m_s}\left(t - \frac{t^2}{2t_0}\right) \, for \, 0 \le t \le t_0 \\ v(t) = \frac{a_0 t_o}{2} = \frac{P_k t_0}{2m_s} \, for \, t > t_0 \end{array}\right\} \quad (14)$$

$$\left.\begin{array}{c} \overline{E}_k = \frac{m_s \overline{v}^2}{2} = \frac{P_k^2}{2m_s}\left(t - \frac{t^2}{2t_o}\right)^2 \, for \, 0 \le t \le t_0 \\ \overline{E}_k = \frac{m_s \overline{v}^2}{2} = \frac{(P_k t_0)^2}{8m_s} \, for \, t > t_0 \end{array}\right\} \quad (15)$$

Since the amount of available information on the detailed characteristics of the burst processes, particularly at their the very beginning phase, is relatively scarce, one may determine roughly the value of t_o based on the velocity value observed in situ. Therefore:

$$t_0 = \frac{2vm_s}{P_k} \quad (16)$$

where v_o—6 m/s averaged approximation acc. to [41,43,44]. Compliance with Equation (1) can be obtained for the value of the range of horizontal movement D = 3.82 m.

Acceleration and velocity values of detached wedge movement over time are presented in Figure 11.

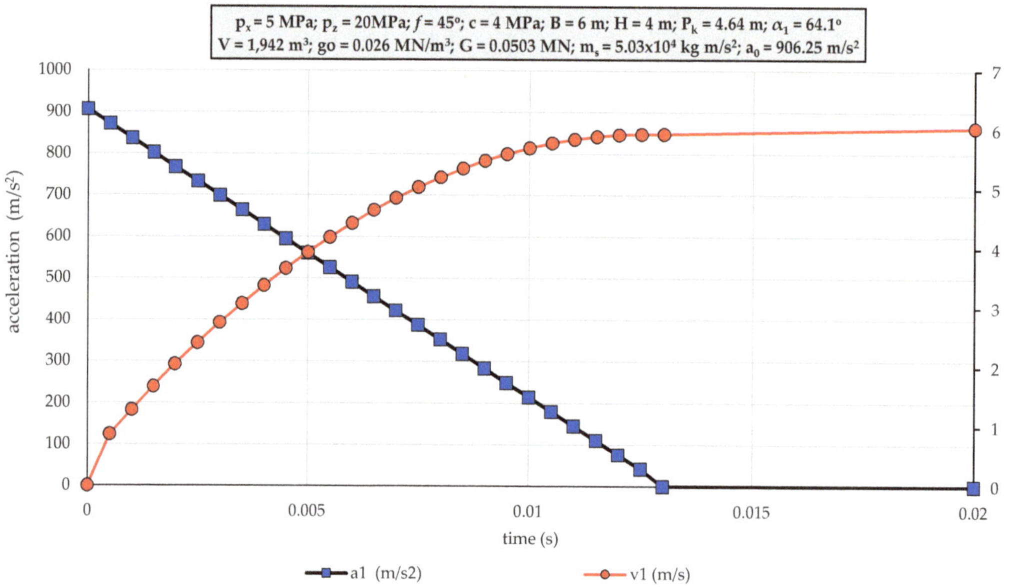

Figure 11. Predicted changes in the acceleration (**left**) and velocity (**right**) of the detached rock wedge movement (t_0 = 0.0131 s).

4.4. Identification of Rockburst-Prone Mine Areas

Knowing the characteristics of the rockburst mechanism, the location of areas prone to instability was determined. Therefore, the general equation relating the rock's strength parameters ($c \neq 0$ and $\Phi \neq 0$) to the load that is acting in conditions where the limit equilibrium may be obtained only in one location is presented below:

$$sin\alpha[(cos\alpha tan\phi + sin\alpha)\int_{\frac{B}{2}}^{\frac{B}{2}+d} \sigma_z(x)dx + (sin\alpha tan\phi - cos\alpha)\int_{\frac{B}{2}}^{\frac{B}{2}+d} \sigma_x(x)dx] + \frac{cH}{2} = 0 \quad (17)$$

where ϕ—the angle of internal friction; α—the angle of inclination of the external plane of the block of solid rock ejection; B—opening width; H—opening height; $d = \frac{h}{2tan\alpha}$; and c—cohesion. Gradually increasing the value of $p_z = H_0\gamma_{0,av}$ (where H_o—overburden thickness, $\gamma_{0,av} = 0.026\frac{MN}{m^3}$)—the average bulk weight of the rock mass), the value of the critical depth H_{cr}, at which the beginning of the rockburst process may be observed, has been assessed within the parameters' space and is characterized in Section 3. Simultaneously, searching for the critical value F = 1, the angle α value has been scanned from zero to $\frac{\pi}{2}$.

Based on the available literature concerned with the strain–strength parameters of the rock mass in the area of Polish copper mines [38], the values of critical depths, from the point of view of rockburst hazards, have been determined. The laboratory-obtained values of rock mass parameters have been scaled down to in situ (rock mass) values using the approach given by [45] (Table 4).

Table 4. Averaged values of the UCS laboratory test-based data for rocks within the copper ore body.

Type of Rock		Sandstones	Anhydrites	Grainy Dolomites and Limestones
UCS (MPa)		71–142	100–141	151–224
GSI		45–70	60–75	75–90
Angle of internal friction (°)	Triaxial tests	64–67	61–65	64–69
	Reduced using GSI	31.5–44.0	36.4–44.2	46.2–62.7
Cohesion c (MPa)	Triaxial tests	17.5–20.5	15.8–18.8	18.0–22.5
	Reduced using GSI	2.8–11.6	5.5–11.4	13.2–30.1
Stress ratio		0.14–0.60	0.27–0.80	0.33–1.39
Horizontal stress p_x (MPa)		4–12	6–30	8–40

These estimates permitted the identification of the critical depths from which a serious rockburst may occur. This knowledge has proved that copper ore exploitation at depths greater than 1100 m will be associated with an increased level of rockburst risk, indicating that at this depth, the extensive use of ductile rock bolts is highly recommended as basic ground support, at least in the long-life workings such as machine chambers, etc. In each case of well-known mining-geology conditions, the development of individual preliminary analyses using data shown in Figure 12 is strongly recommended.

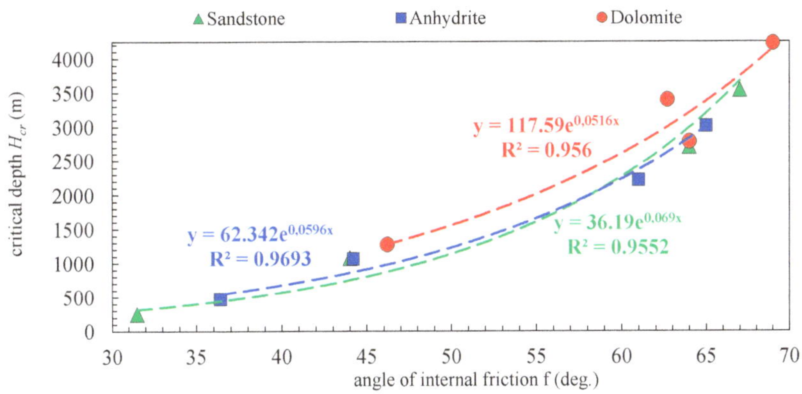

Figure 12. Critical depth for different types of rocks within a drift's sidewall in selected mining-geologic conditions (B = 6 m, H = 3 m, p_x = 10 MPa).

As a result, the rockburst factor of safety over the entire mine may be determined according to the following formula:

$$F_{srb} = \frac{H_{cr}(x,y)}{H_o(x,y)} \tag{18}$$

where H_{cr}—critical depth; H_o—depth of mining.

This type of map may be associated with the x–y maps of ductile sidewall support distribution with detailed data concerning the unbalanced force P_k as well as the number, length, and capacity of the utilized rock bolts. To illustrate the advantages of the proposed approach, the rockburst factor of safety distribution (F_{srb}) over one of the mining panels of the Polish copper mine has been analyzed. The values of angles of internal friction and cohesion that are adequate for different types of rock have been assessed using the method based on tangents to the parabolic envelope of Mohr's circles [41]. At the same time, the H_{cr} value has been calculated, which permitted us to find the critical rock wedge that is associated with a stability index F equal to 1. This required minimizing the index F value by scanning all possible shapes of wedges while increasing the depth of mining ($H_o = p_z/\gamma_0$). The resulting map (Figure 13) of the rockburst safety factor proved the usefulness of the proposed approach for burst hazard general recognition.

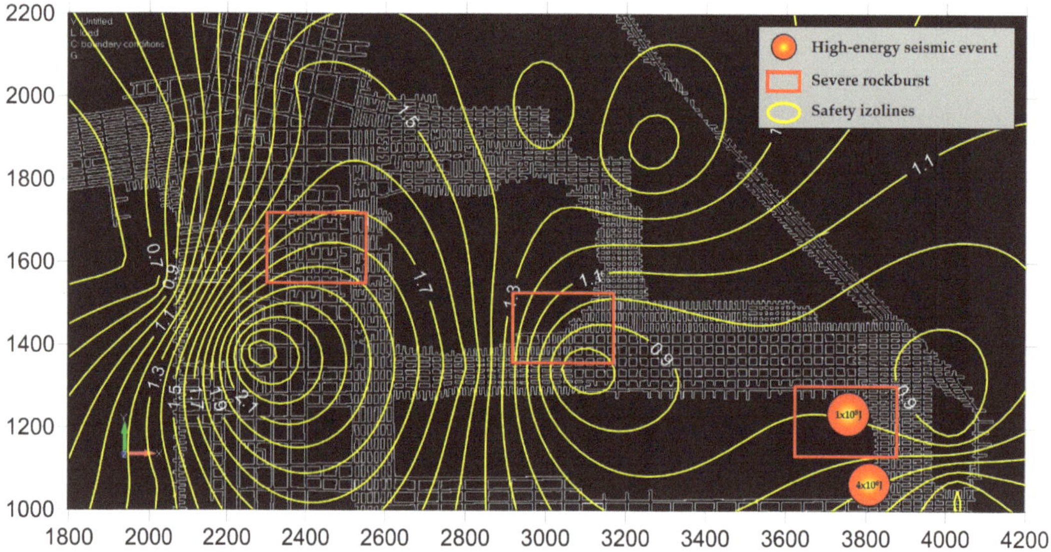

Figure 13. Rockburst safety factor distribution over one of the analyzed regions in the Rudna mine.

The orange circles shown in Figure 10 indicate the areas of severe rockbursts associated with the M_L = 3.4 seismic events which occurred in November 2016 in one of the mining panels [46]. These areas are strongly correlated with safety factor F_{srb} values of about 0.9–1.0, which prove the usefulness of the proposed method for the preliminary rockburst risk analysis. Bearing in mind that sandstone's share in the thickness of the ore body profile is usually relatively large, serious, and frequent rockbursts from the sidewalls due to the low values of rock strain–strength parameters may be expected. Therefore, one may conclude that in situ stress measurement and validating the proposed hypotheses are necessary.

5. Conclusions

The developed 2-dimensional model of rockbursts from an underground excavation's sidewalls, based on the LEM, is a relatively simple and sufficiently general analytical approach useful for formulating practical conclusions and recommendations with respect to selecting the best methods of ground support, particularly in the geologic and mining conditions of deep mines. The model presented herein may be applied to continuous types of rocks as well as to rocks characterized by more than one joint/discontinuity system. Also, the rock heterogeneity within sidewalls does not disturb the theoretical background of the developed models but requires additional analytical work.

Furthermore, the developed approach may indicate the appropriate reinforcement of sidewalls or, if this is not achievable, it allows us to determine the trajectory of the motion of a detached rock wedge. This is the main advantage and novelty of the developed approach over the FEM. It is also worth noting that the developed set of charts allowed us to determine if it is effective to use a rigid type of ground reinforcement in a given set of mining/geologic conditions or if the ductile ground reinforcement is able to dissipate the kinetic energy of the ejected rock wedge within the frame of justified economical limits.

The adequate assumptions concerned with the form as a function of time, within which the rock wedge's debonding process develops with the associated acceleration of the initial value a(0) = α_o and the final value $\alpha(t_o) = 0$, is one of the most important problems of the considered model. Although the initial and final values of acceleration should not be too controversial, the utilized form of the function a(t) may be disputable. A linear function, as used in this study, may be accepted preliminarily with no additional research work. However, other serious doubts are associated with the time t_o (in this study t_o = 0.013 s), which describes the moment when acceleration reaches the value equal to zero. Since this argument is of the highest importance because it significantly influences the movement velocity, it is necessary to compare the obtained velocities with those cited in the literature. Therefore, it is strongly recommended to launch research projects that should permit one to relate the time of acceleration fading out to the most important parameters that characterize the local mining–geologic conditions.

Finally, one may notice that the results of calculation obtained based on numerical modeling (FEM) have proved their sufficient mutual similarity to LEM-based results.

Author Contributions: Conceptualization, W.P. and K.F.; methodology, W.P. and K.F.; software, W.P. and K.F.; validation, B.P.-W. and P.M.; formal analysis, K.F.; investigation, W.P., K.F. and B.P.-W.; resources, W.P., K.F., B.P.-W. and P.M.; data curation, K.F. and B.P.-W.; writing—original draft preparation, W.P. and K.F.; writing—review and editing, K.F., B.P.-W. and P.M.; visualization, K.F. and P.M.; supervision, W.P.; project administration, K.F.; funding acquisition, K.F. All authors have read and agreed to the published version of the manuscript.

Funding: This paper was prepared within the framework of the AGEMERA project "Agile Exploration and Geo-modelling for European Critical Raw materials". The project received funding from the European Union's Horizon Europe Research and Innovation Program under grant agreement No. 101058178.

Institutional Review Board Statement: Not applicable.

Informed Consent Statement: Not applicable.

Data Availability Statement: Not applicable.

Conflicts of Interest: The authors declare no conflict of interest.

References

1. Ghorbani, M.; Shahriar, K.; Sharifzadeh, M.; Masoudi, R. A Critical Review on the Developments of Rock Support Systems in High Stress Ground Conditions. *Int. J. Min. Sci. Technol.* **2020**, *30*, 555–572. [CrossRef]
2. Fuławka, K.; Pytel, W.; Mertuszka, P. The Effect of Selected Rockburst Prevention Measures on Seismic Activity—Case Study from the Rudna Copper Mine. *J. Sustain. Min.* **2021**, *17*, 1–10. [CrossRef]
3. Ptak, M.; Podolska, P.; Podolski, R. Challenges for Science: The Exploitation of Deep Deposits. *E3S Web Conf.* **2018**, *71*, 00008. [CrossRef]
4. Cichy, T.; Prusek, S.; Świątek, J.; Apel, D.B.; Pu, Y. Use of Neural Networks to Forecast Seismic Hazard Expressed by Number of Tremors per Unit of Surface. *Pure Appl. Geophys.* **2020**, *177*, 5713–5722. [CrossRef]
5. Pu, Y.; Apel, D.B.; Prusek, S.; Walentek, A.; Cichy, T. Back-Analysis for Initial Ground Stress Field at a Diamond Mine Using Machine Learning Approaches. *Nat. Hazards* **2021**, *105*, 191–203. [CrossRef]
6. Blake, W.; Hedley, D.G.F. *Rockbursts: Case Studies from North American Hard-Rock Mines*; Society for Mining, Metallurgy, and Exploration: Littleton, CO, USA, 2003; ISBN 9780873352321.
7. Mark, C. Coal Bursts in the Deep Longwall Mines of the United States. *Int. J. Coal Sci. Technol.* **2016**, *3*, 1–9. [CrossRef]
8. Zhang, W.; Feng, X.-T.; Xiao, Y.-X.; Feng, G.-L.; Yao, Z.-B.; Hu, L.; Niu, W.-J. A Rockburst Intensity Criterion Based on the Geological Strength Index, Experiences Learned from a Deep Tunnel. *Bull. Eng. Geol. Environ.* **2020**, *79*, 3585–3603. [CrossRef]
9. Pan, J.; Ren, F.; Cai, M. Effect of Joint Density on Rockburst Proneness of the Elastic-Brittle-Plastic Rock Mass. *Shock Vib.* **2021**, *2021*, 5574325. [CrossRef]
10. Wang, J.; Apel, D.B.; Pu, Y.; Hall, R.; Wei, C.; Sepehri, M. Numerical Modeling for Rockbursts: A State-of-the-Art Review. *J. Rock Mech. Geotech. Eng.* **2021**, *13*, 457–478. [CrossRef]
11. Małkowski, P.; Niedbalski, Z. A Comprehensive Geomechanical Method for the Assessment of Rockburst Hazards in Underground Mining. *Int. J. Min. Sci. Technol.* **2020**, *30*, 345–355. [CrossRef]
12. Zhou, J.; Li, X.; Mitri, H.S. Evaluation Method of Rockburst: State-of-the-Art Literature Review. *Tunn. Undergr. Space Technol.* **2018**, *81*, 632–659. [CrossRef]

13. Farhadian, H. A New Empirical Chart for Rockburst Analysis in Tunnelling: Tunnel Rockburst Classification (TRC). *Int. J. Min. Sci. Technol.* **2021**, *31*, 603–610. [CrossRef]
14. Wojtecki, Ł.; Mendecki, M.J.; Gołda, I.; Zuberek, W.M. The Seismic Source Parameters of Tremors Provoked by Long-Hole Destress Blasting Executed During the Longwall Mining of a Coal Seam Under Variable Stress Conditions. *Pure Appl. Geophys.* **2020**, *177*, 5723–5739. [CrossRef]
15. Wojtecki, Ł.; Konicek, P. Estimation of Active Rockburst Prevention Effectiveness during Longwall Mining under Disadvantageous Geological and Mining Conditions. *J. Sustain. Min.* **2016**, *15*, 1–7. [CrossRef]
16. Tannant, D.D.; McDowell, G.M.; Brummer, R.K.; Kaiser, P.K. Ejection velocities measured during a rock burst simulation experiment. In *Rockbursts and Seismicity in Mines 93, Proceedings of the 3rd International Symposium, Kingston, ON, Canada, 16–18 August 1993*; Balkema: Rotterdam, The Netherlands, 1993; pp. 129–133.
17. He, M.; Cheng, T.; Qiao, Y.; Li, H. A Review of Rockburst: Experiments, Theories, and Simulations. *J. Rock Mech. Geotech. Eng.* **2023**, *15*, 1312–1353. [CrossRef]
18. Qiu, S.; Feng, X.; Zhang, C.; Xiang, T. Estimation of Rockburst Wall-Rock Velocity Invoked by Slab Flexure Sources in Deep Tunnels. *Can. Geotech. J.* **2014**, *51*, 520–539. [CrossRef]
19. Barton, N. The Shear Strength of Rock and Rock Joints. *Int. J. Rock Mech. Min. Sci. Geomech. Abstr.* **1976**, *13*, 255–279. [CrossRef]
20. Esterhuizen, G.S. Extending Empirical Evidence through Numerical Modelling in Rock Engineering Design. *J. South. Afr. Inst. Min. Metall.* **2014**, *114*, 755–764.
21. Terzaghi, K. *Theoretical Soil Mechanics*; John Wiley & Sons, Inc.: Hoboken, NJ, USA, 1943; ISBN 9780470172766/9780471853053.
22. Malan, D.; Napier, J. A Limit Equilibrium Fracture Zone Model to Investigate Seismicity in Coal Mines. *Int. J. Min. Sci. Technol.* **2018**, *28*, 745–753. [CrossRef]
23. Sengani, F.; Mulenga, F. Application of Limit Equilibrium Analysis and Numerical Modeling in a Case of Slope Instability. *Sustainability* **2020**, *12*, 8870. [CrossRef]
24. Ohnishi, Y.; Nishiyama, S.; Sasaki, T. Development and application of discontinuous deformation analysis. In *Rock Mechanics in Underground Construction*; World Scientific: Singapore, 2006; pp. 59–70.
25. Zhao, J.; Xiao, M.; Chen, J.; Li, D. Explicit Dynamic DDA Method Considering Dynamic Contact Force. *Shock Vib.* **2016**, *2016*, 7431245. [CrossRef]
26. Fan, H.; Huang, D.; Wang, G.; Wang, J. Discontinuous Deformation Analysis for SH-Body. *Comput. Geotech.* **2020**, *117*, 103234. [CrossRef]
27. Wu, W.; Zhu, H.; Lin, J.-S.; Zhuang, X.; Ma, G. Tunnel Stability Assessment by 3D DDA-Key Block Analysis. *Tunn. Undergr. Space Technol.* **2018**, *71*, 210–214. [CrossRef]
28. Lin, J.; Zuo, Y.; Wang, J.; Zheng, L.; Chen, B.; Sun, W.; Liu, H. Stability Analysis of Underground Surrounding Rock Mass Based on Block Theory. *J. Cent. South Univ.* **2020**, *27*, 3040–3052. [CrossRef]
29. Hoek, E. Practical Rock Engineering. RocScience. 2007. Available online: http://www.rocscience.com/hoek/PracticalRockEngineering.asp (accessed on 21 July 2023).
30. Billaux, D.; Detournay, C.; Hart, R.; Rachez, X. (Eds.) *FLAC and Numerical Modeling in Geomechanics 2001*; CRC Press: Boca Raton, FL, USA, 2020; ISBN 9781000099607.
31. Khabbaz, H.; Fatahi, B.; Nucifora, C. Finite Element Methods against Limit Equilibrium Approaches for Slope Stability Analysis. In *Proceedings of the 11th Australia—New Zealand Conference on Geomechanics—Ground Engineering in a Changing World (ANZ 2012), Melbourne, Australia, 14–18 July 2012*; Geomechanical Society: Sydney, Australia; New Zealand Geotechnical Society: Wellington, New Zealand, 2012; ISBN 9780646543017.
32. Blake, W. Rockburst mechanics. *Q. Colo. Sch. Mines* **1972**, *67*, 1–64.
33. Tajduś, A.; Cieślik, J.; Tajduś, K. Rockburst Hazard Assessment in Bedded Rock Mass: Laboratory Tests of Rock Samples and Numerical Calculations. *Arch. Min. Sci.* **2014**, *59*, 591–608. [CrossRef]
34. Fulawka, K.; Szumny, M.; Pytel, W.; Mertuszka, P. Conceptual Design of Underground Laboratory under Dynamic Load Condition in Deep Copper Mine. In Proceedings of the 22nd EGU General Assembly, Online, 4–8 May 2020.
35. Fulawka, K.; Pytel, W.; Mertuszka, P.; Szumny, M. Finite Element Method-Based Geomechanical Risk Assessment of Underground Laboratory Located in the Deep Copper Mine. In Proceedings of the 23rd EGU General Assembly, Online, 19–30 April 2021.
36. Maugis, D. Stresses and Displacements around Cracks and Elliptical Cavities: Exact Solutions. *Eng. Fract. Mech.* **1992**, *43*, 217–255. [CrossRef]
37. Exadaktylos, G.E.; Stavropoulou, M.C. A closed-form elastic solution for stresses and displacements around tunnels. *Int. J. Rock Mech. Min. Sci.* **2002**, *39*, 905–916. [CrossRef]
38. Kargar, A.R.; Rahmannejad, R.; Hajabasi, M.A. A semi-analytical elastic solution for stress field of lined non-circular tunnels at great depth using complex variable mrthod. *Int. J. Solids Struct.* **2014**, *51*, 1475–1482. [CrossRef]
39. Ye, Z.K.; Ai, Z.Y. A novel complex variable solution for the stress and displacement fields around a shallow non-circular tunnel. *Comput. Geotech.* **2023**, *164*, 105797. [CrossRef]
40. RocSciecnce. RS2 Finite Element Analysis for Excavations—Rocscience. 2016. Available online: https://www.rocscience.com/rocscience/products/rs2 (accessed on 21 September 2023).
41. Hoek, E.; Diederichs, M.S. Empirical Estimation of Rock Mass Modulus. *Int. J. Rock Mech. Min. Sci.* **2006**, *43*, 203–215. [CrossRef]
42. NEi Software. *NEi Nastran Reference Manual*, Version 9.2; Trafford Publishing: Bloomington, IN, USA, 2009.

43. Ortlepp, W.D. The Design of Support for the Containment of Rockburst Damage in Tunnels—An Engineering Approach. In Proceedings of the International Symposium on Rock Support in Mining and Underground Construction, Sudbury, ON, Canada, 16–19 June 1992.
44. Nussbaumer, M.M. A Comprehensive Review on Rock Burst. Ph.D. Thesis, Massachusetts Institute of Technology, Cambridge, MA, USA, 2000.
45. Pytel, W. *Geomechanical Problems of Ground Support Selection for Mine Excavations*; KGHM Cuprum: Wrocław, Poland, 2012; 423p. (In Polish)
46. Lasocki, S.; Orlecka-Sikora, B.; Mutke, G.; Pytel, W.; Rudzinski, L.; Markowski, P.; Piasecki, P. A catastrophic event in Rudna copper-ore mine in Poland on 29 November, 2016: What, how and why. In Proceedings of the Ninth International Symposium on Rockbursts and Seismicity in Mines, RaSiM9, Santiago, Chile, 15–17 November 2017.

applied sciences

MDPI

Article

Experimental Study on the Relationship between the Degree of Surrounding Rock Fragmentation and the Adaptability of Anchor Support

Shuai Wang [1], Lianguo Wang [1,*], Furong Tang [2,*], Ke Ding [3], Zhaolin Li [3], Bo Ren [1], Chongyang Jiang [1] and Jiaxing Guo [1]

1 State Key Laboratory for Geomechanics and Deep Underground Engineering, China University of Mining and Technology, Xuzhou 221116, China; swang@cumt.edu.cn (S.W.); 02140873@cumt.edu.cn (B.R.); chongyang_jiang@cumt.edu.cn (C.J.); guo1jiaxing@126.com (J.G.)
2 School of Materials Science and Physics, China University of Mining and Technology, Xuzhou 221116, China
3 State Key Laboratory of Mining Response and Disaster Prevention and Control in Deep Coal Mines, Anhui University of Science and Technology, Huainan 232001, China; dingke@cumt.edu.cn (K.D.); lzhlcumt@163.com (Z.L.)
* Correspondence: cumt_lgwang@163.com (L.W.); tfr96@cumt.edu.cn (F.T.)

Abstract: Taking the roadway peripheral rock anchoring unit as the research object, the rock compression test containing the anchor solid was carried out to analyze the influence of the degree of peripheral rock fragmentation and the anchor support method on the mechanical properties of the rock body. The test results showed that the smaller the size of the structural surface, the more a greater number of anchor rods were needed, which in turn provided better support. With the increase in the size of the structural surface, the uniaxial compressive strength and modulus of elasticity of the specimen showed a gradual decrease. Numerical tests of the uniaxial compression of rock containing cohesive units showed that the deformation of the specimen near the anchor bar was significantly reduced, while the main rupture surface was blocked, and an obvious reinforcement zone was formed near the anchor bar. Under the double-anchor condition, the anchor tension stress was more obvious, the reinforcement zone was wider, and the rock rupture surface was strongly blocked, all of which made its reinforcement effect the more obvious. This double-anchor condition showed that the anchoring effect of the anchor rods on the specimens was reflected in two aspects of reinforcement and crack stopping. The denser the anchor rods, the wider the reinforcement zone and hence the more likely that the superposition effect will occur, which allowed the anchor rods to play a greater supporting role in stabilizing the rock. The research results can provide a theoretical basis for the design of anchor support and early warning prediction of destabilization damage in the fractured surrounding rock of coal mine roadways.

Keywords: anchor support; fractured perimeter rock; similar material simulation; numerical simulation; support mechanism

1. Introduction

At this stage, coal is the main energy source in China, with an output of about 4.45 billion tons in 2022, and coal resources account for more than 70% of one-time energy consumption. According to incomplete statistics, more than 90% of China's coal production comes from shaft mining, which requires a large number of roadways to be excavated underground, and it is important to keep the roadways open and the perimeter rock stable for the construction and production of coal mines. The length of newly dug roadways in China is about 12,000 km per year. As a result of the continuous improvement of mining depth, width, and intensity, the depth of the roadway is increasing year by year, and the geological conditions are becoming more and more complicated. Additionally, the complicated and difficult conditions such as high geostress roadways, roadways with strong mining influence, soft and broken perimeter rock roadways, and extra-large section roadways and refuge are becoming more common, which significantly increases the difficulty of the roadway support [1].

As the mining depth increases, the ground stress rises significantly, and the mechanical properties of the deep rock body show different characteristics from those of the shallow rock body [2–6]. Due to the complex geological structure [7], the rock body of underground engineering is usually rich in defects such as joints, faults, and discontinuous surfaces, which can cause different degrees of fragmentation of the surrounding rock and have an important impact on the safe construction of underground engineering and the stability of the surrounding rock [8–11]. Due to the existence of geological structural surfaces in the rock mass, the mechanical strength of the rock mass is greatly reduced, and most engineering practices have proved that the structural surfaces in the rock mass play a controlling role in the deformation and destructive strength of the rock mass. Therefore, an accurate understanding of the deformation and damage law of the rock body containing internal joints is crucial for the reinforcement of weak and broken perimeter rock [12]. Jeon et al. [13] showed that the deformation of the tunnel perimeter rock increased significantly due to the existence of the weak face and obvious shear deformation appearing along the weak face through physical tests and numerical simulation. Meanwhile, Moir et al. [14] illustrated the characteristics of the rock body rupture evolution of the fracture network under the action of the load through numerical calculations and showed that the geometrical distribution of the fracture network was very small. The results of their study showed that the fracture network was not only a good solution, but that it had the effect of the deformation of the rock body as well. The results show that the geometric distribution of the fissure network and the local stress field have an important influence on the development of the rupture zone of the surrounding rock.

In terms of surrounding rock reinforcement technology, anchor support technology, as one of the most common rock reinforcement means, plays an indispensable role in mining engineering, slope reinforcement, and other fields [15,16]. Anchor support significantly improves the tunnel support effect, reduces the cost of tunnel support, and reduces the labor intensity of workers. What is more, anchor support greatly simplifies the end support and overhead support process of the coal mining face, ensures safe production, and creates good conditions for the rapid advancement of the coal mining face and the substantial increase in coal production. At present, anchor support technology has been commonly used at home and abroad and is one of the key technologies essential for coal mines to realize high productivity and high efficiency. It has achieved good results in complex and difficult conditions such as high geostress roadways, soft rock roadways, roadways affected by strong dynamic pressure, and roadways digging and staying along the empty roadways in kilometer-deep wells [17–22]. Its working principle is to fully utilize and improve the bearing capacity of the rock body, which has a good reinforcement effect and economic cost [23–25]. In recent years, many scholars at home and abroad have explored and researched the action mechanism of anchors from various aspects, including some classical theories such as suspension theory, combined beam theory, combined arch theory, perimeter rock loosening circle support theory, and perimeter rock strength reinforcement theory [26,27]. The research methods are mainly divided into indoor tests, model tests, theoretical analysis, and numerical simulation. Li et al. [28] investigated the reinforcing effect of anchors on rock bodies containing penetrating fissures under uniaxial tensile conditions through indoor experiments, and the results showed that the anchored specimens showed plastic damage characteristics and the anchors increased the deformation modulus and uniaxial tensile strength of the jointed rock body. Wong and Chau (1998) and Wong et al. (2001) conducted experiments on rock-like materials containing two and three fissures (rectangular specimen dimensions: $60 \times 120 \times 25$ mm) subjected to uniaxial compression. Their findings revealed that the peak strength of the specimens was not contingent upon the initial crack density but rather on the actual number of pre-existing flaws involved in the coalescence process. Chen et al. [29] carried out tensile-shear tests on the anchored specimens by developing a new test method to characterize the strain distribution of the surface of the anchors under the tensile-shear loading and, at the same time revealed that the angle of the installation of the anchors. In a separate study, Sagong and Bobet [30] examined a series of specimens (rectangular specimen dimensions: 101.6 × 203.2 × 30 mm) composed of gypsum and featuring three and sixteen fissures, which were subjected to uniaxial compression. Their research focused on elucidating the influence of continuity and the ligament influence of continuity and ligament length on the stress associated with crack coalescence. Su et al. [31] simulated the formation process of rock bodies in fault-fracture zones by a novel test method and prefabricated model specimens of anchors anchored by uniaxial. The effect of the anchorage form on the bearing capacity and damage mode of the fault fractured rock body was investigated by compression test. Jing et al. [32] investigated the effect of the nodal angle and anchor density on the anchorage strength, deformation behavior, and axial force evolution characteristics of the anchorage in the model specimen under the action of normal stress by prefabricating a large-scale nodal model specimen (500 mm × 500 mm × 480 mm). With the continuous development of advanced testing techniques and computer simulation software, unprecedented progress has been made in recent years in the study of anchor support principles. Deb and Das [33] introduced a doubly enriched finite element (DEFE) procedure designed for simulating grouted bolts passing through rock joints. Nie et al. [34] incorporated a rock bolt element, grounded in analytically derived interface behavior, into a two-dimensional discontinuous deformation analysis (DDA) program to assess the effectiveness of the rock reinforcement system, yielding valuable insights into the role of DDA in the design of rock reinforcement systems. Lin et al. [35] constructed a three-dimensional bolt-reinforced model within rock joints using the FLAC3D numerical calculation program. They investigated the influence of bolt inclination angles on both flat and undulating joint surfaces.

In the actual project, due to the influence of primary joints, fissures, and other structural surfaces and weathering and erosion, the surface of the roadways of the soft surrounding rock was highly fragmented, forming irregular bulk block structures of different sizes [36,37]. The surrounding rock had large deformation and strong rheological properties. The bearing capacity and anchoring mechanical behavior of this fractured rock mass were significantly different compared with the intact rock mass [38]. Throughout the literature at home and abroad, experimental studies on the relationship between the degree of surrounding rock fragmentation and the adaptability of anchor support are rarely reported. In view of this, this study, through the development of model-similar materials and a specialized anchor addition scheme, carried out compression tests on the model specimens of similar materials under the action of anchor anchoring to investigate the influence law of aggregate particle size and anchor anchoring method on the strength, deformation, and fracture characteristics of the specimens. And the reinforcing effect of the anchor rods on the soft and broken surrounding rock was analyzed by numerical simulation.

2. Test Program

The traditional combined arch theory believes that if the anchor spacing is relatively close, this double conical compression zone between the anchors overlaps and the combination is arch-shaped, and as a result the extrusion combination action can be formed (see Figure 1). In Figure 1, Region A is one of the anchor units in the combined arch bearing structure, and Region B represents the double-anchor bar anchorage unit. This test is based on the force state of the anchor solid in the surrounding rock. When combined with the effect of the different degrees of fragmentation on the anchoring effect of the anchor rods, the anchor solid can be investigated.

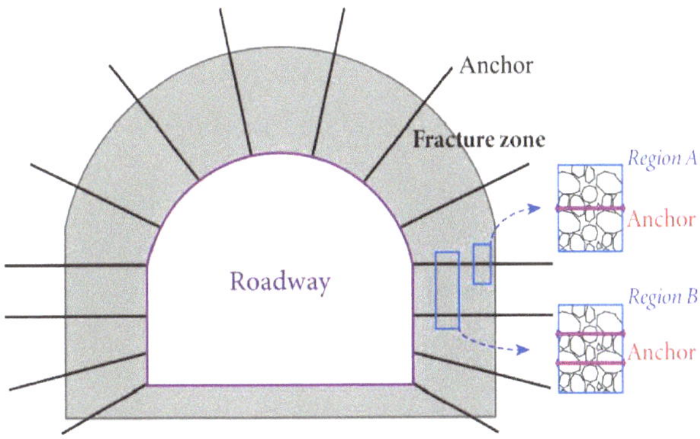

Figure 1. Schematic diagram of the combined arch of the surrounding rocks.

2.1. Modeling

Considering the comparability of the test results, it is recommended to use gypsum and quartz sand as the raw materials to configure the specimens [39]. The specimen size was set to 50 mm × 50 mm × 100 mm with reference to the international standards of rock mechanical testing and the results of previous research [40–43], where gypsum was used as a binder [44]. In order to simulate the large number of fissures and other structural surfaces that exist in the rock mass, three kinds of quartz sand with different grain sizes (Grain size A: 2–4 mm, Grain size B: 4–8 mm, and Grain size C: 10–15 mm) of quartz sand as the aggregate were used (Figure 2). The specimens containing quartz sand and gypsum with different grain sizes were used to simulate rock bodies with different degrees of microfractures and structural surfaces. The smaller grain sizes of quartz sand and gypsum had smaller contact surfaces, forming structural surfaces of smaller sizes inside the specimens, and vice versa.

Figure 2. Quartz sand with different grain size.

The steps for preparing the specimens (Figure 3) are as follows:

1. Brush the mold release agent on the inside of 50 mm × 50 mm × 100 mm mold, weigh quartz sand, gypsum, and water separately with a balance and set aside;
2. Put gypsum and quartz sand in a container, add water evenly and mix it quickly, and put it into the mold for vibration (the ratio of materials is water: gypsum: quartz sand = 1:1.5:3);
3. Smooth the upper surface after shaking evenly, and remove the mold 24 h after the specimen is finished;
4. Put the mixture into the drying box at 60 °C for maintenance, after about 10 days of maintenance drying;
5. After the specimen is completely dry, polish the specimen and measure with vernier calipers to ensure that the error is within the allowable range.

Figure 3. Specimen production process.

After reviewing a large number of studies and the mechanical properties of the materials, based on the similarity theorem, the selection of the anchor rods required that their mechanical properties are basically the same as those used in the project. After comparing the mechanical properties of relevant carbon steel materials, it was decided to select the steel wire processed from No. 45 steel as the anchor rod for this test. The mechanical properties parameters are shown in Table 1.

Table 1. Mechanical Parameters of Common Anchor and Selected Screws.

Materials	Tensile Strength /MPa	Shear Strength /MPa	E/GPa	Anchoring Force /MPa	Elongation/%
General Anchor	200~600	260~600	200	≥50	≥16
Selection of screws	600	400	210	20~40	≥16

The geometric similarity ratio of the anchor parameters selected for this test is 10:1, which can be adjusted according to the actual conditions in the laboratory, and the basic parameters are shown in Table 2.

Table 2. Geometric parameters of specimen anchoring.

Typology	Drilling Diameter /mm	Anchor Diameter /mm	Spacing /mm
Engineering Prototypes	30	20	800
Anchoring specimen	3	2	80

Drill anchor holes in the horizontal direction of the test piece using a bench drill. Screw the nut and spacer on one end of the anchor rod and insert it into the drilled anchor hole on the test piece. Add the spacer on the other end and tighten it with the nut while applying a certain amount of preload.

It was decided to select the double-ended steel wire screw processed from No. 45 steel as the reinforcing anchor rod for this test with the use of fastening nuts and circular spacers.

The tensile strength of the anchor was 600 MPa, the modulus of elasticity was 210 GPa, and the elongation was ≥16%. The holes were drilled horizontally in the specimen with a diameter of 2 mm using a bench drill, one end of the anchor rod was screwed with a nut and a spacer and inserted into the drilled holes in the specimen, and the spacer was added and fastened with a nut on the other side, while applying a certain amount of pre-tensioning force. The specimens made of each grain size aggregate were arranged with 0, 1, and 2 anchors, respectively, and the finished fabricated plus anchor body is shown in Figure 4.

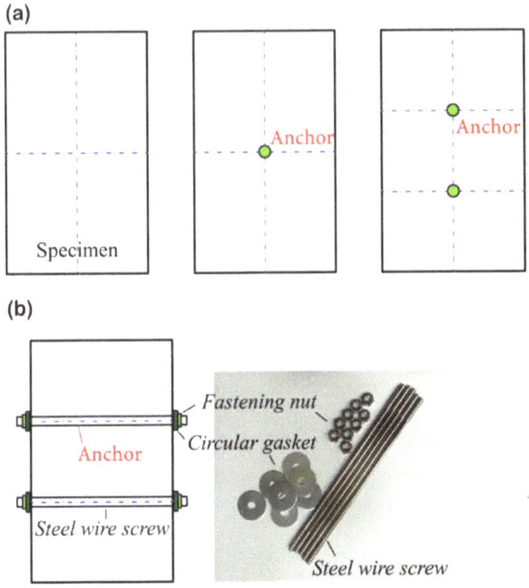

Figure 4. (**a**) Anchoring specimen layout; (**b**) Anchoring specimen.

2.2. Test Methods

A RLJW-2000 microcomputer-controlled rheological experimental machine was used to carry out the rock unidirectional compression test, as shown in Figure 5. The data of the stress can be measured by the experimental machine by placing two micrometers at the diagonal of the chassis for measuring the axial deformation of the specimen. This test was loaded at the rate of 0.25 mm/min until the specimen completely loses the load-bearing capacity, and the results of the experiment were compared to analyze the effect of the anchor on the uniaxial compressive strength of the specimens of different types and the damage characteristics of the experimental program, as shown in Table 3.

Table 3. Test program.

Group/Type of Specimen	Anchor Type	Specimen Number
Grain size A: 2–4 mm aggregate	0 anchor	1-0-1~1-0-4
	1 anchor	1-1-1~1-1-4
	2 anchors	1-2-1~1-2-4
Grain size B: 4–8 mmaggregate	0 anchor	2-0-1~2-0-4
	1 anchor	2-1-1~2-1-4
	2 anchors	2-2-1~2-2-4
Grain size B: 10–15 mmaggregate	0 anchor	3-0-1~3-0-4
	1 anchor	3-1-1~3-1-4
	2 anchors	3-2-1~3-2-4

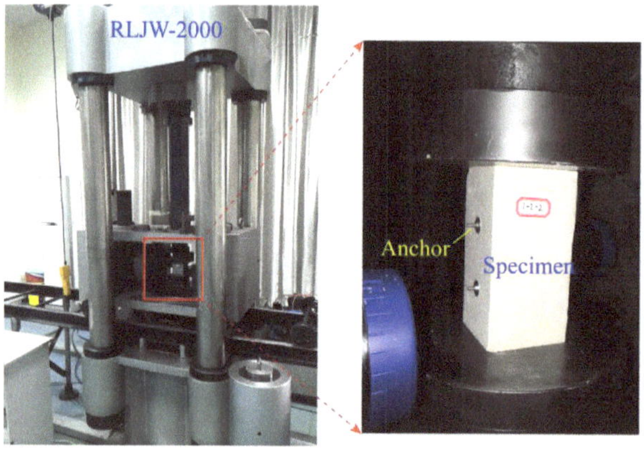

Figure 5. Test setup and specimen loading.

3. Results

3.1. Stress–Strain Curve of Anchor Solid

The axial stress–strain curves during uniaxial compression of the specimens with different grain sizes are shown in Figure 6. Compared with the unanchored specimen, it can be seen that the residual strength value of the anchored specimen in the late stage of damage had a greater increase, but the residual strength value in the compression process of the 2–4 mm grain size aggregate specimen was higher than that of the 4–8 mm and 10–15 mm grain size aggregate specimens. Consequently, the trend of the stress–strain curves of the larger aggregate specimens close to the horizontal direction was more obvious. This indicates that under uniaxial compression conditions, anchors can increase the compressive strength and elastic modulus of specimens. They can also improve the residual strength of the rock mass, but the ability of the anchor bar to improve the residual strength of the rock mass gradually decreases with the increase in the specimen crushing degree.

Comparing the stress–strain curves of the different types of anchored specimens and unanchored specimens revealed that the anchors change the deformation characteristics of the surrounding rock body. Under the no-anchor condition, the brittle characteristics of the rock-like body specimens were more obvious, and when the specimen reached the peak strength, i.e., when the specimen reached the limit of its ability to resist the external load, the rate of decrease in the stress–strain postpeak curves was faster, and the postpeak modulus was larger. For the anchored specimen, its brittle characteristics were not obvious, and the deformation of the anchored rock body showed plastic characteristics. When the anchored rock sample reached the peak compressive strength, the peak stress–strain curve began to decline. Compared to the unanchored specimen, the stress decline rate was reduced, and when the stress dropped to a certain value, the stress–strain curve of the specimen had a small fluctuation, that is, the specimen stress decline rate was small with regard to the anchor at this time. The role of the anchor was more obvious. There was no-anchor specimen in the load loading process. When the specimen reached the peak compressive strength, the rate of stress decline was faster. When the anchor specimen reached the peak strength in the rupture process, the anchor bar prevented internal cracks from expanding and penetrating inside the specimen, thus slowing down the rupture rate.

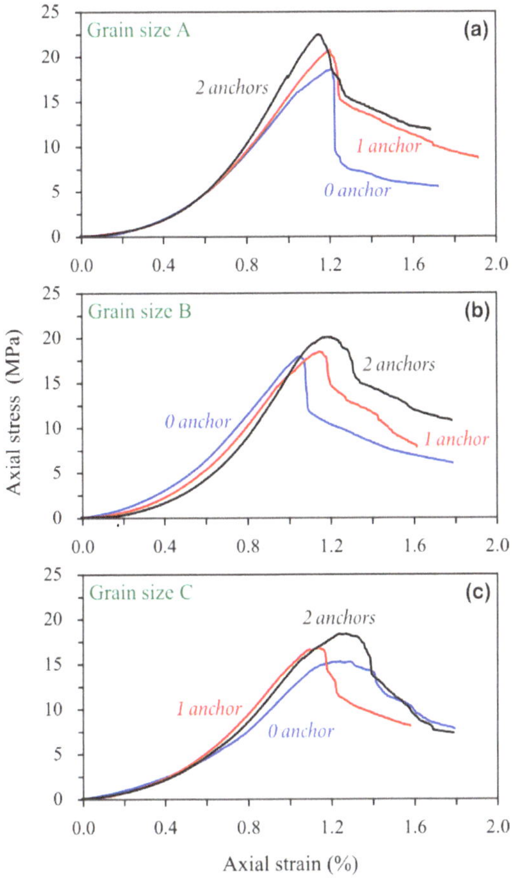

Figure 6. Stress–strain curve of anchor solid. (**a**–**c**) represent anchor solids composed of particle sizes of 2–4 mm, 4–8 mm, and 10–15 mm, respectively.

The uniaxial compressive strength of the specimens with quartz sand of different grain sizes as the aggregate showed significant differences. These specimens had a smaller surface area for the smaller grain size quartz sand aggregate, a smaller contact surface for gypsum, and the formation of structural surfaces of smaller sizes inside the specimen with a smaller degree of fragmentation, while they also had a larger surface area for the larger grain size quartz sand aggregate, a larger contact surface for gypsum, and the formation of structural surfaces of larger sizes inside the specimen with a higher degree of fragmentation. The change characteristics of the compressive strength and elastic modulus of the rock mass under the conditions of the different grain sizes and anchoring methods are shown in Figure 7. It is worth explaining that the modulus of elasticity of the rock mass in the test is the slope of the approximate straight line segment of the axial stress–strain curve, i.e., the generalized modulus of elasticity, as can be seen from the diagram:

1. The uniaxial compressive strength and modulus of elasticity of the specimens showed the following pattern: small-size aggregate specimen > medium-size aggregate specimen > large-size aggregate specimen. The average compressive strength of the unanchored specimens with 2–4 mm aggregate (Grain size A) was 19.75 MPa, that of the unanchored specimens with 4–8 mm aggregate (Grain size B) was 18.30 MPa, and that of the unanchored specimens with 10–15 mm aggregate (Grain size C) was

16.94 MPa. Uniaxial compressive strength decreased by 7.34% and 14.23%, respectively. The average modulus of elasticity of the unanchored specimens with 2–4 mm aggregate was 3.074 GPa, the average modulus of elasticity of the unanchored specimens with 4–8 mm aggregate was 2.884 GPa, and that of the unanchored specimens with 10–15 mm aggregate was 2.549 GPa, with the modulus of elasticity decreasing by 6.18% and 17.06%, respectively.

2. The anchoring effect of the anchors on the specimens of different grain sizes varied. For the specimens with different grain sizes of the aggregates with the different anchoring methods, the uniaxial compressive strength and elastic modulus of the specimens followed this pattern: no anchor < single anchor < double anchor. Among them, the uniaxial compressive strength of the specimens with 2–4 mm aggregate size increased by 12.90% and 22.81%, and the modulus of elasticity increased by 16.39% and 23.34%, respectively, compared with that without the anchor in the single-anchored and double-anchored specimens; the uniaxial compressive strength of the specimens with 4–8 mm aggregate size increased by 4.16% and 11.35%, and the modulus of elasticity increased by 3.33% and 14.34%, respectively, compared with that without the anchor in the single-anchored and double-anchored specimens, which increased by 3.33% and 14.25%; the specimens with 10–15 mm particle size aggregate increased the uniaxial compressive strength by 4.88% and 6.02% and the modulus of elasticity by 7.19% and 9.16%, respectively, when the single and double anchored were compared to the unanchored.
3. With the increase in the specimen aggregate particle size, the degree of improvement of the mechanical properties of the anchor rods on the specimen gradually decreased, due to the formation of large-size aggregate and the gypsum size of the structural surface. Because the specimen crushing degree was larger, the anchoring effect of the anchor rods could not be given full play in the crushed rock, and the anchor rods on the crushed rock support adaptability was reduced. Therefore, under the condition of roadways surrounded by rock with a high degree of crushing, auxiliary support measures need to be taken to form a joint coordinated support form to control the phenomenon of a poor anchor support effect of the crushed surrounding rock, the large deformation of roadways surrounded by rock, and other phenomena affecting the safe production of the mine.

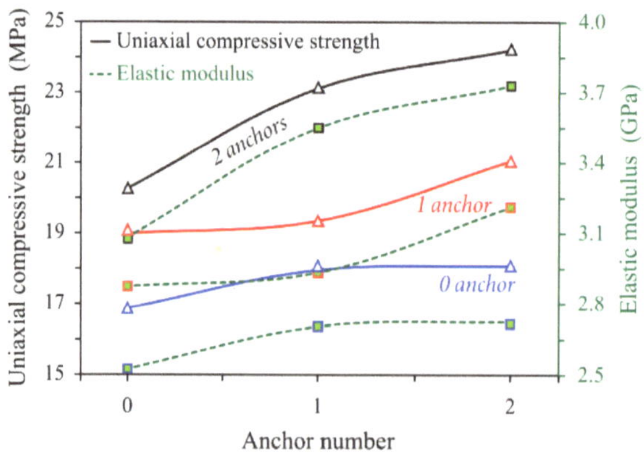

Figure 7. Influence of the degree of surrounding rock fragmentation and the number of anchors added on the mechanical parameters of the rock mass.

3.2. Rupture Characteristics

The final damage pattern of the rock mass under the conditions of different grain sizes and anchoring methods is shown in Figure 8. As can be seen from the graph,

1. Under the condition of no anchor (0 anchor), the uniaxial compression of the specimen formulated with large-grained aggregate was more serious than that of the small-grained aggregate specimen. The specimen as a whole showed diagonal shear damage, but the secondary fissures showing splitting damage developed to a high degree, and different sizes of block spalling phenomenon appeared. When the aggregate diameter was 2–4 mm (Grain size A), the specimen split along the axial direction, i.e., the morphology of the rupture surface was parallel to the loading axial direction, and no shear rupture occurred; when the aggregate diameter was 4–8 mm (Grain size B), the shear rupture was relatively slight, and the shear rupture surface was not completely through the upper and lower parts of the rupture surface; and when the aggregate diameter was 10–15 mm (Grain size C), obvious shear damage occurred on the free surface, and a large number of block spalling phenomenon occurred. This shows the strength difference of the specimens made of the aggregates with different grain size ratios.
2. Under the action of the single anchor (1 anchor), the rock body along the anchor direction was subjected to a certain support resistance constraint, and the rupture of the specimen was relatively homogenized. The rupture crack direction of the rock body changed, the main rupture cracks in the aggregate diameter of 2–4 mm and 4–8 mm rupture cracks in the anchor position rupture crack direction changed at a certain angle, compared with the no-anchor condition rupture degree, which was small; 10–15 mm aggregate single-anchor specimens produced intensive rupture cracks, appearing to a certain extent of the block exfoliation. Combined with the rupture surface of the specimen under the anchored condition in Figure 8, the anchoring effect of the anchor on the specimen was reflected in the two aspects of reinforcement and crack stopping. Under the pre-stressing effect of the anchor, when the crack developed to the range of the anchor's action, the anchor played the role of stopping the crack from developing or changing the direction. The role of the tray along the anchor axial force was applied, and when the tray was in contact with the formation of conical stress concentration within the rock mass, the rock mass played a reinforcing effect.
3. Under the double-anchor condition (2 anchors), the damage pattern of the anchored rock body, when compared with single anchor with an aggregate diameter of 2–4 mm, had less developed specimen secondary cracks; when the aggregate diameter was 4–8 mm, the main rupture cracks were distributed along the direction of the axial stress, and the specimen underwent splitting damage, and the secondary cracks were obviously reduced. When the aggregate diameter was 10–15 mm, the rupture cracks produced by the single-anchor specimen are greater. When the aggregate diameter was 10–15 mm, the rupture cracks produced by the single-anchor specimens were intensive, and those produced by the double-anchor specimens were relatively scattered. This shows that the increase in anchors effectively enhanced the cementation between the anchors and the rock body, which can augment the supporting role of anchors, thus alleviating the degree of rupture of the rock body and improving the overall bearing capacity.

The anchoring effect of the anchor on the specimen was reflected in the two aspects of reinforcement and crack stopping. Under the prestressing reinforcement of the anchor, when the crack developed into the reinforcement range of the anchor, the anchor played the role of stopping the crack from developing or changing its direction. The action of the tray exerting force along the axial direction of the anchor rods created a conical stress concentration within the rock mass in contact with the tray, which acted as a reinforcement to the rock mass. As shown in Figure 9, after the rupture of the specimen, an obvious reinforcement zone was formed at the anchor reinforcement, and this extrusion reinforcement effect improved the integrity and bearing capacity of the rock mass.

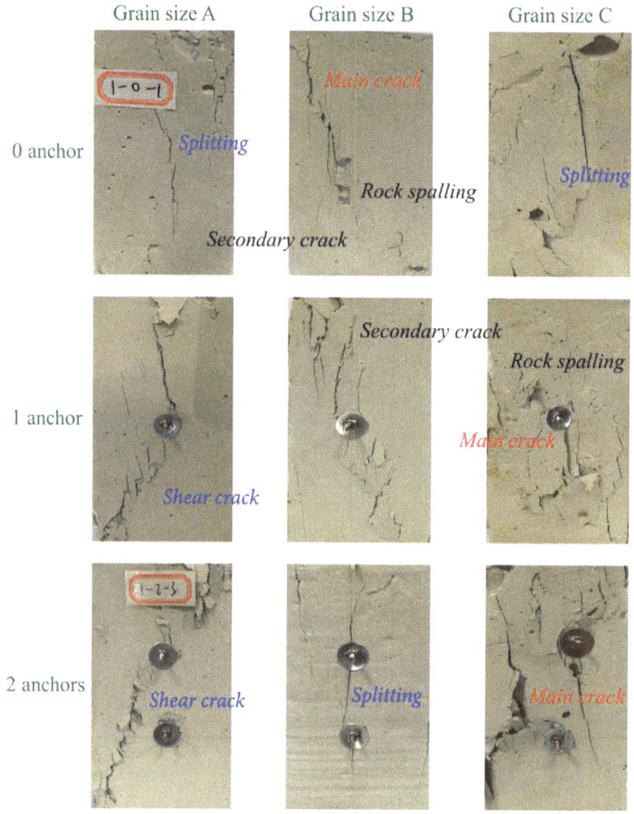

Figure 8. Rupture morphology of specimens with different grain sizes.

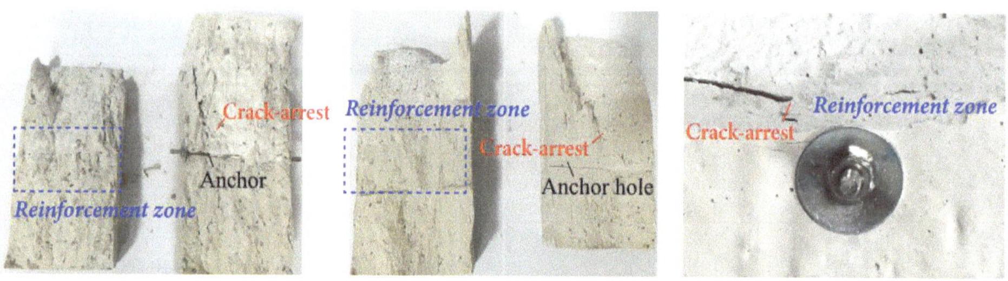

Figure 9. Anchor solid specimen rupture pattern.

4. Numerical Modeling

In order to further analyze the protective effect of the anchor rods on the rock, this paper used the finite element software ABAQUS [45–47] to establish a numerical model of uniaxial compression of the rock. The boundary conditions of the model were similar to the boundary conditions of the test, and the bottom boundary of the model was fixed to the vertical displacement, and the top boundary gradually applied the axial load. The test was divided into three cases: (1) no anchor; (2) embedded 1 anchor; (3) embedded 2 anchors (Figure 10). In the model, cohesive units were inserted between the rock units based on the damage mechanics theory. The rock matrix material parameters were a modulus of elasticity of 10 GPa, a Poisson's ratio of 0.25, and a density of 1700 kg/m^3. The cohesive unit

adopted the traction-separation theorem, the fracture energy of the unit was 0.1 N/mm, and it was assumed that there is no friction on the contact surface once the unit is cracked. The anchor material parameters were a modulus of elasticity of 210 GPa and a Poisson's ratio of 0.3.

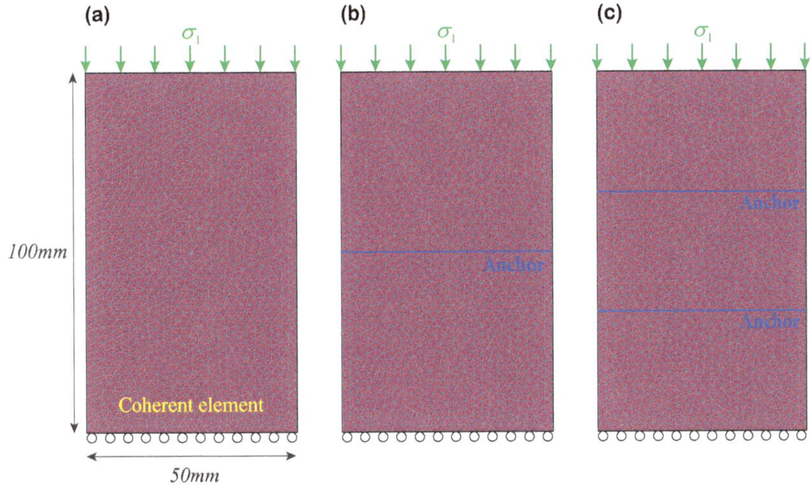

Figure 10. Numerical model of anchor solid. (**a**–**c**) denote the anchored solid consisting of no anchor specimen, 1 anchor specimen, and 2 anchor specimens, respectively.

5. Discussion of the Results

Figure 11 shows the evolution of rock deformation and damage under the different working conditions. For the no-anchor condition (0 anchor), the specimen showed typical oblique shear damage characteristics, and at the same time, in the region near the top and bottom boundaries, a block spalling phenomenon occurred, and with the continuous increase in the axial load, the main macroscopic rupture surface running through the specimen was presented eventually. Under the single-anchor action (1 anchor), the deformation of the specimen near the anchor was significantly lower than that in other areas due to the limiting effect of the anchor. At the same time, the main rupture surface of the rock was blocked, and it can be seen from the figure that an obvious reinforcement zone was formed near the anchor. Under the condition of double anchors (2 anchors), the restriction effect of anchors was more obvious, the range of strengthened area was wider, and the rock rupture surface was strongly blocked. This indicates that the anchors played a reinforcing and fracture stopping role on the rock, and the more anchors (or the more intensive) meant a more obvious reinforcing effect.

Figure 12 shows the maximum stress distribution characteristics of the anchor solid. From the figure, it can be seen that due to the embedded anchor in the specimen, the lateral expansion of the specimen occurred after the external load was applied, and the tensioning effect of the anchor was obvious. The specimen gradually produced oblique shear damage near the top-bottom boundary. However, due to the tensile action of the anchor rods, the crack expansion was inhibited, and the tension stress of the anchor rods near the potential rupture area (ellipse area) in the figure was more obvious (the anchor rods are in the red color, which represents a higher stress), which indicates that the anchor rods played a role of reinforcing and stopping cracks in the rock, and the more densely the anchors were installed, the more obvious their reinforcing effect was.

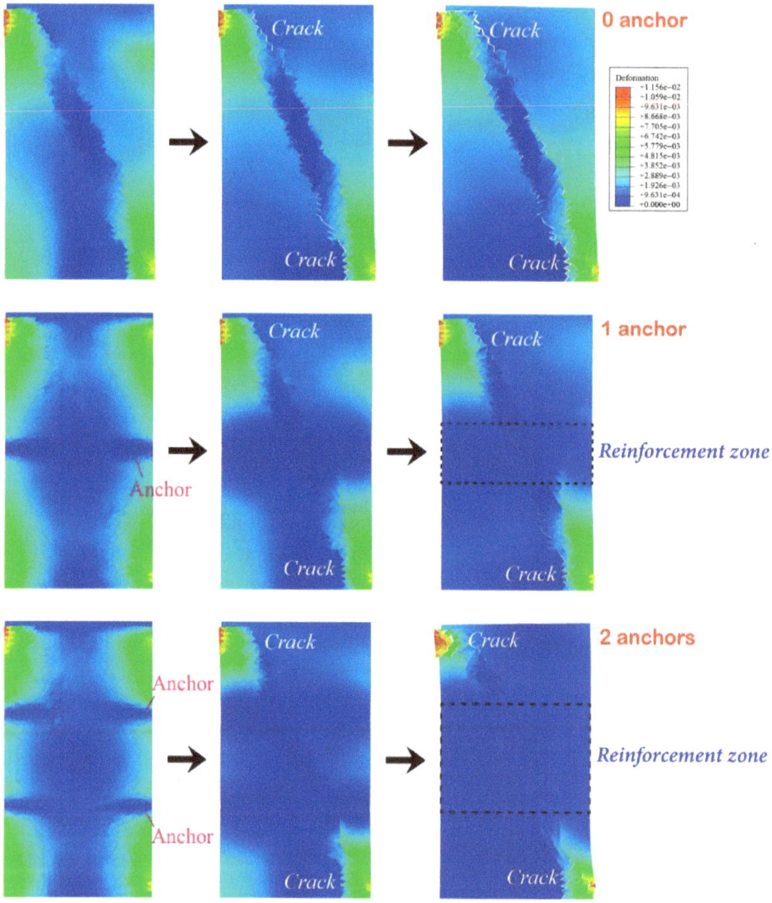

Figure 11. Deformation damage cloud diagram of different specimens.

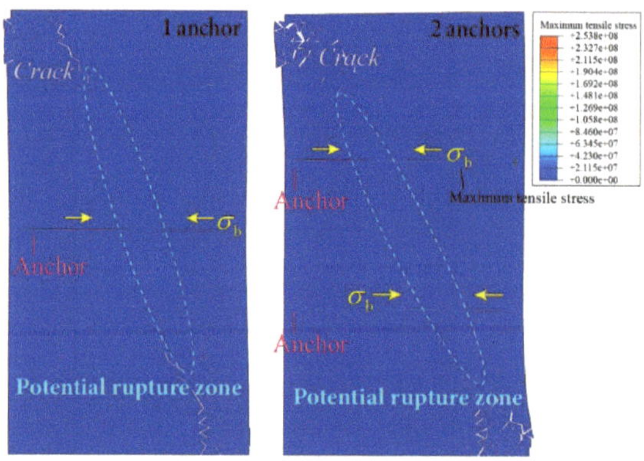

Figure 12. Characterization of the maximum stress distribution in the anchor solid.

Figure 13 shows a schematic diagram of the reinforcing action of the anchor rods, from which it can be seen that the anchor rods limited the deformation of the specimen and formed an anchor support anchorage force, similar to a small lateral circumferential

pressure. In the area of anchor cementation and action, a reinforcement zone was formed, which can greatly limit the development of internal cracks as well as the lateral expansion of the rock and provide protection to the specimen. Compared with the limited scope of the reinforced zone under the single-anchor condition, the reinforced zone under the double-anchor condition had a wider scope, and there was a superposition effect, which effectively strengthened the bonding effect of the anchor and the rock body. This, as a result, gave a better supporting effect to the anchor, which greatly alleviated the rock's rupture development, effectively harmonized its overall deformation, and played a key role in stabilizing it.

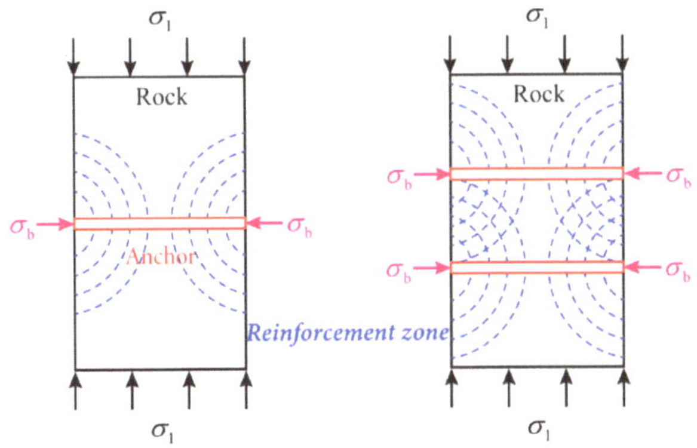

Figure 13. Schematic diagram of the reinforcing effect of anchors. Σ_b represents maximum tensile stress.

6. Conclusions

1. The uniaxial compressive strength of the specimens with different grain size quartz sand as the aggregate showed significant differences, with a smaller surface area for the smaller grain size quartz sand aggregate, a smaller contact surface for gypsum, which formed smaller size structural surfaces inside the specimen and had a smaller degree of fragmentation, whereas the larger grain size quartz sand aggregate formed larger size structural surfaces inside the specimen and a higher degree of fragmentation. The uniaxial compressive strength and modulus of elasticity of the specimens showed the following pattern: small-size aggregate specimen > medium-size aggregate specimen > large-size aggregate specimen.
2. Compared with the specimen without the anchor, the residual strength value of the anchored specimen in the late stage of damage had a greater increase, but the residual strength value was higher than that of the large aggregate specimen in the compression process of the small aggregate specimen, and the trend of the stress–strain curve close to the horizontal direction was more obvious in the specimen of the larger aggregate size. This shows that under uniaxial compression, the anchor can not only improve the compressive strength and elastic modulus of the rock mass but it can also improve the residual strength of the rock mass. However, with the increase in specimen crushing degree, the ability of the anchor to improve the residual strength of the rock mass gradually decreased.
3. The anchoring effect of the anchors on the specimens with different grain sizes varies was crucial. When different anchoring methods are used for the specimens with the different grain sizes of the aggregates, the uniaxial compressive strength and elastic modulus of the specimens follow the following pattern: no anchor<single anchor<double anchor. With the increase in the specimen aggregate size, the degree of improvement of the mechanical properties of the specimens by the anchors decreases gradually.

4. Numerical tests on the uniaxial compression of the rock containing cohesive units were carried out. It was found that under the no-anchor condition, the specimen showed typical oblique shear damage characteristics, and under the single-anchor action, the deformation of the specimen near the anchor bar was obviously reduced. By contrast, the main rupture surface was blocked, and an obvious reinforcement zone was formed near the anchor bar. Under the double-anchor condition, the anchor tension stress was more obvious, the reinforcement zone was wider, and the rock rupture surface was strongly blocked, all of which made its reinforcement effect the more obvious.
5. The anchoring effect of the anchor on the specimen was reflected in the two aspects of reinforcement and crack stopping. Under the prestressing effect of the anchor, when the crack developed to the range of the anchor's action, the anchor played the role of stopping the crack from developing or changing its direction. After the specimen rupture, an obvious conical reinforcement zone was formed at the anchor reinforcement, and this extrusion reinforcement effect improved the integrity and bearing capacity of the rock mass. Moreover, the denser the anchors, the wider the reinforcement zone, and the superposition effect occurred, which better utilized the supporting role of the anchors and played a key role in stabilizing the rock.

Author Contributions: Conceptualization, L.W.; methodology, Z.L.; software, Z.L. and C.J.; validation, B.R.; formal analysis, B.R. and S.W.; investigation, S.W., K.D. and J.G.; resources, F.T.; data curation, S.W. and K.D.; writing—original draft preparation, S.W.; writing—review and editing, S.W.; visualization, C.J.; supervision, F.T. and J.G.; project administration, F.T.; funding acquisition, L.W. All authors have read and agreed to the published version of the manuscript.

Funding: This work was supported by the National Key Research and Development Program of China (No. 2017YFC0603004), Scientific Research Foundation for High-level Talents of Anhui University of Science and Technology (No. 2022yjrc37) and Open Research Foundation for State Key Laboratory of Mining Response and Disaster Prevention and Control in Deep Coal Mines (No. SKLMRDPC22KF05).

Institutional Review Board Statement: Not applicable.

Informed Consent Statement: Not applicable.

Data Availability Statement: Not applicable.

Conflicts of Interest: The authors declare no conflict of interest.

References

1. Kang, H. Support technologies for deep and complex roadways in underground coal mines: A review. *Int. J. Coal Sci. Technol.* **2014**, *1*, 261–277. [CrossRef]
2. He, M.C. Rock mechanics and hazard control in deep mining engineering in China. *Rock Mech. Undergr. Constr.* **2006**. [CrossRef]
3. Sun, J.; Wang, S. Rock mechanics and rock engineering in China: Developments and current state-of-the-art. *Int. J. Rock Mech. Min. Sci.* **2000**, *37*, 447–465. [CrossRef]
4. Xie, H.; Lu, J.; Li, C.; Li, M.; Gao, M. Experimental study on the mechanical and failure behaviors of deep rock subjected to true triaxial stress: A review. *Int. J. Min. Sci. Technol.* **2022**, *32*, 915–950. [CrossRef]
5. Xie, H.; Gao, F.; Ju, Y. Research and development of rock mechanics in deep ground engineering. *Chin. J. Rock Mech. Eng.* **2015**, *34*, 2161–2177.
6. Chai, Y.; Dou, L.; Cai, W.; Małkowski, P.; Li, X.; Gong, S.; Bai, J.; Cao, J. Experimental investigation into damage and failure process of coal-rock composite structures with different roof lithologies under mining-induced stress loading. *Int. J. Rock Mech. Min. Sci.* **2023**, *170*, 105479. [CrossRef]
7. Yoon, S.; Lee, H.; Kim, J. The modeling of fault activation, slip, and induced seismicity for geological CO_2 storage at a pilot-scale site in the Janggi Basin, South Korea. *Int. J. Rock Mech. Min. Sci.* **2023**, *170*, 105441. [CrossRef]
8. Ju, Y.; Zhang, Q.; Yang, Y.; Xie, H.; Gao, F.; Wang, H. An experimental investigation on the mechanism of fluid flow through single rough fracture of rock. *Sci. China Technol. Sci.* **2013**, *56*, 2070–2080. [CrossRef]
9. Yin, Q.; Jing, H.; Zhu, T. Mechanical behavior and failure analysis of granite specimens containing two orthogonal fissures under uniaxial compression. *Arab. J. Geosci.* **2015**, *9*, 31. [CrossRef]

10. Lee, H.; Jeon, S. An experimental and numerical study of fracture coalescence in pre-cracked specimens under uniaxial compression. *Int. J. Solids Struct.* **2011**, *48*, 979–999. [CrossRef]
11. Zhang, B.; Li, S.; Yang, X.; Zhang, D.; Shao, C.; Yang, W. Uniaxial compression tests on mechanical properties of rock mass similar material with cross-cracks. *Rock Soil Mech.* **2012**, *33*, 3674–3679.
12. Wu, W.; Gong, F.; Ren, L.; He, L. Strain rockburst failure characteristics and mechanism of high stress circular hard rock tunnel triggered by dynamic impact load. *Int. J. Rock Mech. Min. Sci.* **2023**, *171*, 105575. [CrossRef]
13. Jeon, S.; Kim, J.; Seo, Y.; Hong, C. Effect of a fault and weak plane on the stability of a tunnel in rock—A scaled model test and numerical analysis. *Int. J. Rock Mech. Min. Sci.* **2004**, *41*, 658–663. [CrossRef]
14. Moir, H.; Lunn, R.J.; Shipton, Z.K.; Kirkpatrick, J.D. Simulating brittle fault evolution from networks of pre-existing joints within crystalline rock. *J. Struct. Geol.* **2010**, *32*, 1742–1753. [CrossRef]
15. Wang, J.; Apel, D.B.; Xu, H.; Wei, C.; Skrzypkowski, K. Evaluation of the Effects of Yielding Rockbolts on Controlling Self-Initiated Strainbursts: A Numerical Study. *Energies* **2022**, *15*, 2574. [CrossRef]
16. Wang, Q.R.; Xie, L.X.; Song, E.X.; Kong, F.L.; Fan, J.Q.; Yu, L.Y.; Xu, J.M.; Shi, X.Y. Model tests on dynamic responses of surrounding rock and support structure on underground tunnel under combined dynamic and static loading. *Int. J. Rock Mech. Min. Sci.* **2023**, *171*, 105572. [CrossRef]
17. Li, C.C.; Stjern, G.; Myrvang, A. A review on the performance of conventional and energy-absorbing rockbolts. *J. Rock Mech. Geotech. Eng.* **2014**, *6*, 315–327. [CrossRef]
18. Kang, H.; Wu, Y.; Gao, F.; Jiang, P.; Cheng, P.; Meng, X.; Li, Z. Mechanical performances and stress states of rock bolts under varying loading conditions. *Tunn. Undergr. Space Technol.* **2016**, *52*, 138–146. [CrossRef]
19. Huang, Z.; Broch, E.; Lu, M. Cavern roof stability—Mechanism of arching and stabilization by rockbolting. *Tunn. Undergr. Space Technol.* **2002**, *17*, 249–261. [CrossRef]
20. Skrzypkowski, K.; Korzeniowski, W.; Zagórski, K.; Dudek, P. Application of Long Expansion Rock Bolt Support in the Underground Mines of Legnica–Głogów Copper District. *Stud. Geotech. Et Mech.* **2017**, *39*, 47–57. [CrossRef]
21. Cao, R.; Cao, P.; Lin, H. Support technology of deep roadway under high stress and its application. *Int. J. Min. Sci. Technol.* **2016**, *26*, 787–793. [CrossRef]
22. Chen, Y.; Meng, Q.; Xu, G.; Wu, H.; Zhang, G. Bolt-grouting combined support technology in deep soft rock roadway. *Int. J. Min. Sci. Technol.* **2016**, *26*, 777–785. [CrossRef]
23. Chen, S.H.; Yang, Z.M.; Wang, W.M.; Shahrour, I. Study on Rock Bolt Reinforcement for a Gravity Dam Foundation. *Rock Mech. Rock Eng.* **2012**, *45*, 75–87. [CrossRef]
24. Wang, J.; Li, S.-C.; Li, L.-P.; Zhu, W.; Zhang, Q.-Q.; Song, S.-G. Study on anchorage effect on fractured rock. *Steel Compos. Struct.* **2014**, *17*, 791–801. [CrossRef]
25. Zhang, B.; Li, S.; Xia, K.; Yang, X.; Zhang, D.; Wang, S.; Zhu, J. Reinforcement of rock mass with cross-flaws using rock bolt. *Tunn. Undergr. Space Technol.* **2016**, *51*, 346–353. [CrossRef]
26. Wang, J. New development of rock bolting technology for coal roadway in China. *J. China Coal Soc.* **2007**, *32*, 113–118.
27. Liu, Q.; Lei, G.; Peng, X. Advance and review on the anchoring mechanism in deep fractured rock mass. *Chin. J. Rock Mech. Eng.* **2016**, *35*, 312–332.
28. Li, S.; Zhang, N.; Lu, A.; Li, M.; Yang, L. Experimental study of anchoring effect of discontinuous jointed rock mass under uniaxial tension. *Chin. J. Rock Mech. Eng.* **2011**, *30*, 1579–1586.
29. Chen, Y. Experimental study and stress analysis of rock bolt anchorage performance. *J. Rock Mech. Geotech. Eng.* **2014**, *6*, 428–437. [CrossRef]
30. Sagong, M.; Bobet, A. Coalescence of multiple flaws in a rock-model material in uniaxial compression. *Int. J. Rock Mech. Min. Sci.* **2002**, *39*, 229–241. [CrossRef]
31. Su, H.; Jing, H.; Zhao, H.; Yu, L.; Wang, Y. Strength degradation and anchoring behavior of rock mass in the fault fracture zone. *Environ. Earth Sci.* **2017**, *76*, 179. [CrossRef]
32. Jing, H.W.; Yang, S.Q.; Zhang, M.L.; Xu, G.A.; Chen, K.F. An experimental study on anchorage strength and deformation behavior of large-scale jointed rock mass. *Tunn. Undergr. Space Technol.* **2014**, *43*, 184–197. [CrossRef]
33. Deb, D.; Das, K.C. A new doubly enriched finite element for modelling grouted bolt crossed by rock joint. *Int. J. Rock Mech. Min. Sci.* **2014**, *70*, 47–58. [CrossRef]
34. Nie, W.; Zhao, Z.; Ning, Y.; Sun, J. Development of rock bolt elements in two-dimensional discontinuous deformation analysis. *Rock Mech. Rock Eng.* **2014**, *47*, 2157–2170. [CrossRef]
35. Lin, H.; Xiong, Z.; Liu, T.; Cao, R.; Cao, P. Numerical simulations of the effect of bolt inclination on the shear strength of rock joints. *Int. J. Rock Mech. Min. Sci.* **2014**, *66*, 49–56. [CrossRef]
36. Grasselli, G. Manuel Rocha Medal Recipient Shear Strength of Rock Joints Based on Quantified Surface Description. *Rock Mech. Rock Eng.* **2006**, *39*, 295. [CrossRef]
37. Lee, J.S.; Bang, C.S.; Mok, Y.J.; Joh, S.H. Numerical and experimental analysis of penetration grouting in jointed rock masses. *Int. J. Rock Mech. Min. Sci.* **2000**, *37*, 1027–1037. [CrossRef]
38. Fan, L.; Liu, S. A conceptual model to characterize and model compaction behavior and permeability evolution of broken rock mass in coal mine gobs. *Int. J. Coal Geol.* **2017**, *172*, 60–70. [CrossRef]

39. Li, S.C.; Wang, Q.; Wang, H.T.; Jiang, B.; Wang, D.C.; Zhang, B.; Li, Y.; Ruan, G.Q. Model test study on surrounding rock deformation and failure mechanisms of deep roadways with thick top coal. *Tunn. Undergr. Space Technol.* **2015**, *47*, 52–63. [CrossRef]
40. Brown, E.T. (Ed.) *Rock Characterization, Testing & Monitoring: ISRM Suggested Methods*; International Society for Rock Mechanics, Pergamon Press: Oxford, UK, 1981.
41. Fairhurst, C.E.; Hudson, J.A. Draft ISRM suggested method for the complete stress-strain curve for intact rock in uniaxial compression. *Int. J. Rock Mech. Min. Sci.* **1999**, *36*, 279–289.
42. Wang, S.; Wang, L.G.; Tian, J.S.; Fan, H.; Jiang, C.Y.; Ding, K. An Experimental Study on the Effects of True Triaxial Loading and Unloading Stress Paths on the Mechanical Properties of Red Sandstone. *Minerals* **2022**, *12*, 204. [CrossRef]
43. Wang, S.; Wang, L.; Ren, B.; Ding, K.; Jiang, C.; Guo, J. Study of the mechanical characteristics of coal-serial sandstone after high temperature treatment under true triaxial loading. *Sci. Rep.* **2023**, *13*, 13036. [CrossRef]
44. Wang, X.; Liu, X.; Wang, E.; Li, X.; Zhang, X.; Zhang, C.; Kong, B. Experimental research of the AE responses and fracture evolution characteristics for sand-paraffin similar material. *Constr. Build. Mater.* **2017**, *132*, 446–456. [CrossRef]
45. Huang, X.; Bie, Z.; Wang, L.; Jin, Y.; Liu, X.; Su, G.; He, X. Finite element method of bond-based peridynamics and its ABAQUS implementation. *Eng. Fract. Mech.* **2019**, *206*, 408–426. [CrossRef]
46. Hibbitt; Sorensen, K. *ABAQUS: Theory Manual*; Hibbitt, Karlsson & Sorensen: Providence, RI, USA, 1997; Volume 2.
47. Tahmasebinia, F.; Yang, A.; Feghali, P.; Skrzypkowski, K. A Numerical Investigation to Calculate Ultimate Limit State Capacity of Cable Bolts Subjected to Impact Loading. *Appl. Sci.* **2023**, *13*, 15. [CrossRef]

applied sciences

MDPI

Article

Investigation into Mining Economic Evaluation Approaches Based on the Rosenblueth Point Estimate Method

Jiaoqun Li [1], Tong Wu [1], Zengxiang Lu [1,*] and Saisai Wu [2,*]

[1] School of Mining Engineering, Liaoning University of Science and Technology Liaoning, Anshan 114051, China; ljq_as@ustl.edu.cn (J.L.); 2604452468@ustl.edu.cn (T.W.)
[2] School of Resources Engineering, Shanxi Key Laboratory of Geotechnical and Underground Space Engineering XAUAT, Xi'an 710055, China
* Correspondence: zengxiang_lu@sohu.com (Z.L.); saisai.wu@xauat.edu.cn (S.W.)

Abstract: Conducting technical and economic evaluations is important for mining investment and mining operation decision-making. Traditional economic evaluation methods rarely address the issue of evaluation reliability and usually require complex calculations to obtain the optimal solution. In this study, the Rosenblueth point estimate method for reliability evaluation of engineering project schemes is introduced. Combined with the cash flow method for economic evaluation of mines, the Rosenblueth point estimate method for evaluating the reliability of mining economy is established. Based on the technical and economic index of the case mine, taking the ore grade as a sensitivity indicator, empirical research on established models and methods was carried out. The results of the economic reliability evaluation and the variation rules obtained using the Rosenblueth point estimate method model were basically consistent with the actual production and operation rules of mining enterprise. The similar results also proved that the proposed model has good applicability and reliability for mining economic evaluation. Using the proposed RPEM economic reliability model, the economic reliability of a certain iron mine in Liaoning Province was calculated to be 99.95, which was a huge improvement compared with the traditional evaluation method. Additionally, the calculation process of the proposed model for economic reliability evaluation is simple and the accuracy is controllable. The economic reliability of the project can be calculated based on changes in sensitivity indicators, and the value range of sensitivity indicators can also be calculated through the required reliability. The obtained results and the proposed evaluation model provide a decision-making basis for mining investment projects and operation management.

Keywords: Rosenblueth method; economic benefit; reliability analysis; economic index

1. Introduction

In present times, China's national economy has entered a stage of rapid development. Mineral resources, as an important foundation for promoting social development, must be managed with long-term and stable development paths to ensure the economic benefits of mining enterprises [1–3]. With the primary goal of obtaining economic benefits, mining project development and construction require production processes and modes that can control production costs, improve production efficiency, ensure engineering safety, and maximize profits [4–6]. Economic evaluation can improve operations and management and increase economic benefits. Conducting economic evaluation of mining project development, scientifically predicting the cost level of mining production, improving cost prediction accuracy, discovering the rule of cost changes, and identifying factors that affect costs and effectively controlling them can improve the cost management level and core competitiveness of enterprises [7–9]. Therefore, mining enterprises have a strong focus on how to evaluate their economic benefits.

Currently, the widely used economic evaluation methods include the cash flow method, discounted cash flow method, financial internal rate of return analysis method, net present value and breakeven analysis method, etc. [10]. For example, Zhou Ling [11] used the APRP model for economic evaluation of the green mining evaluation index system and proposed suggestions for the improvement of the growth stage of green mining development in a certain region. Wang Quansheng [12] discovered the problems in using the cash flow method for project economic evaluation and mining rights evaluation and proposed suggestions for the improvement of the determination of relevant parameter selection. In order to improve the utilization rate of resources, Shi Qingbao et al. [13] adopted the internal rate of the return analysis method to evaluate the internal rate of return and the development and utilization value of resource development according to the reserves and industrial indicators in a mining rights project. They proposed the methods and measures for optimizing the internal rate of return. Chen Wenjun et al. [14] utilized the method of combining full lifecycle theory with economic benefit evaluation to evaluate the rationality of mine technical renovation and expansion from aspects such as investment cash flow and financial viability. Based on the traditional capital asset pricing model, Xu Yixin et al. [15] introduced the systematic risk correction coefficient, established an improved capital asset pricing model, and used it to evaluate the value of mining rights. Yang Huaimin et al. [16] suggested that dynamic breakeven analysis can consider the time value of funds, and the evaluation results obtained were more reasonable. Dariusz et al. [17] pointed out that when using the breakeven point to calculate the profitability of mines, the sensitivity of factors should be adjusted based on the impact of each indicator on revenue, taking into account the specific situation of the mine. Wen Wei et al. [18] used TOPSIS and grey correlation method to construct an economic risk evaluation model and evaluated the economic risks of different countries from a macro perspective. Li Guoqing et al. [19] used 0–1 Integer Programming to establish the operation plan optimization model, which could be used to optimize the mining plan, and improve the utilization rate of mineral resources. Kong Wenyuan et al. [20] pointed out that considering infrastructure investment when optimizing mining plans can improve economic benefits and make investment plans more reasonable. Based on the four principles of system comprehensiveness, main factors, operability, and scientific objectivity, Yuliduzi Sidike et al. [21] constructed a mining economic competitiveness evaluation index system with four secondary indicators and 18 tertiary indicators. They used principal component analysis to evaluate the mining economic competitiveness of 10 major mining provinces in China and proposed directions for improving the mining economic competitiveness of Heilongjiang Province.

Various existing evaluation methods have played important roles in mining economic evaluation, but in this process, the calculation procedures of each evaluation method may be cumbersome or simple, and the reliability of evaluation results may be high or low. Seeking a mining economic evaluation method that can simplify the evaluation calculation process and improve the reliability of evaluation results can effectively improve the efficiency and accuracy of evaluation work. The Rosenblueth point estimate method [22] (RPEM) is a method of obtaining reliability and failure probability by calculating the mean–variance of the objective function, which is widely used in engineering reliability analysis [23–25]. RPEM only needs to calculate the objective function at the selected weighted points, which is easy to implement and does not require numerical iteration [24]. However, there is currently no precedent for using this method for economic evaluation. In terms of engineering reliability analysis, Yang Kui [26] used RPEM for slope stability analysis of hydropower stations and compared the calculation results with the results obtained by Monte Carlo method. It was found that the probability of slope failure obtained by these two methods was similar. Cai Qian et al. [27] used RPEM for real-time analysis of the probability of earth-rock dam failure, concluding that RPEM can balance data accuracy and efficiency. Yang Yang et al. [28] applied RPEM to the reliability analysis of tunnel bolt support structures and obtained results that were basically consistent with the actual situation on site, improving computational efficiency. Kahe Maryam Sadat et al. [29] used

RPEM to evaluate the uncertainty of groundwater model parameters, believing that it has the advantages of simplicity and efficiency compared to Monte Carlo methods. Using TRIGRS landslide prediction mode and adopting RPEM to solve the uncertainty problem of parameters, Xu Zenghui et al. [30] studied the relationship between rainfall events and landslide occurrence.

Based on the widespread application of RPEM in engineering reliability evaluation [25], the major objective of this study was to introduce RPEM into mining economic evaluation to shorten the cycle of economic evaluation and improve the reliability of evaluation. The development and production processes of mining projects are complex and involve many technical and economic indicators. In order to study the applicability of RPEM in mining economic evaluation, on the basis of establishing an RPEM economic evaluation method, we studied the probability of achieving benefits with different geological grades, conducted empirical research to verify the rationality of the results, and explored the feasibility of using RPEM for mining economic evaluation.

2. RPEM Economic Evaluation Method

2.1. The Basic Principles of RPEM

RPEM, also known as moment estimation method, is often used to purposefully select points composed of special values under the condition of unclear distribution of random variables. The mean and variance are used to find the first four moments of the functions point by point and to calculate the reliability of the evaluated project. This method has a simple principle, fewer calculations, and does not need to consider the probability distribution of basic input random variables in advance. The processing of related variables is simple, while the requirements of engineering accuracy can be met [31]. The calculation process of RPEM is shown in Figure 1.

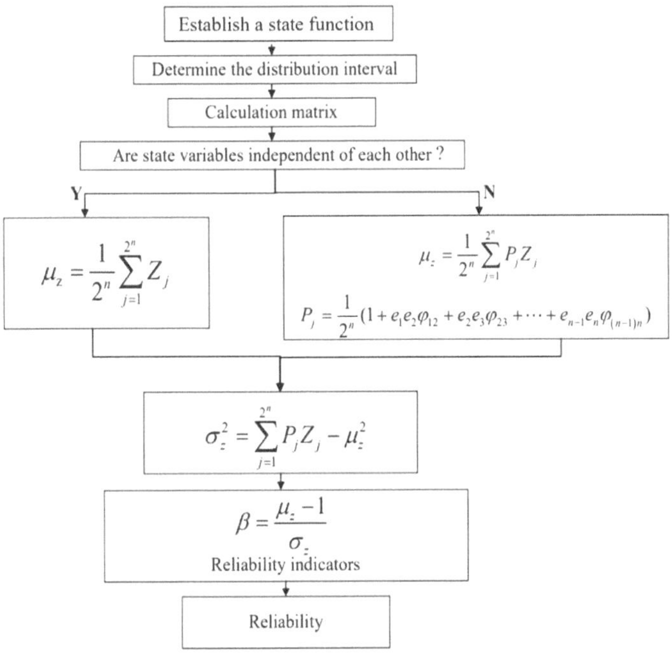

Figure 1. Flow chart of RPEM.

When using RPEM for mine technical and economic evaluation, the target state function is first established: $Z = Z\ (x_1, x_2, \ldots, x_i, \ldots, x_n)$, among which, x_i represents a random variable subject to normal distribution, that is, mining and dressing cost, ore geological grade, concentrate sales price, dressing recovery rate, and other indicators; next, the value points are selected, and within the interval (x_i min, x_i max), two value points are symmetrically seleceted (such as the positive and negative standard deviation of the mean of the state variable). Using μ, σ representing the mean and standard deviation of variables, there are:

$$x_{i1} = \mu_{xi} + \sigma_{xi} \tag{1}$$

$$x_{i2} = \mu_{xi} - \sigma_{xi} \tag{2}$$

At this point, there are 2^n value points and 2^n combinations of corresponding value points for n random variables. When the probabilities of these 2^n combinations occurring are equal and independent of each other, the first moment of the mean function is:

$$\mu_z = \frac{1}{2^n}\sum_{j=1}^{2^n} Z_j \tag{3}$$

When a group of variables is related and the probability of occurrence is not equal, the probability of occurrence P_j for each group of variables is related to its correlation coefficient:

$$P_j = \frac{1}{2^n}(1 + e_1e_2\varphi_{12} + e_2e_3\varphi_{23} + \cdots + e_{(n-1)}e_n\varphi_{(n-1)n}) \tag{4}$$

where $j = 1, 2, 3, \ldots, n$. When $x = x_{i1}$, $e_i = 1$. When $x = x_{i2}$, $e_i = -1$; φ_{ij} is the correlation coefficient of a random variable. The first moment of the state function is:

$$\mu_z = \frac{1}{2^n}\sum_{j=1}^{2^n} P_jZ_j \tag{5}$$

The second moment is:

$$\sigma_z^2 = \sum_{j=1}^{2^n} P_jZ_j^2 - \mu_z^2 \tag{6}$$

Reliability indicators β can be established as shown in Equation (7):

$$\beta = \frac{\mu_z - 1}{\sigma_z} \tag{7}$$

Then, the calculation of reliability R can be performed using:

$$R = \phi(\beta) \tag{8}$$

2.2. *Construction of RPEM Economic Evaluation Model*

2.2.1. Evaluation Index System

There are numerous and complex indicators that affect the economy of mining projects [32], which can be divided into three secondary indicators: reserve indicators, production indicators, and operational indicators. Then, the indicators are classified according to their attributes to form a three-level indicator. The established technical and economic evaluation index system is shown in Figure 2.

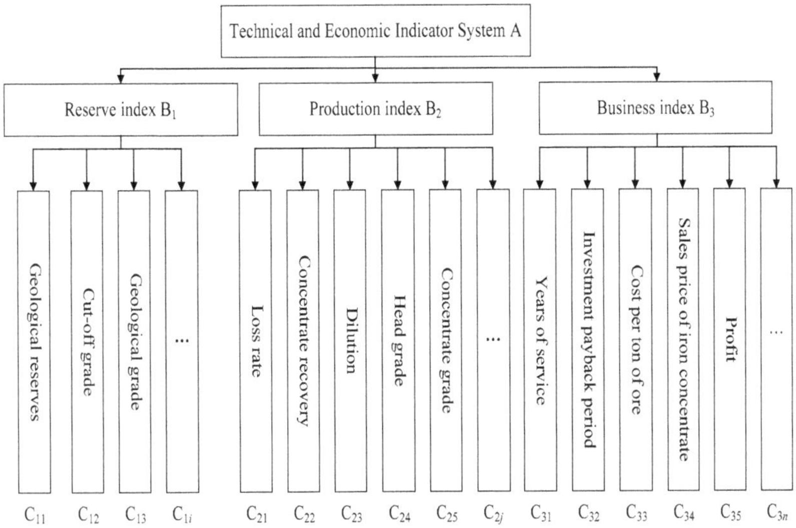

Figure 2. Technical and economic index system.

2.2.2. Solving the Reliability of RPEM Economic Evaluation

To study the applicability of RPEM to the economic evaluation of mining, taking the mining and selection production process of underground mines as an example, the commonly used technical and economic indicators are selected for mining economic evaluation and are used to establish a reliability model for economic evaluation of RPEM. The main indicators include mining cost, beneficiation cost, ore dilution rate, ore grade, concentrate grade, beneficiation recovery rate, concentrate sales price, etc. These indicators are all included in the indicator system shown in Figure 2.

When establishing an RPEM economic evaluation reliability model based on the cash flow method, only the costs and benefits directly generated during the mining production process are considered, and the impact of its fluctuations on costs and profits is analyzed using the ore grade as a sensitivity indicator. When the mining and selection production balance and concentrate production and sales balance are achieved, the specific calculations are as follows:

(1) Cash inflows from mines (only when calculating concentrate sales revenue) F:

$$F = A \cdot \frac{g \cdot \varepsilon \cdot W}{\gamma} \tag{9}$$

where A is the annual mining output of the mine, 10,000 t/year; G is the ore grade (i.e., the ore grade after beneficiation), %; ε is the recovery rate of mineral processing, %; W is the sales price of concentrate, CNY/t; γ is the concentrate grade, %.

(2) Annual cash outflow from mines P:

When the production and sales of concentrate produced by the mine are balanced, the annual cash outflow P of the mine is:

$$P = A \cdot \frac{g \cdot \varepsilon}{\gamma} \cdot C_c + C_1 + C_2 \tag{10}$$

where Cc is the production cost of ton of ore concentrate, CNY/t; C_1 represents various taxes and fees, in CNY 10,000; C_2 refers to the annual average fixed assets investment and current asset investment of the enterprise, CNY 10,000.

(3) Net cash flow of mines I:

$$I = F - P = A \cdot \frac{g \cdot \varepsilon}{\gamma} \cdot (W - C_c) - C_1 - C_2 \tag{11}$$

(4) Construction of Objective State Function for Economic Reliability Evaluation

Based on the basic principles of RPEM, the cash flow method is combined with RPEM to construct the objective state function of the RPEM economic reliability evaluation model, as shown in Equation (12).

$$Z = \frac{F}{P} = \frac{A \cdot g \cdot \varepsilon \cdot W}{A \cdot g \cdot \varepsilon C_c + \gamma(C_1 + C_2)} \tag{12}$$

The economic reliability evaluation target state function represented by Equation (12) reflects the level of benefits that can be achieved by mining projects. The higher the economic reliability, the higher the mining benefits, and vice versa.

From Equation (12), it can also be seen that when the Z value is 1, it indicates that the cash inflow and outflow of the project are equal. At this point, the breakeven point of a certain technical and economic indicator can be determined by formula transformation.

(5) Solution of Economic Reliability Based on RPEM

First, Equation (12) is used as the objective state function, according to the RPEM calculation process shown in Figure 1; the sensitivity indicators for evaluation and their possible range of values are determined. Secondly, within the range of sensitivity indicators, the mean μz and variance $\sigma_z{}^2$ of sensitivity indicators is calculated according to Equations (1)–(6), and Equation (7) is reused to solve reliability index β. Finally, the reliability R is calculated according to Equation (7). By solving the reliability R under different values of sensitivity indicators and comparing them, the economic reliability of the project under these sensitivity indicators can be analyzed.

The range of values for the economic reliability R obtained from this is 0–1. The larger the R value, the higher the economic reliability, which indicates that the project is more reliable.

3. Empirical Study

3.1. Case Mine Overview

A certain iron mine in Liaoning Province, China was selected as an example for conducting applied research. The mining area is located in the Liaodong region, with a flat terrain and an inclination angle of 62° to 87° for the ore body, with a thickness of 6.0 to 43.2 meters. The ore body occurs at an elevation between +200 m and −300 m. The main minable ore bodies are Fe10, Fe11-1, and Fe11-2. The main useful mineral is magnetite with a TFe content of 25~35%, which belongs to magnetite lean ore.

The mining of this deposit adopted open-pit multi-step mining operation and a highway transportation method in the early stage. After nearly a decade of open-pit mining, the service life of open-pit mining was only 2.38 years due to the combined constraints of the comprehensive cost of deposit mining and surface terrain conditions. In order to ensure the production efficiency of the enterprise and achieve the full exploitation and utilization of the mining resources, according to the provisions of the "Code for Geological Exploration of Iron, Manganese, and Chromium Mines" (DZ/T0200-2002), the retained reserves of the mine were evaluated. Based on the resource estimation results, as of the evaluation date, the three main minable ore bodies had a total reserve of 17,178.39 kt (122 b + 333), of which the ore bodies below +70 m were mined underground, and the usable reserve of ore was 13,243.82 kt. The mining and beneficiation production capacity of the mine was 1300 kt/a. The main production technical and economic indicators of iron mine in Liaoning Province are shown in Table 1 [33].

Table 1. The main production technical and economic indicators.

Index	Symbol	Underground Mining
Average geological grade of recoverable reserves (%)	g	26.3%
Dip angle of ore body (°)	α	62~87
Thickness of ore body (m)	m	20~40
Actual mining recovery rate (%)	η	88
Actual dilution rate in mining (%)	ρ	20
Raw ore (ore extraction) grade (%)	g′	21.09
Concentrate grade (%)	γ	65
Beneficiation recovery (%)	ε	77.2
Concentrate yield (t/t)	K = G·ε	0.25
Sales price of iron concentrate (CNY/t)	W	744
Underground mining cost (CNY/t)	Cp	78.31
Mineral processing cost (CNY/t)	Cu	36.28
Production cost of ton of ore concentrate (CNY)	Cc	114.59

3.2. *Empirical Study on the Reliability of RPEM Economic Evaluation*

3.2.1. Determine Model Variables

Using ore grade as a sensitivity indicator, its impact on economic benefits is studied and a state function with ore grade as a variable is established, as shown in Equation (13).

$$Z = \frac{F}{P} = Z(g) \tag{13}$$

3.2.2. Determine the Range of Ore Extraction Grade Values for the Mine

The average geological grade of the mine is 26.3% (as shown in Table 1). When using the non-pillar sublevel caving method for mining, if the ore dilution rate fluctuates between 15% and 20%, the change in the average ore grade g of the mine can be calculated by Equation (14):

$$\begin{aligned} g &= g_1 \cdot (1-\rho) \\ &= 26.3\% \times [1-(15\% \sim 20\%)] \\ &= 21.04\% \sim 22.36\% \end{aligned} \tag{14}$$

In the formula, g_1 is the average geological grade of the deposit, %; ρ is the ore dilution rate, %.

Therefore, when the mine adopts the non-pillar sublevel caving method for mining, the variation range of ore grade in the mine is 21.04~22.36%.

3.2.3. Mine Cash Flow Calculation

According to Equation (14), the variation range of mining grade is obtained, and the economic benefits within this range are discussed. The coefficient of variation of the extracted ore grade is 0.1, and 11 sets of data are taken within the variation range of the ore grade (21.04%, 22.36%). Based on the mining and beneficiation production capacity of the mine, if other indicators remain unchanged, the corresponding concentrate output and concentrates sales revenue for the 11 sets of ore grades can be calculated.

The designed annual mine output of the underground mining project is 1300 kt, the mine ore grade is 21.04%, the production cost of concentrate per ton of ore is CNY 114.59 (excluding tax), and the annual average fixed assets investment and current asset investment is CNY 9.9686 million. The taxes and fees paid by enterprises are subject to changes in sales revenue, as changes in ore grade affect concentrate production, which in turn affects changes in sales revenue. When the ore grade is 21.04%, the annual tax paid by the enterprise is CNY 39.6797 million. Taking a mining grade of 21.04% as an example, the annual cash flow of the mine is calculated, and the specific steps are as follows:

(1) Annual cash inflows from mines F:

$$F = A\frac{g\cdot\varepsilon\cdot w}{\gamma} \begin{aligned} &= 130 \times \tfrac{21.04\% \times 77.2\% \times 744}{65\%} \\ &= 241.6941(\text{million Yuan}) \end{aligned} \tag{15}$$

(2) Annual cash outflow from mines P:

$$\begin{aligned} P &= A \cdot \tfrac{g\cdot\varepsilon}{\gamma} \cdot C_c + C_1 + C_2 \\ &= 130 \times \tfrac{21.04\% \times 77.2\%}{65\%} \times 114.59 + 3967.97 + 996.86 \\ &= 198.6153(\text{million Yuan}) \end{aligned} \tag{16}$$

(3) Annual net cash flow of the mine I:

$$I = F - P \begin{aligned} &= 241.6941 - 198.6153 \\ &= 43.0788(\text{million Yuan}) \end{aligned} \tag{17}$$

3.2.4. Calculation of Economic Reliability of Mines

(1) Establish target state function

Taking the ore extraction grade of 21.04% as an example and referring to Equation (13), a target state function with the ore extraction grade as a variable is established, as shown in Equation (18):

$$Z = \frac{F}{P} = Z(21.04\%) \tag{18}$$

(2) Solution to Economic Reliability

According to the basic principles and methods of RPEM in Section 1, with its coefficient of variation of 0.1, two symmetrical value points for an ore grade of 21.04% can be obtained according to Equations (1) and (2),

$$x_{i1} = \mu_{xi} + \sigma_{xi} = 21.04\% + 2.1\% = 23.14\% \tag{19}$$

$$x_{i2} = \mu_{xi} - \sigma_{xi} = 21.04\% - 2.1\% = 18.94\% \tag{20}$$

By using the cash flow calculation method for mines, the cash flows for ore grades of 23.14% and 18.94% are calculated, resulting in Z_1 = 1.2814 and Z_2 = 1.1464. Its mean μ_z and variance ${\sigma_z}^2$ of the state function for ore grade variables within the interval of 21.04% and 22.36% can be calculated, as shown in Equations (21) and (22):

$$\mu_z = \frac{1}{2}(Z_1 + Z_2) = \frac{1}{2} \times (1.2814 + 1.1464) = 1.2139 \tag{21}$$

$$\sigma_z^2 = \sum_{j=1}^{2^n} P_j Z_j^2 - \mu_z^2 = 0.0091 \tag{22}$$

According to the calculation results of Equations (21) and (22), Equation (7) is used to solve the reliability index of the state function β, as shown in Equation (23).

$$\beta = \frac{\mu_z - 1}{\sigma_z} = \frac{0.2139}{0.0954} = 2.2421 \tag{23}$$

At this point, when the ore extraction grade is 21.04%, the economic reliability R of the mine is:

$$R = \phi(2.2421) = 98.75\% \tag{24}$$

Within the variation range of ore grade [21.04%, 22.36%], with a coefficient of variation of 0.1, taking 11 sets of data, the cash flow, reliability indicators, and economic reliability of each group of data are calculated separately according to the above steps. The results are shown in Table 2.

Table 2. Reliability calculation results.

Ore Grade /%	Concentrate Production /t	Income /CNY 10,000	Net Cash Flow /CNY 10,000	Reliability Indicators	Economic Reliability /%
21.04	4.21 × 105	2.42 × 104	0.43 × 104	2.2421	98.75
21.17	4.23 × 105	2.43 × 104	0.44 × 104	2.2759	98.84
21.30	4.26 × 105	2.45 × 104	0.45 × 104	2.3226	98.98
21.43	4.29 × 105	2.46 × 104	0.46 × 104	2.3647	99.09
21.57	4.31 × 105	2.48 × 104	0.47 × 104	2.4031	99.18
21.70	4.34 × 105	2.49 × 104	0.48 × 104	2.4473	99.27
21.83	4.37 × 105	2.51 × 104	0.49 × 104	2.4924	99.36
21.96	4.33 × 105	2.52 × 104	0.50 × 104	2.5292	99.41
22.09	4.42 × 105	2.54 × 104	0.51 × 104	2.5682	99.49
22.22	4.44 × 105	2.55 × 104	0.52 × 104	2.6115	99.55

Repeating the above process, the cash flow, reliability indicators, and economic reliability of its variables when they change within its value range can be obtained.

3.3. Analysis of Calculation Results

According to Table 2, when the mining grade increased from 21.04% to 22.36%, the company's revenue increased from CNY 2.42 × 104 ten thousand to CNY 2.57 × 104 ten thousand, an increase of 10.33%; net cash flow increased by CNY 0.1 × 104 ten thousand, an increase of 23.26%; and the economic reliability calculated using the RPEM-based economic reliability model increased from 98.75% to 99.95%, an increase of 1.2%.

It can be seen that as the ore grade increases, the concentrate production increases, and the profits and net cash flow of the mine also increase. The economic reliability calculated based on the RPEM economic reliability evaluation model is also continuously increasing. This feature is completely consistent with the actual production and operation situation of mining enterprises. This indicates that the economic reliability evaluation model established based on RPEM has its applicability for the economic evaluation of mining projects.

To more intuitively predict the trend of changes in reliability and revenue, the annual revenue and economic reliability data of enterprises in Table 2 are converted into a curve chart, as shown in Figure 3.

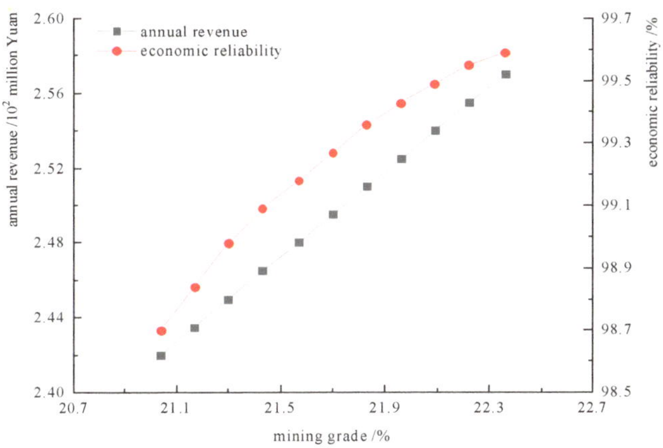

Figure 3. Relationship curve between annual revenue and economic reliability with changes in ore grade.

From Figure 3, it can be seen that as the ore grade increases, the mine income curve and economic reliability curve also change and increase, indicating that while the enterprise income increases, the economic reliability also increases. The trend of curve changes is in line with the objective facts of mine production and operation.

The empirical research results indicate that the mining economic reliability evaluation model established based on the RPEM method has good applicability, and the evaluation results comply with the basic laws of changes in mining production and operation.

4. Conclusions

The RPEM method for studying the economic reliability of mines was introduced. Combined with the cash flow method, a reliability model for mining economic evaluation was proposed. Based on the actual production of empirical mines and the impact of dilution rate on ore grade, the impact of changes in ore grade on mine revenue and economic reliability was studied using ore grade as a sensitivity indicator. The established RPEM economic reliability evaluation model was validated. The empirical research results indicate that the economic reliability results and their variation patterns obtained by using the RPEM evaluation model are basically consistent with the actual prodHanuction and operation laws of mining enterprises, indicating that the established RPEM economic evaluation method has good applicability and reliability for mining economic evaluation. By analyzing the changes in economic reliability when different sensitive technical and economic indicators change, the problem of mining economic evaluation can be solved and the reasonable range of relevant parameters can be quickly found. The RPEM calculation process is simple and can solve the economic reliability problem of the project based on changes in sensitivity indicators, and can also calculate the value range of sensitivity indicators based on the required reliability. The obtained results and the proposed evaluation model provide a decision-making basis for the production and operation of mining enterprises. Future study involving the dynamic evaluation indicators in mining economic evaluation, expenditures and profits outside the mining production enterprises is also suggested.

Author Contributions: Conceptualization and methodology, J.L.; validation and formal analysis, T.W.; resources, data curation and supervision, Z.L.; writing—review and editing, S.W. All authors have read and agreed to the published version of the manuscript.

Funding: The work was supported by the National Nature Science Foundation of China (No. 51774176), Shannxi Province Key Research and Development Program (2023-GHYB-06).

Institutional Review Board Statement: Not applicable.

Informed Consent Statement: Not applicable.

Data Availability Statement: The detailed data is available upon request.

Acknowledgments: Not applicable.

Conflicts of Interest: The authors declare no conflict of interest.

References

1. Wang, Q.; Song, C.; Li, W.; Han, W. Analysis of the Development Process and Current Situation of International Cooperation in China's Mining Industry. *Geol. Explor.* **2022**, *58*, 229–238.
2. Chen, J.; Liu, L.; Zeng, B.; Tao, K.; Zhang, C.; Zhao, H.; Li, D.; Zhang, J. A constitutive model to reveal the anchorage mechanism of fully bonded bolts. *Rock Mech. Rock Eng.* **2023**, *56*, 1739–1757. [CrossRef]
3. Wu, S.; Hao, W.; Yao, Y.; Li, D. Investigation into durability degradation and fracture of cable bolts through laboratorial tests and hydrogeochemical modelling in underground conditions. *Tunn. Undergr. Space Technol.* **2023**, *138C*, 105198. [CrossRef]
4. Li, Q.; Zhang, C.; Yang, Y.; Ansari, U.; Han, Y.; Li, X.; Cheng, Y. Preliminary experimental investigation on long-term fracture conductivity for evaluating the feasibility and efficiency of fracturing operation in offshore hydrate-bearing sediments. *Ocean. Eng.* **2023**, *281*, 114949. [CrossRef]
5. Li, Q.; Zhao, D.; Yin, J.; Zhou, X.; Li, Y.; Chi, P.; Han, Y.; Ansari, U.; Cheng, Y. Sediment Instability Caused by Gas Production from Hydrate-bearing Sediment in Northern South China Sea by Horizontal Wellbore: Evolution and Mechanism. *Nat. Resour. Res.* **2023**, *32*, 1595–1620. [CrossRef]

6. Wang, F.; Liu, X.; Jiang, B.; Zhuo, H.; Chen, W.; Chen, Y.; Li, X. Low-loading Pt nanoparticles combined with the atomically dispersed FeN4 sites supported by FeSA-N-C for improved activity and stability towards oxygen reduction reaction/hydrogen evolution reaction in acid and alkaline media. *J. Colloid Interface Sci.* **2023**, *635*, 514–523. [CrossRef]
7. Chen, J.; Tao, K.; Zeng, B.; Liu, L.; Zhao, H.; Zhang, J.; Li, D. Experimental and numerical investigation of the tensile performance and failure process of a modified Portland cement. *Int. J. Concr. Struct. Mater.* **2022**, *16*, 61. [CrossRef]
8. Wu, S.; Zhang, Z.; Chen, J.; Yao, Y.; Li, D. Characterisation of stress corrosion durability and time-dependent performance of cable bolts in underground mine environments. *Eng. Fail. Anal.* **2023**, *150*, 107292. [CrossRef]
9. Li, G.; Wu, B.; Hou, J.; Wang, H.; Wang, J.; Chen, L.; Fan, C. Prediction Model for Mining Cost in Underground Metal Mines. *Met. Mines* **2022**, *551*, 62–69.
10. The Third Group of Shanxi Provincial Management Accounting Expert Team for Enterprise Service; Xin, M. Economic Evaluation of the Ekou Iron Mine Openpit to Underground Mining and Dressing Project. *Financ. Account.* **2019**, *588*, 28–30.
11. Zhou, L. Construction and Demonstration of Green Mining Economic Evaluation Index System. *Stat. Decis. Mak.* **2018**, *34*, 68–70.
12. Wang, Q. Application of cash flow method in project economic evaluation and mining rights evaluation. *Coal Econ. Res.* **2006**, *12*, 45–47.
13. Shi, Q.; Liu, A.; Li, C.; Cai, J.; Huang, S. Technical and economic evaluation of mining right (deposit) project based on internal rate of return method. *Mod. Min.* **2021**, *37*, 17–18.
14. Chen, W.; Xiao, X.; Peng, Z. Research on the Full Life Cycle Economic Benefit Evaluation of Mining Technical Renovation and Expansion Projects. *Min. Res. Dev.* **2020**, *40*, 164–168.
15. Xu, Y.; Wang, X.; Lu, Z. Improved mining rights evaluation method based on CAPM model. *Non Ferr. Met. Sci. Eng.* **2022**, *13*, 108–113.
16. Yang, H.; Wang, Z. Economic Evaluation of Coal Mines Based on Dynamic Breakeven Analysis. *Min. Res. Dev.* **2017**, *37*, 109–111.
17. Fuksa, D.; Trzaskuś-Żak, B.; Gołaś, Z.; Utrata, A. An evaluation of practical applicability of multi-assortment production break-even analysis based on mining companies. *Arch. Min. Sci.* **2017**, *62*, 33–44. [CrossRef]
18. Wen, W.; Zheng, M. Macroeconomic Risk Assessment of Overseas Mining Project Mergers and Acquisitions. *Non Ferr. Met. Sci. Eng.* **2016**, *7*, 110–116.
19. Li, G.; Li, B.; Hu, N.; Hou, J.; Xiu, G. Optimization Model for Mining Operation Plan of Underground Metal Mines. *J. Eng. Sci.* **2017**, *39*, 342–348.
20. Kong, W.; Gu, X.; Wang, Q.; Xu, X.; Wang, R. Optimization of open-pit coal mining plans considering infrastructure investment. *Met. Mines* **2021**, *540*, 188–193.
21. Yuliduzi, S.; Peng, J. Evaluation of the Economic Competitiveness of Heilongjiang Mining Industry: A Comparison Based on 10 Provinces (Regions). *China Land Resour. Econ.* **2017**, *30*, 44–48.
22. Rosenblueth, E. Point estimates for probability moments. *Proc. Natl. Acad. Sci. USA* **1975**, *72*, 3812–3814. [CrossRef] [PubMed]
23. Conciatori, D.; Bruhwiler, E.; Morgenthaler, S. Calculation of reinforced concrete corrosion initiation probabilities using the Rosenblueth method. *Int. J. Reliab. Saf.* **2009**, *3*, 345–362. [CrossRef]
24. Chen, Z.; Du, J.; Yan, J.; Sun, P.; Li, K.; Li, Y. Point estimation method: Validation, efficiency improvement, and application to embankment slope stability reliability analysis. *Eng. Geol.* **2019**, *263*, 521–544. [CrossRef]
25. Napa-García, G.F.; Beck, A.T.; Celestino, T.B. Reliability analyses of underground openings with the point estimate method. *Tunn. Undergr. Space Technol.* **2017**, *64*, 154–163. [CrossRef]
26. Yang, K. Reliability analysis of hydropower station slopes based on improved Rosenblueth method. *Water Conserv. Sci. Technol. Econ.* **2018**, *24*, 69–72.
27. Cai, W.; Wang, Z.; Li, D. Real time analysis of the probability of earth-rock dam failure based on Rosenblueth method. *Hydroelectr. Energy Sci.* **2018**, *36*, 61–65.
28. Yang, Y.; Xu, M.; Cheng, A.; Wang, P.; Lu, J.; Tan, T. Reliability analysis of roadway roof bolt support structure based on Rosenblueth method. *Min. Res. Dev.* **2020**, *40*, 86–90.
29. Kahe, M.S.; Javadi, S.; Roozbahani, A.; Mohammadi, K. Parametric uncertainty analysis on hydrodynamic coefficients in groundwater numerical models using Monte Carlo method and RPEM. *Environ. Dev. Sustain.* **2021**, *23*, 11583–11606. [CrossRef]
30. Xu, Z.; Jin, J.; Cai, Y.; Yang, T. Simulation Study on the Impact of Climate Change on Shallow Landslides on the Loess Plateau: A Case Study of Baota District in Yan'an. *Res. Soil Water Conserv.* **2021**, *28*, 387–393.
31. Wang, J.-P.; Huang, D. RosenPoint: A Microsoft Excel-based program for the Rosenblueth point estimate method and an application in slope stability analysis. *Comput. Geosci.* **2012**, *48*, 239–243. [CrossRef]
32. Lin, Y. Construction of Technical and Economic Evaluation Index System for Bauxite Resource Development. *Min. Res. Dev.* **2020**, *40*, 160–163.
33. Lu, Z.; Cai, M. *Key Technologies for Transition from Open Pit to Underground Mining in Metal Mines*; Metallurgical Industry Press: Beijing, China, 2019.

applied sciences

MDPI

Article

Reasonable Coal Pillar Width and Surrounding Rock Control of Gob-Side Entry Driving in Inclined Short-Distance Coal Seams

Fulian He [1,2], Wenli Zhai [1,*], Jiayu Song [1], Xuhui Xu [1], Deqiu Wang [1] and Yanhao Wu [1]

1 School of Energy and Mining Engineering, China University of Mining and Technology (Beijing), Beijing 100083, China; flhe1966@126.com (F.H.); bqt2100101001t@student.cumtb.edu.cn (J.S.)
2 Beijing Key Laboratory for Precise Mining of Intergrown Energy and Resources, China University of Mining and Technology (Beijing), Beijing 100083, China
* Correspondence: wlzhai8@163.com

Abstract: During gob-side entry driving under complex conditions in inclined short-distance coal seams, the roadway loses stability and deforms seriously, which affects the safety and efficiency of mine production. In this study, a reasonable coal pillar width was explored by means of on-site investigation, theoretical analysis, numerical simulation, and engineering tests. The following research results were obtained: (1) In selecting a reasonable coal pillar width, the influences of the position of residual coal pillars, stratum spacing, main roof breakage, roadway section in the upper coal seam should be considered. From established mechanical models of inclined gob-side roadways, the maximum floor failure depth is 27 m and the concentrated influence range of the #1 coal pillars is 11 m. (2) The stress states of coal pillars with different widths were analyzed by numerical simulation. As the coal pillar width increases, the peak value of the stress increases first and then decreases. Based on the site geological conditions, the optimum coal pillar width was determined to be 8 m, which is consistent with the theoretical calculation results. (3) A new pressure-yield support technology was proposed, and its on-site application confirmed its notable roadway control effect. Our research can provide theoretical support for the control of roadways surrounding rock under similar engineering background conditions.

Keywords: inclined coal seams; gob-side entry driving; numerical simulation; reasonable coal pillar width; yield support

1. Introduction

As the depth of coal mining rises, an increasing number of coal mining areas are faced with the technical problem of short-distance multi-seam mining [1–4]. During the mining of short-distance coal seams, the gob and coal pillars left from the upper coal seams greatly affect the selection and control of the roadway location in the lower coal seams [5–7]. An unreasonable roadway location in the lower coal seams often causes extreme roadway deformation, which greatly affects normal recovery of the working face [8–10]. Therefore, gob-side entry driving in lower coal seams should not only consider factors influencing lower-coal-seam mining, but also avoid the influence of the high stress concentration of residual coal pillars in the upper coal seams, which is of great significance for safe and efficient coal mining [11].

Experts have conducted extensive research on the stress distribution characteristics of short-distance coal seams, the reasonable coal pillar width in gob-side entry driving, and the surrounding rock control [12–14]. Ref [15] analyzed the correlation between the deformation failure of bottom coal in the gob-side entry and coal pillar width by theoretical analysis, numerical calculation, and similar simulation experiments. Ref. [16] studied the mechanical behavior of coal pillar load on the floor rock strata in multi-seam downward mining and discovered that stress was highly concentrated in the floor rock strata near coal pillars, with the stress concentration degree in floor rock strata positively correlated with the distance between the surrounding rock and coal pillars. Quang et al. (2020) [17] adopted numerical simulation software and determined a reasonable design for roadway supporting pillars. Gao (2004) [18] used the method of similarity simulation experiments to study the laws of roof activity. Ghasemi et al. (2014) [19] applied fuzzy logic to predict sate pillar sizing in room and pillar coal mines. Hu et al. (2012) [20] summarized the main factors causing roadway damage in adjacent gobs and found that roadway deformation can be effectively controlled using high-preload and dense-bolt (cable) support systems. With respect to problems such as serious asymmetric deformation and support failure of gob-side roadways, He et al. (2020) [21] proposed a hydraulic fracture technique to pre-fracture the two hard roofs, and the validity of the above roof treatment was verified via field application. Meanwhile, some experts have improved the geological and mechanical conditions of the coal mining process through backfilling technology [22,23].

Based on the engineering background of the 4602 head roadway in the Honghui Coal Mine, this study establishes a theoretical model for floor failure in short-distance coal seams and analyzed the floor failure range of the residual coal pillars. In addition, it establishes a mechanical model of inclined gob-side roadways and deduces an expression for the influence range of low stress areas and its influencing factors. Moreover, the stress distribution characteristics of coal pillars with different coal pillar widths in gob-side entry driving were analyzed by numerical simulation; a reasonable coal pillar width was comprehensively determined, and the yield-anchor mesh-cable support system was proposed. The proposed support system effectively controls the integrity of the roadway surrounding rock and realizes the safety and stability of the roadway support.

2. Engineering Background

2.1. Engineering Profile

The #2 coal seam of the 4602 working face was mined with a dip length of 116 m, a design strike length of 915 m, and the following coal seam parameters: buried depth 354~284 m, average dip angle 17°, average thickness 3.0 m, firmness coefficient f 1.8~2.2, and medium-level bedding development. The average thickness of coal seam #1 is 8.8 m, the average dip angle of the coal seam is 15°, the width of the remaining coal pillar is 10 m, and the vertical distance from coal seam #2 is approximately 19 m. The shallow and deep parts of the 4602 working face are the 4502 gob and the 4702 gob, respectively. The 4602 working face is an island face whose location is illustrated in Figure 1. A mudstone false roof that is 1 m thick exists in the #2 coal seam roof. The immediate roof, whose thickness is unstable and averages 19 m, comprises fine sandstone and siltstone, with gray and argillaceous cementation and developed fractures. The immediate floor, whose average thickness is 2.45 m, comprises argillaceous siltstone with gray white and argillaceous cementation and poor hardness, and it softens when encountering water. The lithologies of the coal seam roof and floor are shown in Figure 2.

2.2. Originally Adopted Support Scheme and Deformation and Failure Characteristics

In the originally adopted gob-side entry driving, the roadway has a rectangular section with a net width of 4000 mm and a net height of 3000 mm. The roadway roof and sides adopt Φ 20 mm × 2400 mm high-strength bolt supports with a row spacing of 800 mm × 800 mm. The anchor cables are Φ 18.9 mm × 5000 mm mining anchor cables; each row has 3 cables with a spacing of 1200 mm, and the row spacing is 1600 mm, with the bolts and anchors fixed in sections, and resin cartridges were used. After the above support scheme is adopted, the roadway surrounding rock deforms markedly during recovery, especially the gob-side roof and sides of the roadway, where the bolts move outward with the deformation of the surrounding rock, with a maximum inward movement of 700 mm (Figure 3a). At the same time, local floor heave occurs in the roadway, with the bottom heave of the roadway floor occurring for a maximum of 800 mm (Figure 3b). The recovery requires repeated side expansion and floor reshaping. If the situation becomes too serious, grouting and re-building metal U-shaped steel shed supports will be needed after the repair. In view of the above roadway deformation characteristics of gob-side entry driving, coal pillar width and support parameters require further optimization.

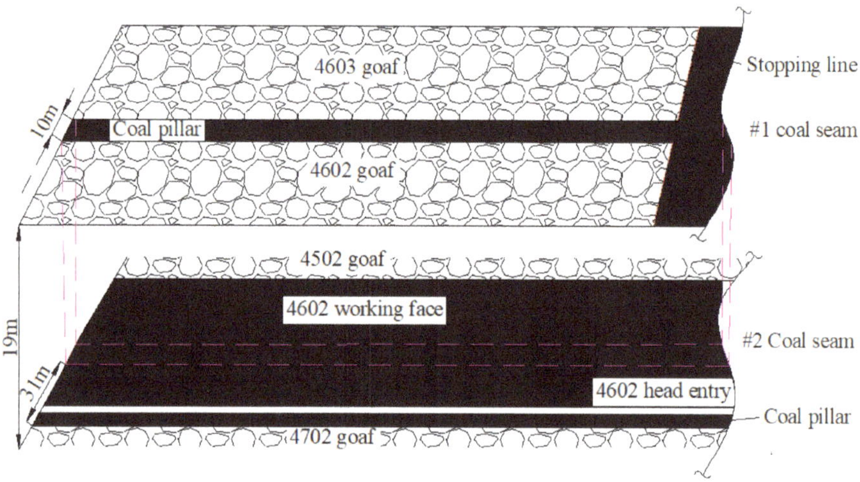

Figure 1. Spatial position diagram of the working face.

Histogram		Rock stratum	Thickness/m	Lithology description
		coarse sandstone	16.5	gray white, coarse-grained structure, massive structure, mainly composed of quartz
		No.1 coal seam	8.8	black, massive, with pyrite film, Semi bright briquette
		siltstone	12.0	dark gray, dense, brittle, developed fissures
		fine sandstone	6.0	dark gray, dense and hard, containing a large number of muscovite fragments, mixed with coal line
		mudstone	1.0	grayish black, argillaceous cementation, softened in case of water
		No.2 coal seam	3.0	black, massive, with pyrite film, Semi-bright briquette
		pelitic siltstone	2.45	dark gray, dense and hard, argillaceous cementation, with coal line
		siltstone	6.5	dark gray, dense, brittle, developed fissures

Figure 2. Rock stratum histogram of working face.

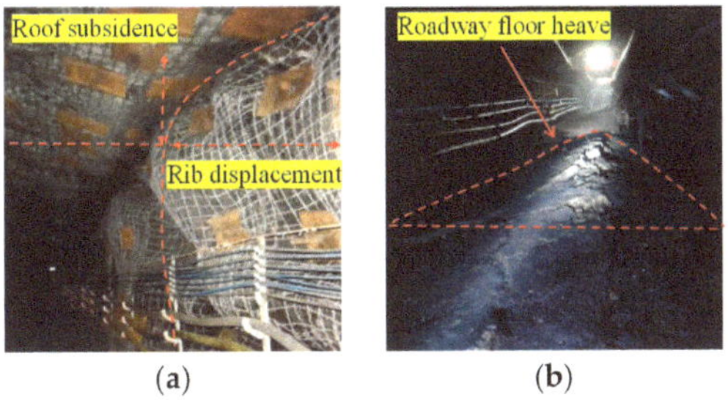

Figure 3. Deformation and failure of a gob-side coal roadway. (**a**) Deformation of the roof and rip. (**b**) Roadway floor heave.

3. Calculation of Coal Pillar Width of Gob-Side Entry Driving

3.1. Study on the Floor Failure Range of Residual Coal Pillars

Residual coal pillars exist after mining the #1 coal seam. The floor failure depth resulting from the residual coal pillars can directly reveal whether the lower-coal-seam mining is affected. The coal pillar load acts on the floor and causes damage (Figure 4).

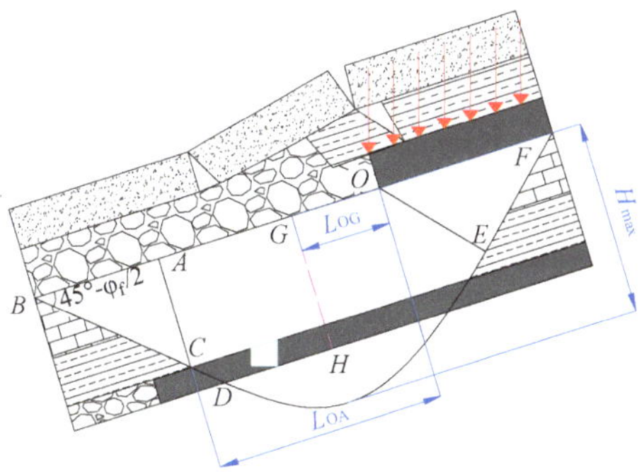

Figure 4. Floor failure model.

The *ED* curve is a logarithmic spiral, and its formula is (Xu et al., 2021 [24]):

$$r = r_0 e^{\alpha \cdot \tan \varphi_f} \tag{1}$$

where r is the spiral radius with O as the origin at an angle of α with r_0; r_0 is the length of OE; α is the angle between r and r_0; and φ_f is the internal friction angle of rock strata.

Since $\angle OFE = \angle EOF = \pi/4 + \varphi/2$,

$$r_0 = B/2\cos\left(\pi/4 + \varphi_f/2\right) \tag{2}$$

where B is the width of the coal pillar acting on the floor.

When the floor reaches the maximum yield failure depth, α can be calculated by:

$$\alpha = \pi/4 + \varphi_f/2 \tag{3}$$

Then the maximum floor failure depth can be expressed as:

$$H_{\max} = \frac{B\cos\varphi_f}{2\cos\left(\frac{\pi}{4} + \frac{\varphi_f}{2}\right)} e^{\left(\frac{\pi}{4} + \frac{\varphi_f}{2}\right)\tan\varphi_f} \tag{4}$$

According to Equation (4), the maximum floor failure depth under the influence of residual coal pillars is related to the coal pillar width and floor lithology. Based on the actual geological conditions of the coal and the rock strata, the internal friction angle φ_f is 38° and the width of the coal pillars B is 10 m. They are substituted into Equation (4): Hmax = 27 m, which is larger than the average distance of 19 m between the two coal seams. Therefore, the failure range of residual coal pillars in the upper coal seams has a significant influence on the lower-coal-seam mining and the mining roadway layout.

Based on the above analysis, the #2 coal seam roadway can be arranged at the position where the residual coal pillar stress is less concentrated on the floor, which is conducive to roadway stability. The influence range of the high floor stress concentration of residual coal pillars in the #1 coal seam L_{OG} is shown in Figure 4,

$$L_{OG} = (h_1 + h_2)\tan\varphi \tag{5}$$

where h_1 is the thickness between the #1 and #2 coal seams, m, calculated as 19 m; h_2 is the #2 coal seam thickness, m, calculated as 3 m; and φ is the propagation angle of stress influence, calculated as 25°.

Based on the calculation of L_{OG}, that is, at the #1 coal seam floor, the concentrated influence range of the coal pillars is 11 m. With reference to the calculation results, if the #2 coal roadway is arranged over 11 m away from the residual coal pillars of the #1 coal seam, the roadway will be less affected by the residual coal pillars of the #1 coal seam.

3.2. Study on the Reasonable Coal Pillar Width

After the recovery of the working face in the upper section, the gob-edge immediate roof collapses, and the main roof rotates. The gangue supports the rock block B of the main roof; the rock block B and adjacent blocks squeeze each other to form a relatively stable structure. The fracture of the main roof divides the lateral abutment pressure into the obvious internal stress field S_1 (stress falling area) and the external stress field S_2 (stress rising area). The location of the main roof fracture is the boundary of the two stress fields (Figure 5a). To provide a good stress environment for the roadway and to facilitate roadway maintenance, roadways are generally arranged in the stress falling area in coal pillar width selection (Song et al., 1986 [25]).

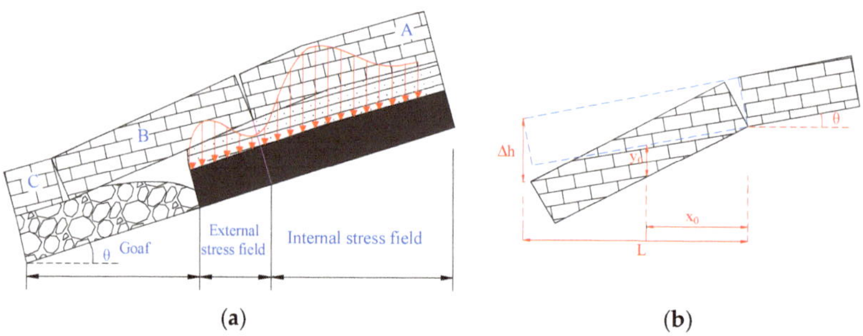

Figure 5. Mechanical model. (**a**) Distribution diagram of the internal and external stress fields. (**b**) Breaking structure model of the main roof.

In the internal stress field, the abutment pressure x m away from the coal wall can be expressed as:

$$\sigma_x = G_x y_x \tag{6}$$

where G_x is the stiffness of the coal body x m away from the coal wall, Pa; y_x is the deformation of the coal body x away from the coal wall, m.

As the abutment pressure develops into the deep part of the coal wall, the stiffness of the coal body gradually rises while the compression gradually decreases. This is simplified as a process of linear variation. According to the geometric relationship, then:

$$y_x = \frac{y_0}{x_0}(x_0 - x), G_x = \frac{G_0}{x_0}x \tag{7}$$

where G_0 is the maximum stiffness of the coal body within the internal stress field, Pa; y_0 is the compression of the gob-edge coal body, m; x_0 is the distribution range of the internal stress field, m. According to Equations (6) and (7), the abutment pressure of the internal stress field is

$$\int_0^{x_0} \sigma_v dx = \frac{G_0 y_0 x_0}{6} \tag{8}$$

According to the theory of internal and external stress fields, the abutment pressure within the internal stress field on the coal body around the stope is equal to the self-weight of the main roof strata during the first weighting of the working face; thus,

$$\frac{G_0 y_0 x_0}{6} = abC_0\gamma \tag{9}$$

where a is the inclined length of the working face, m; C_0 is the first weighting step distance of the adjacent working face, m; b is the thickness of the main roof strata, m; and γ is the bulk density of the main roof rock, kN/m^3. According to Figure 5b, then:

$$\frac{y_0}{x_0} = \frac{\Delta h}{L} = \frac{M - h(K_c - 1)}{L} \tag{10}$$

$$y_0 = \frac{x_0}{L}[M - h(K_c - 1)] \tag{11}$$

where M is the mining thickness of the coal seam, m; h is the thickness of the immediate roof rock strata, m; Δh is the maximum subsidence of the key block, m; L is the span of the hinged rock beams, which is nearly equal to the periodic weighting step distance of the fully mechanized caving face, m; and K_c is the dilatancy coefficient of the immediate roof rock.

According to the inclusion theory, the stiffness of the coal body G_0 (Zhang et al., 2016 [26]) in the plastic state is

$$G_0 = \frac{E}{2(1+v)\xi} \tag{12}$$

where E is the elastic modulus of the coal body, Pa; v is Poisson's ratio; and ξ is the influence coefficient, which is related to the development of fractures in the coal body.

The range of the internal stress field x_0 can be obtained based on Equations (9)~(12):

$$x_0 = \sqrt{\frac{12abC_0\gamma\xi L(1+v)}{E[M - h(K_c - 1)]}} \tag{13}$$

According to geological data from the mine and on-site measurements, $a = 120$ m, $b = 12$ m, $C_0 = 35$ m, $\gamma = 27$ kN/m^3, $\xi = 0.85$, $L = 16$ m, $v = 0.3$, $E = 24$ GPa, $M = 3$ m, $h = 7$ m, and $K_c = 1.32$. These values are substituted in Equation (13) to obtain the internal field range $x_0 = 12.58$ m. The constraints of the reserved coal pillars in gob-side entry driving is to arrange roadways within the internal stress field. The 4602 head roadway is designed 4.6 m wide, and the width of coal pillars should not exceed 8 m.

A reasonable width should ensure not only the stability of the roadway surrounding the rock during excavation, but also the safety and stability of the working face during recovery. Although the 4602 head roadway can be arranged in a low stress environment by reserving 3~5 m small coal pillars, the 4602 working face belongs to an island face which is also affected by the residual coal from the #1 coal seam. Hence, the 4602 working face is likely to experience large-scale instability and damage. Internal fractures of coal pillars may develop, leading to poor stability and air leakage in the gob, unfavorable for fire prevention. Meanwhile, according to the above calculation, the roadway should also be arranged outside the influence range of the coal pillar stress concentration in the upper coal seam, so the coal pillar width is preliminarily determined to be in the range of 6~8 m. Due to the limitations of theoretical model calculation, the final coal pillar width is determined by calculating the stress distribution characteristics of coal pillars with different widths through numerical simulation.

4. Numerical Simulation of Coal Pillar Width of Gob-Side Entry Driving in Inclined Coal Seams

4.1. Establishment of the Numerical Model

According to the mining relationship and the roof and floor strata conditions of the 4602 working face, a numerical calculation model that is 470 m long, 200 m wide, and 200 m high is established (Figure 6). In order to simulate the overburden load, the upper boundary stress of the model is set as 7.5 MPa. The displacement on the left, right, and bottom boundaries is fixed to 0. The lateral pressure coefficient is set as 1.2, and the coal and rock masses are calculated by the Mohr–Coulumb model. The horizontal distance between the #1 coal seam residual pillars and the #2 coal seam is 31 m; therefore, the above calculated stress concentration influence range of residual pillars is 11 m and the range of the internal stress field is 8 m. The reasonable pillar width can be further determined by simulating the stress distribution characteristics of the roadway surrounding the rock with pillar widths of 4 m, 6 m, 8 m, 10 m, 12 m, and 14 m, respectively. The calculation process is as follows: model establishment → original rock stress equilibrium → excavation of the #1 coal → excavation of the 4502 and 4702 working faces in the #2 coal seam → excavation of the 4602 head roadway → calculation result analysis. The mechanical parameters of the coal and rock mass in numerical simulation are obtained by conversion on the basis of laboratory test data. Table 1 gives the mechanical parameters of the coal rock mass.

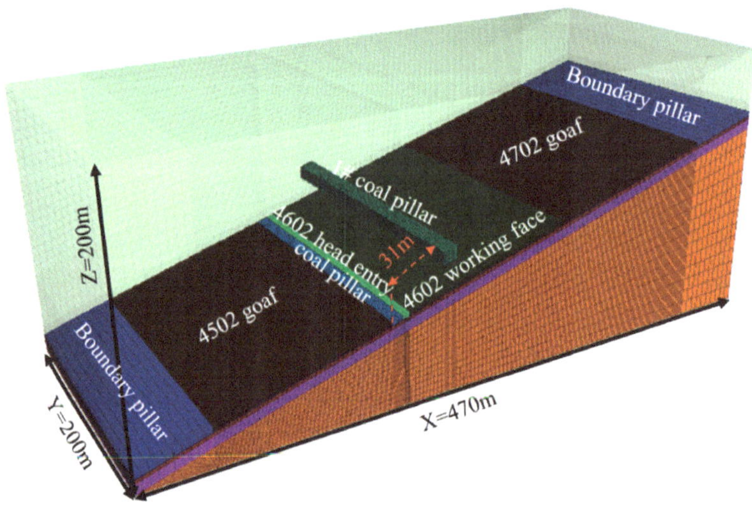

Figure 6. Three-dimensional numerical model.

Table 1. Mechanical parameters of the coal rock mass.

Lithology	ρ (kg/m^3)	E_i (GPa)	c_i (MPa)	φ_i (deg.)	σ_{ti} (MPa)	v_i
Coarse sandstone	2556	10.25	7.64	29.31	1.86	0.20
No.1 coal seam	1335	2.46	2.91	23.59	1.35	0.25
Siltstone	2548	7.25	4.59	28.42	1.62	0.18
Fine sandstone	2552	10.63	6.90	31.74	1.72	0.19
Mudstone	1220	3.58	1.50	15.53	0.80	0.27
No.2 coal seam	1329	2.97	2.35	21.86	1.10	0.23

4.2. Analysis of the Model Results

The analysis of the influence range of the residual coal pillar can provide a certain basis for the selection of the width of the coal pillar during the mining of the lower coal seam. From the numerical simulation results (Figure 7), it can be seen that the peak value of the advance abutment pressure during the mining of the #2 coal seam is 16.9 MPa. The residual coal pillar has a certain impact on the mining of the #2 coal seam, and the influence range of the stress concentration is 10.5 m (Figure 7b), which is almost identical to the theoretical calculated value of 11 m for the concentrated influence range of the remaining coal pillar of the coal seam; thus, the accuracy of theoretical calculation is further verified. Therefore, when roadways are driven along the goaf in coal seam #2, the roadways should be arranged to avoid the influence of the #1 residual coal pillar as much as possible.

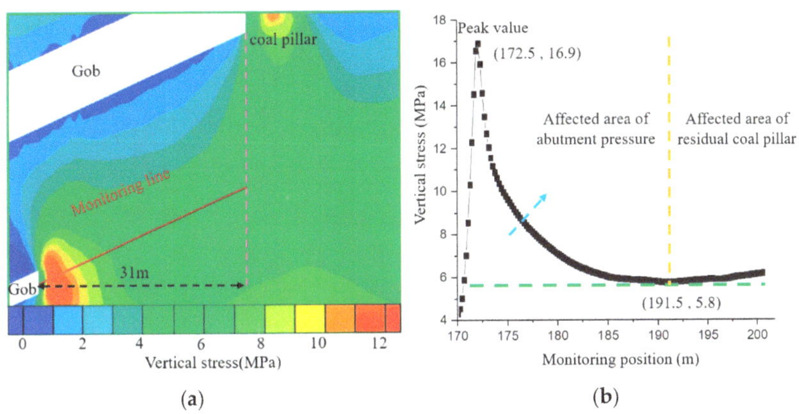

Figure 7. Influence range of residual coal pillar stress. (**a**) Stress distribution. (**b**) Stress monitoring curve.

When the coal and rock masses boast a strong bearing capacity, stress concentration is likely to occur in a certain range. Stress is re-distributed after roadway excavation. The simulation and analysis of the stress distribution in the coal pillars is of certain guiding significance for determining the reasonable coal pillar width. The stress nephograms and distribution curves of different coal pillar widths were obtained by numerical calculation results (Figure 8).

According to Figure 8, overall, the stress distributions of different coal pillar widths show the following rules:

(1) At coal pillar widths of 8~14 m, the internal stress of coal pillars presents an asymmetric bimodal distribution with high stress on the gob sides and low stress on the roadway sides. When the width is below 6 m, the internal stress presents a single peak distribution.

(2) After the roadway excavation, the vertical stress is re-distributed. At a small coal pillar width, stress is concentrated on the right side of the roadway. As the width increases, the position of the stress concentration gradually shifts to the gob side.
(3) At widths of 4 m and 6 m, fractures develop in the coal pillars, leading to serious damage and poor bearing capacity, which fails to form stress concentration. However, as the width continuously increases to 8 m, the mechanical properties of the coal and rock mass improve, and the peak value of the stress nucleus rises. When the width reaches 10 m, the maximum stress peak value is 11.30 MPa. When the width exceeds a certain value and reaches 12 m, the abutment pressure increases and surpasses the ultimate strength of the coal body. In this case, the stress in the coal pillar is relieved, and the peak value of the stress nucleus decreases. The peak stress value rises first and then decreases in turn in the following order: 1.67 MPa, 7.22 MPa, 10.67 MPa, 11.30 MPa, 10.27 MPa, and 10.25 MPa.
(4) When the width of the coal pillar reaches 8 m, the coal pillar boasts an overall good bearing capacity. An increase in the width improves the integrity of the coal pillar, but this will cause a significant waste of coal resources. However, a width below 8 m corresponds to poor integrity and difficult maintenance; at the same time, it is easy to cause air leakage in the coal pillar, which is not conducive to fire prevention management in the goaf. Therefore, a width of 8 m can ensure the stability of coal pillars through appropriate measures and maximize the recovery of coal resources.

In summary, when the width of the coal pillar is 8 m, coal pillars have an overall good bearing capacity. Meanwhile, the roadway is within the internal stress field, and can also avoid the impact of high stress caused by the residual coal pillars in the upper coal seams. Under strengthened support, the roadway can fully meet the requirements for basic transportation, ventilation, and side walking during recovery. Moreover, strengthened support can maximize the recovery of coal resources and ensure the safety and stability of roadway during recovery. As a result, it is determined that the coal pillar width of gob-side entry driving should be set to 8 m through theoretical analysis and on-site actual situation.

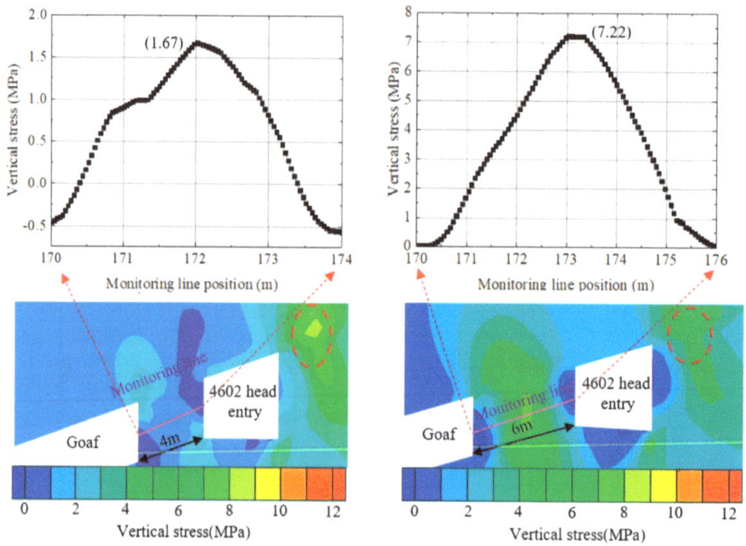

Figure 8. *Cont.*

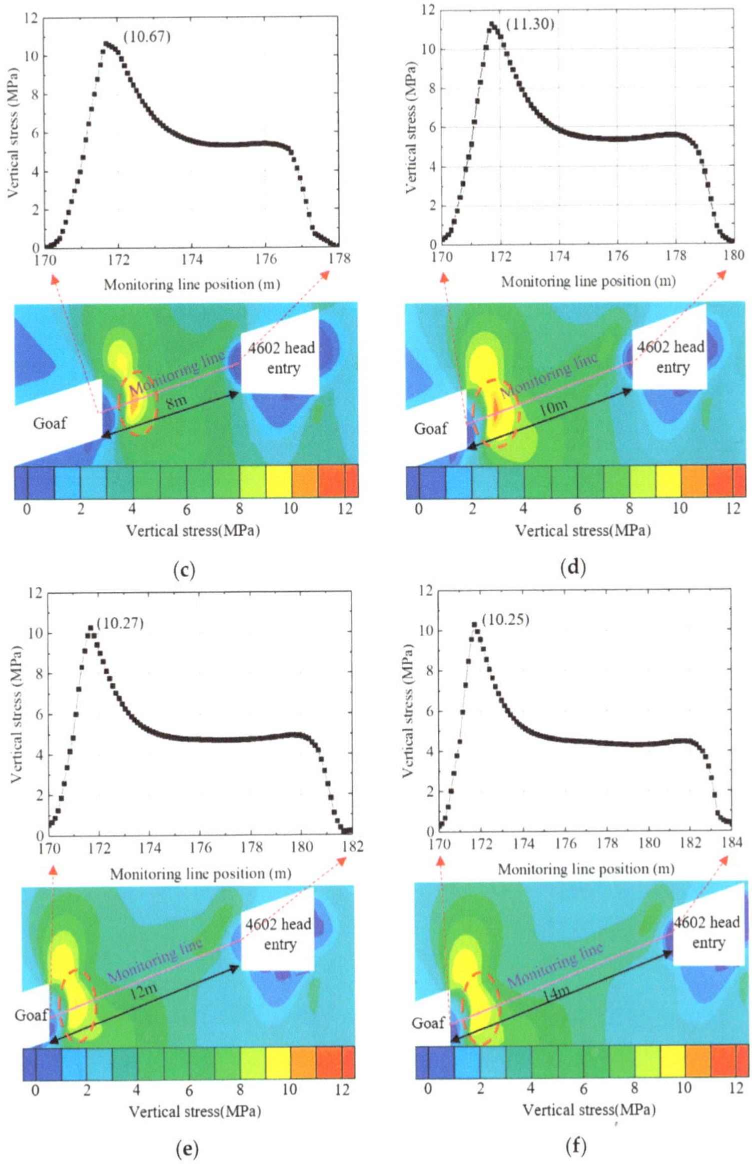

Figure 8. Stress distribution of different coal pillar widths. (**a**) Stress variation of a 4 m coal pillar. (**b**) Stress variation of a 6 m coal pillar. (**c**) Stress variation of an 8 m coal pillar. (**d**) Stress variation of a 10 m coal pillar. (**e**) Stress variation of a 12 m coal pillar. (**f**) Stress variation of a 14 m coal pillar.

5. Determination of the 4602 Head Roadway Support Scheme

The 4602 head roadway is in a high stress environment for gob-side entry driving. Hence, targeted design and improvements in anchor net cable support are required. The stability of a roadway surrounding rock can be controlled using yield support technology [27–31]. The principle of "yield before resist, a combination of yield and resist" was adopted to control the deformation of the roadway surrounding rock. "Pressure yielding" by the use of pressure-yield bolts can lead to a certain plastic deformation of roadways surrounding rock, so that the deformation of surrounding rock can be released to a certain extent. "Pressure resisting" by high-strength bolts can resolutely resist the deformation of roadways surrounding rock. "Pressure resisting" does not mean simply improving the bolt strength. When the bolt strength increases to a certain extent, the mechanical parameter ratio of the roadway surrounding rock varies slightly. Meanwhile, when the bolt has a higher strength, its elongation is lower, and fails to resist the deformation of the roadway surrounding rock. At this time, the performance of the pressure-yield bolts can be brought into full play.

5.1. Bolt Support System

The parameters of the support system are as follows: Φ 20 mm × 2600 mm high-strength anchor bolts, 150 mm × 150 mm × 10 mm high-strength trays with 260 N·m tightening torque. The anchor bolts are matched with pressure-yield pipes. The pressure-yield distance is 26 mm and the load-yield point is 120~150 kN (Figure 9a). The anchor bolt layout is shown in Figure 10, and surface support is realized by a prepared metal mesh.

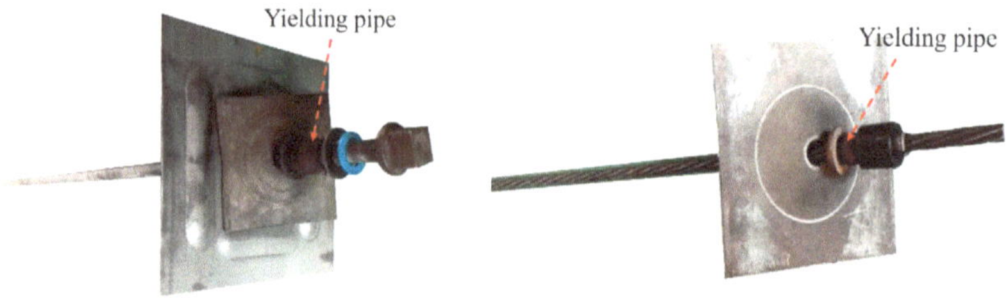

(**a**) (**b**)

Figure 9. Pressure-yield supporting materials. (**a**) Pressure-yield anchor bolt. (**b**) Pressure-yield anchor cable.

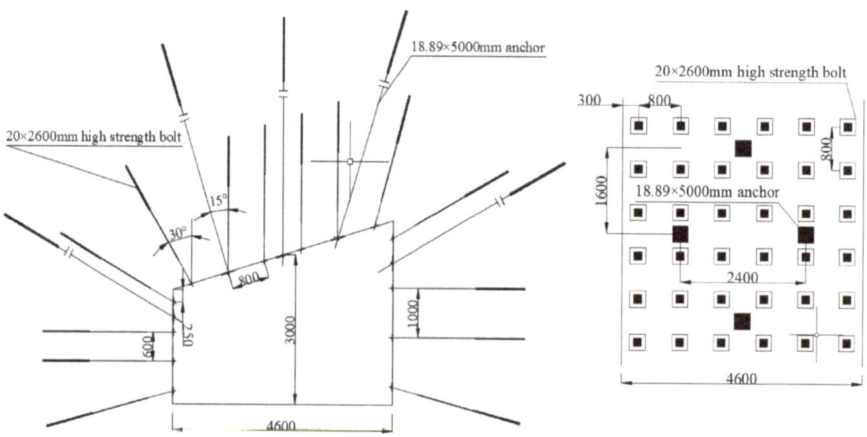

(**a**) (**b**)

Figure 10. Support scheme. (**a**) Front view. (**b**) Top view of roof support.

5.2. Auxiliary Support System

The roadway roof is supported by Φ 18.89 mm × 5000 mm mining anchor cables. The upper parts of the two sides of the roadway are both equipped with a Φ 18.89 mm × 5000 mm anchor cable for reinforcement support. The installation preload is 100 kN, equipped with 300 × 300 × 12 mm high-strength arc trays, and the cable layout is given in Figure 10. The anchor cables are matched with pressure-yield pipes. The pressure-yield distance is 26 mm, and the pressure-yield point is 220~250 kN (Figure 9b).

6. Engineering Experiment

An engineering experiment of the support scheme was conducted in the 4602 head roadway with the coal pillar width being 8 m. During excavation, mine pressure observation points are arranged along the roadway to monitor the roadway support quality, thereby monitoring the reasonableness of the selected roadway support parameters. These provide a basis for the optimization and adjustment of our roadway support scheme. Two monitoring stations were arranged in the 4602 head roadway, and the distance between each monitoring station was 50 m. The monitoring content of each measuring station included the displacement of the roadway surrounding rock and the stress on the anchor bolt (cable), and a specially assigned person monitored and collected the data regularly. The monitoring results were analyzed as follows.

6.1. Roadway Deformation Data and Analysis

Based on the data recorded at the observation points of the roadway surrounding the rock displacement, curves of the roadway displacement variation with working face distance were drawn (Figure 11). The roadway is basically stable when excavation advances to the observation point about 80 m away from the working face. Influenced by the gob= and coal-pillar-concentrated stress in the #1 coal seam as well as the coal pillars in the #2 coal seam, the coal pillar sides deform markedly (up to 82 mm) while the working face sides deform relatively less (up to 65 mm). When excavation advances to the observation point 80 m away from the working face, the roof subsidence curve stabilizes and becomes essentially linear, with a maximum roof subsidence of 95 mm.

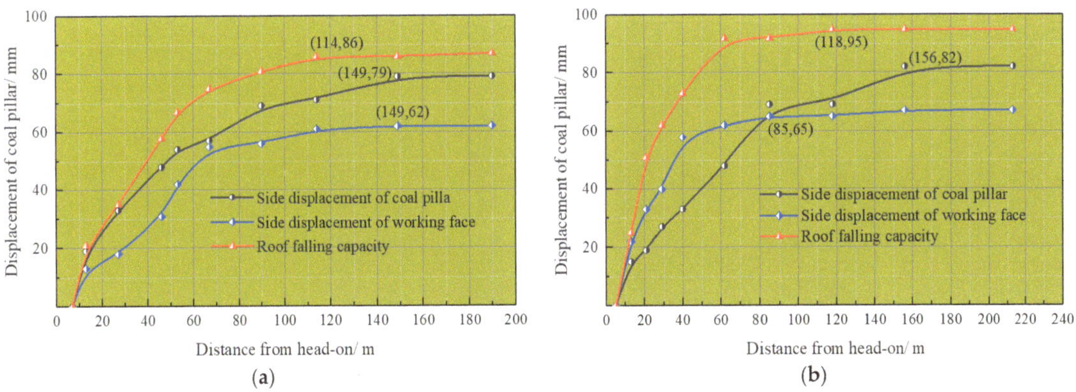

Figure 11. The curves of roadway displacement variation with working face distance. (**a**) Displacement of the #1 measuring station. (**b**) Displacement of the #2 measuring station.

According to the roadway displacement, the displacements of the roof and two sides are well controlled. The coal pillar sides deform quite markedly, and the displacement of the left side accounts for 56% of the deformation of the two sides.

6.2. Observed Data of Bolt Stress and Analysis

According to the data recorded at the two groups of bolt-stress observation points, the curves of bolt stress variation with working face distance were drawn (Figure 12). The bolt stress on the working face sides and roof increases linearly before the distance from the working face exceeds 80 m and tends to stabilize after the distance from the working face exceeds 80 m. Influenced by the coal pillars, the bolt stress on the coal pillar side is likely to cause stress concentration, which leads to high bolt stress on the coal pillar side. The maximum bolt stress is 140 kN, and the maximum bolt stress of the working face side is 113 kN. The coal pillar side bolt is subject to a higher stress than the working face side bolt, and the maximum roof bolt stress is 122 kN. According to the data on anchor bolt stress, bolts are subject to high stress. The bolt stress on the coal pillar side and the roof is close to or even exceeds the load-yield point of the pressure-yield pipe. The pressure-yield device plays a role to make the anchor bolts bear the stress together and prevent the breakage of the local anchor bolt by excessive stress.

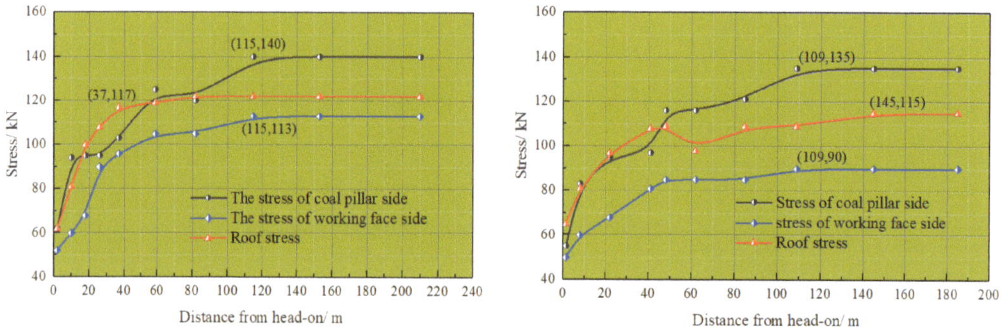

(**a**) (**b**)

Figure 12. The curves of bolt stress variation with working face distance. (**a**) Load of the #1 measuring station. (**b**) Load of the #2 measuring station.

6.3. Observed Data of Anchor Stress and Analysis

According to the recorded data of the two groups of anchor cable stress, the curves of anchor cable stress variation with working face distance were drawn (Figure 13). After the installation of anchor cables, the stress of the anchor cable basically increases linearly and tends to stabilize when the observation point is approximately 100 m away from the working face, the maximum stress of the anchor cable being 210 kN, which is smaller than the cable-breaking strength of 400 kN. Cables can bear the stress together with anchor bolts and play the role of auxiliary strengthened support.

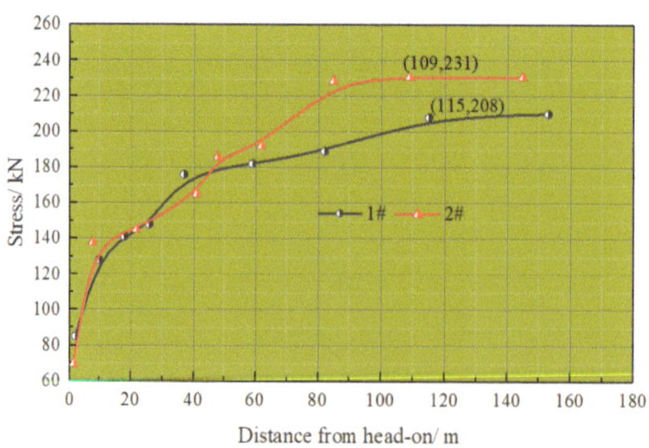

Figure 13. The curves of anchor cable stress variation with working face distance.

According to the above ground pressure observation and analysis, stable support of roadway excavation along the gob in the inclined coal seam is effectively realized by the pressure-anchor mesh-cable support scheme. The on-site support effect is illustrated in Figure 14.

Figure 14. Effect drawing of on-site support.

6.4. Potential Future Research

In the future, we will continuously optimize the width of coal pillars and tunnel support schemes through theoretical and on-site testing methods. At the same time, we will conduct innovative research on anchor rod (cable) support technology to provide more support for safe, efficient, and green mining in mines.

7. Conclusions

(1) A theoretical model of short-distance coal seam floor failure and a mechanical model of inclined gob-side roadway were established. The influence depth of the residual coal pillars in the upper coal seam on the floor is 27 m, and the influence ranges of the stress concentration in coal pillars and the low-stress area of the inclined coal seam are 11 m and 12.6 m, respectively. With reference to the on-site production geological condition, the coal pillar width is preliminarily determined to be in the range of 6~8 m.

(2) Numerical simulation analysis reveals that when the width of coal pillars is smaller than 6 m, the internal stress of coal pillars is distributed in a single-peak shape, and stress is concentrated near the roadway. When the width is 8~14 m, the internal stress concentration position of coal pillars moves to the gob side, showing an asymmetric bimodal distribution of high stress on the gob side and low stress on the roadway side.

(3) Through numerical simulation calculation, the residual coal pillar has a certain impact on the mining of the #1 coal seam, and the influence range of the stress concentration is 10.5 m, which is almost consistent with the theoretical calculation of the concentrated influence range of the remaining coal pillar of the coal seam, which is 11 m, verifying the accuracy of the theoretical calculation. As the coal pillar width rises, the overall peak stress increases first and then decreases. When the coal pillar width reaches 8 m, the coal pillars have an overall strong bearing capacity.

(4) The gob-side entry driving in inclined short-distance coal seams should adopt the technology of pressure-yield anchor net cable support. The technology boasts good overall support, realizes the safety and stability of roadway support and provides a reference for mine roadway support under similar conditions.

Author Contributions: Conceptualization, F.H.; Methodology, W.Z.; Software, J.S.; Formal analysis, X.X.; Data curation, W.Z.; Writing—original draft, W.Z.; Writing—review & editing, D.W. and Y.W.; Supervision, F.H.; Project administration, F.H. All authors have read and agreed to the published version of the manuscript.

Funding: This work was supported by the National Natural Science Foundation of China (No.51974317), and the Fundamental Research Funds for the Central Universities (2022YJSNY09).

Data Availability Statement: The data used to support the findings of this study are included in the article.

Conflicts of Interest: The authors declare no conflict of interest.

References

1. Kuyuk, A.F.; Ghoreishi-Madiseh, S.A.; Sasmito, A.P.; Hassani, F. Designing a Large-Scale Lake Cooling System for an Ultra-Deep Mine: A Canadian Case Study. *Energies* **2019**, *12*, 811. [CrossRef]
2. Palei, S.K.; Das, S.K. Logistic regression model for prediction of roof fall risks in bord and pillar workings in coal mines: An approach. *Saf. Sci.* **2009**, *47*, 88–96. [CrossRef]
3. Yan, H.; Weng, M.Y.; Feng, R.M.; Li, W.K. Layout and support design of a coal roadway in ultra-close multiple-seams. *J. Cent. South Univ.* **2015**, *22*, 114385–114395. [CrossRef]
4. Yavuz, H. An estimation method for cover pressure re-establishment distance and pressure distribution in the goaf of longwall coal mines. *Int. J. Rock Mech. Min. Sci.* **2004**, *41*, 193–205. [CrossRef]
5. Mo, S.; Tutuk, K.; Saydam, S. Management of floor heave at Bulga Underground Operations—A case study. *Int. J. Min. Sci. Technol.* **2019**, *29*, 73–78. [CrossRef]
6. Shang, H.F.; Ning, J.G.; Hu, S.C.; Yang, S.; Qiu, P.Q. Field and numerical investigations of gateroad system failure under an irregular residual coal pillar in close-distance coal seams. *Energy Sci. Eng.* **2019**, *7*, 2720–2740. [CrossRef]
7. Yue, X.Z.; Tu, M.; Li, Y.F.; Zhang, J.S.; Gao, L. Calculation of subsidiary mining stress in floor roadway under the remaining boundary pillar of close coal seam mining. *J. Min. Saf. Eng.* **2021**, *38*, 246–252,259.
8. Das, A.J.; Mandal, P.K.; Sahu, S.P.; Kushwaha, A.; Bhattacharjee, R.; Tewari, S. Evaluation of the Effect of Fault on the Stability of Underground Workings of Coal Mine through DEM and Statistical Analysis. *J. Geol. Soc. India* **2018**, *92*, 732–742. [CrossRef]
9. Islam, M.R.; Shinjo, R. Mining-induced fault reactivation associated with the main conveyor belt roadway and safety of the Barapukuria coal mine in Bangladesh: Constraints from BEM simulations. *Int. J. Coal Geol.* **2009**, *79*, 115–130. [CrossRef]
10. Vu, T.T. Solutions to prevent face spall and roof falling in fully mechanized longwall at underground mines, Vietnam. *Min. Miner. Depos.* **2022**, *16*, 127–134. [CrossRef]
11. Xu, B.; Dong, S.N.; Xu, Y.L. Waterproof pillar size design in steep coal seam. *J. Min. Sci. Technol.* **2017**, *2*, 301–315.
12. Murphy, M.M.; Ellenberger, J.L.; Esterhuizen, G.S.; Miller, T. Analysis of roof and pillar failure associated with weak floor at a limestone mine. *Int. J. Min. Sci. Technol.* **2016**, *26*, 471–476. [CrossRef] [PubMed]
13. Ram, S.; Singh, A.K.; Kumar, R.; Kumar, A.; Kumar, N.; Waclawik, P.; Gautam, A.; Raja, M. Design of rock bolt-based goaf edge support for conventional depillaring with stowing. *Arab. J. Geosci.* **2021**, *14*, 2164. [CrossRef]
14. Yadav, A.; Behera, B.; Sahoo, S.K.; Singh, G.S.P.; Sharma, S.K. An Approach for Numerical Modeling of Gob Compaction Process in Longwall Mining. *Min. Metall. Explor.* **2020**, *37*, 631–649.
15. Wang, K.; Huang, Y.L.; Gao, H.D.; Zhai, W.; Qiao, Y.F.; Li, J.M.; Ouyang, S.Y.; Li, W. Recovery technology of bottom coal in the gob-side entry of thick coal seam based on floor heave induced by narrow coal pillar. *Energies* **2020**, *13*, 3368. [CrossRef]
16. Peng, S.S.; Chandra, U. Getting the most from multiple seam reserves. *Coal Min. Process.* **1980**, *17*, 78–84.
17. Quang, P.L.; Zubov, V.; Duc, T.P. Design a reasonable width of coal pillar using a numerical model a case study of khe cham basin, vietnam. In Proceedings of the Vth International Innovative Mining Symposium, Kemerovo, Russia, 19–21 October 2020; Volume 174, p. 01043.
18. Gao, M.Z. Similarity model test of strata movement with steep seam. *Chin. J. Rock Mech. Eng.* **2004**, *23*, 441–445.
19. Ghasemi, E.; Ataei, M.; Shahriar, K. An intelligent approach to predict pillar sizing in designing room and pillar coal mines. *Int. J. Rock Mech. Min. Sci.* **2014**, *65*, 86–95. [CrossRef]
20. Hu, B.; Kang, H.P.; Lin, J. Comparison and application of high prestress and intensive support system in close, soft and cracked roadway support. *Adv. Mater. Res.* **2012**, *524–527*, 29–35. [CrossRef]
21. He, F.L.; Gao, S.; Zhang, G.C.; Jiang, B.Y. Ground response of a gob-side gateroad suffering mining-induced stress in an extra thick coal seam. *Geomech. Eng.* **2020**, *22*, 1–9.
22. Malashkevych, D.; Petlovanyi, M.; Sai, K.; Zubko, S. Research into the coal quality with a new selective mining technology of the waste rock accumulation in the mined-out area. *Min. Miner. Depos.* **2022**, *16*, 103–114. [CrossRef]
23. Smoliński, A.; Malashkevych, D.; Petlovanyi, M.; Rysbekov, K.; Lozynskyi, V.; Sai, K. Research into Impact of Leaving Waste Rocks in the Mined-Out Space on the Geomechanical State of the Rock Mass Surrounding the Longwall Face. *Energies* **2022**, *15*, 9522. [CrossRef]
24. Xu, X.H.; He, F.L.; Li, X.B.; Lv, K.; He, C.X.; Li, L. Roadway impact mechanism and optimization design under coal pillar disturbance in large mining height working face. *J. Min. Saf. Eng.* **2021**, *38*, 547–555,564.
25. Song, Z.Q. *Mining Pressure and Strata Control*; China University of Mining and Technology Press: Xuzhou, China, 1986.
26. Zhang, G.C.; He, F.L.; Lai, Y.H.; Song, J.W.; Xiao, P. Reasonable width and control technique of segment coal pillar with high-intensity fully-mechanized caving mining. *J. China Coal Soc.* **2016**, *41*, 2188–2194.
27. Chen, J.H.; Liu, L.; Zeng, B.Q.; Tao, K.M.; Zhang, C.; Zhao, H.B.; Li, D.Q.; Zhang, J.W. A constitutive model to reveal the anchorage mechanism of fully bonded bolts. *Rock Mech. Rock Eng.* **2023**, *56*, 1739–1757. [CrossRef]
28. Dyczko, A.; Kamiński, P.; Jarosz, J.; Rak, Z.; Jasiulek, D.; Sinka, T. Monitoring of Roof Bolting as an Element of the Project of the Introduction of Roof Bolting in Polish Coal Mines—Case Study. *Energies* **2021**, *15*, 95. [CrossRef]
29. Krykovskyi, O.; Krykovska, V.; Skipochka, S. Interaction of rock-bolt supports while weak rock reinforcing by means of injection rock bolts. *Min. Miner. Depos.* **2021**, *15*, 8–14. [CrossRef]

30. Wang, J.; Apel, D.B.; Xu, H.W.; Wei, C.; Skrzypkowski, K. Evaluation of the Effects of Yielding Rockbolts on Controlling Self-Initiated Strainbursts: A Numerical Study. *Energies* **2022**, *15*, 2574. [CrossRef]
31. Zuo, J.P.; Liu, J.H.; Liu, D.J.; Wu, L.L.; Sun, Y.J. Control theory of uniform strength support in deep roadway. *J. Min. Sci. Technol.* **2021**, *6*, 148–159.

applied sciences

MDPI

Article

Construction Method of Honeycomb Gas Flow Network Channel: Structure-Based and Engineering Response on Plastic Zone Distribution Form

Chao Yuan [1,2,*], Lian-Xin Li [1], Lei Fan [2], Yue-Yang Xu [1] and Jun-Jie Wu [1]

1 Mining Engineering, School of Resources & Environment and Safety Engineering, Hunan University of Science and Technology, Xiangtan 411201, China

2 Work Safety Key Lab on Prevention and Control of Gas and Roof Disasters for Southern Coal Mines, Hunan University of Science and Technology, Xiangtan 411201, China; fanlei@mail.hnust.edu.cn

* Correspondence: yuanchaozh1@126.com

Abstract: The plastic zone of the surrounding rock of the borehole is a high-quality channel for gas flow. Studying the distribution form of the plastic zone in the surrounding rock of the boreholes and the characteristics of the extended connection network is essential for the optimal layout of the gas drainage boreholes. Considering gas pressure parameters based on the modified Terzaghi effective stress principle and Mohr–Coulomb strength criterion. The implicit equation of the plastic zone boundary of the surrounding rock of the borehole is derived. The influence of two-way confining pressure ratio, coal body strength, gas pressure and hole diameter on the distribution of the surrounding plastic rock zone and the permeability-increasing circle is analyzed, and a honeycomb gas flow network channel structure based on the distribution of plastic zones is constructed. The gas flow and enrichment zone of the coal seam are connected, and the gas drainage rate is improved.

Keywords: plastic zone; gas pressure; permeability-increasing circle; network channel; borehole spacing

1. Introduction

Borehole extraction is the main technical means of gas control at present. Effective extraction radius is an important basis for borehole design, which directly affects the extraction effect. If the hole spacing is too large, it is easy to form a blind area of extraction. If the hole spacing is too small, the effective extraction radius overlaps with each other, which is easy to cause unnecessary waste of manpower and material resources [1,2]. For this reason, researchers have carried out shock waves [3], loose blasting [4,5], hydraulic fracturing [6,7], CO_2 fracturing [8,9] and large borehole [10,11] in improving the permeability-increasing of coal seam and increasing the effective extraction radius. It achieved a certain effect of increasing permeability and increasing flow. In addition, many researchers have carried out a lot of fruitful research work on the theoretical solution of gas flow channels and effective radius of extraction, field test technology and other key issues, and achieved a series of results [12]. Liu et al. [13] studied the evolution law of mining gas channels and its influence on the law of gas-guided flow. They established the judgment method of the gas channel flow state. Zhang et al. [14] divided the gas flow channel into mesoscopic channels and macroscopic channels according to the fracture degree of the coal body. They analyzed their formation mechanism and evolution law of them. Xu et al. [15] studied the unloading damage range of gas channels under different mining heights. Ma et al. [16,17] established the model of the plastic zone of a borehole surrounding a rock and gas permeability-increasing circle. He put forward the concept of a high-quality gas channel. Ma et al. [18] used a similar simulation method to study the formation and evolution mechanism of gas channels in close-distance coal seams.

In summary, the formation of the plastic zone of the surrounding rock of the borehole provides a high-quality channel for gas flow. The distribution pattern of the plastic zone and the characteristics of the connection network are very important for the optimal layout of the gas extraction borehole. However, the current research mainly focuses on the formation and evolution of the plastic zone of the single-hole surrounding rock and the expansion of the plastic zone between the surrounding rock of each borehole. The research on the scientific rationality of the hole spacing is relatively lagging behind, and it also ignores the influence of gas pressure on the plastic zone of the surrounding rock of the borehole. Resulting in a certain blindness and randomness in the actual engineering application, which seriously affects the efficient extraction of gas. Therefore, considering the gas pressure parameter, the implicit equation of the plastic zone boundary of the borehole surrounding rock is derived based on the modified Terzaghi effective stress principle and Mohr-Coulomb strength criterion. The distribution pattern of the plastic zone of the borehole surrounding rock and the variation law of the permeability-increasing circle with the confining pressure ratio, the mechanical parameters of the coal body, the gas pressure and the aperture are analyzed. The distribution pattern of the plastic zone of the borehole surrounding rock is used to establish the honeycomb gas flow network channel structure, which is beneficial to gas extraction. Based on this, the main criterion of the borehole spacing is proposed, and the engineering test is carried out.

2. Mechanical Model of Surrounding Rock for Borehole in Gas Coal Body

The surrounding rock around the gas extraction borehole will form a certain range of plastic zones under the action of in-situ rock stress. A large number of cracks are derived from the plastic zone, and the cracks are the main channels for gas flow. The formation of the plastic zone provides a high-quality channel for gas flow and increases gas permeability. On the basis of predecessors, the mechanical model is established, and the formula is deduced to analyze the influence of various factors on the gas penetrating circle.

The circular borehole is excavated in the coal seam. The radius of the circular borehole is a. The original gas pressure in the coal seam is P_g. The gas pressure at the free surface of the surrounding rock is P_a. The unit long thole axis is determined along the hole axis, which is an axisymmetric plane strain issue without taking body force into account. To assist the study of the stress state, the mechanical model of the borehole surrounding rock is simplified by assuming that the two-way stress field is a uniformly distributed load. The following fundamental assumptions and the basis of some assumptions are made.

(1) The drilling section is a circular hole with radius a and buried depth $H \geq 20a$, which is an infinitely long flat hole; (2) The surrounding rock is an ideal elastic–plastic body, which is an isotropic medium. The coal body with good integrity can be regarded as an isotropic continuous medium; (3) Ignoring the mutual transformation between free gas and adsorbed gas in coal. The adsorption equilibrium of coal is destroyed after pressure relief so that the adsorbed gas becomes free gas. In practice, it must be converted under special circumstances, so it can be ignored; (4) The model's external boundary conditions are vertical principal stress P_1 and horizontal principal stress P_3, both of which are parallel to the rectangular coordinate axis.

On this basis, a mechanical model of a circle with increased permeability is developed, as shown in Figure 1.

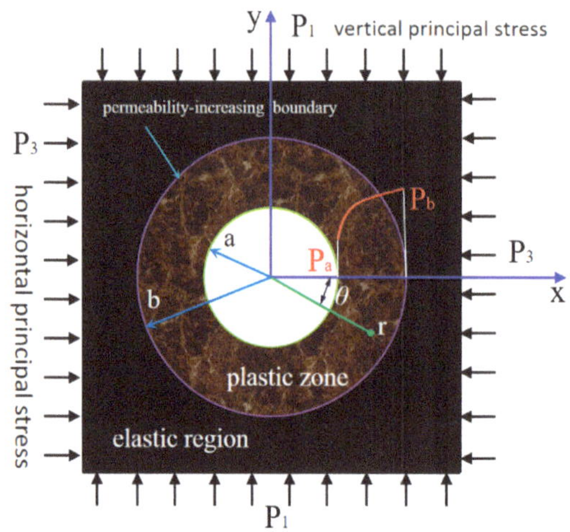

Figure 1. Force model of surrounding rock of borehole.

3. Gas Pressure Distribution of Surrounding Rock in Boreholes

After excavation, the gas flow in the surrounding rock is axisymmetric radial seepage. The cylindrical gas flow Q flows across the radius r per unit length per unit time:

$$Q = -2\pi r K \frac{\mathrm{d}p}{\mathrm{d}r} \tag{1}$$

We may obtain from Equation (1):

$$p(r) = c_1 - \frac{Q}{2\pi K} \ln r \tag{2}$$

Using Equation (2) and the boundary conditions of the coal borehole's surrounding rock, $p(\mathrm{a}) = P_\mathrm{a}$, $p(\mathrm{b}) = P_\mathrm{g}$, the following result is obtained:

$$c_1 = P_\mathrm{a} + \frac{Q}{2\pi K} \ln \mathrm{a} \tag{3a}$$

$$\frac{Q}{2\pi K} = \frac{P_\mathrm{g} - P_\mathrm{a}}{\ln(\mathrm{a}/\mathrm{b})} \tag{3b}$$

Substituting Equations (3a) and (3b) into Equation (2) yields the gas pressure distribution law:

$$p(r) = P_\mathrm{a} + \frac{(P_\mathrm{g} - P_\mathrm{a}) \ln(\mathrm{a}/r)}{\ln(\mathrm{a}/\mathrm{b})} \tag{4a}$$

$$\frac{\mathrm{d}p}{\mathrm{d}r} = -\frac{(P_\mathrm{g} - P_\mathrm{a})}{r \ln(\mathrm{a}/\mathrm{b})} \tag{4b}$$

4. Theoretical Solutions of the Surround Rock Plastic Zone of the Borehole

The mechanical model of the borehole surrounding rock in Figure 1 is decomposed into two parts by the action of two-way ground stress and gas pressure. One part is the uniform ground stress field condition under gas pressure, as shown in Figure 2a. The other part is to bear the same pressure and tensile stress in the vertical and horizontal directions, respectively, without taking gas pressure into account, as shown in Figure 2b.

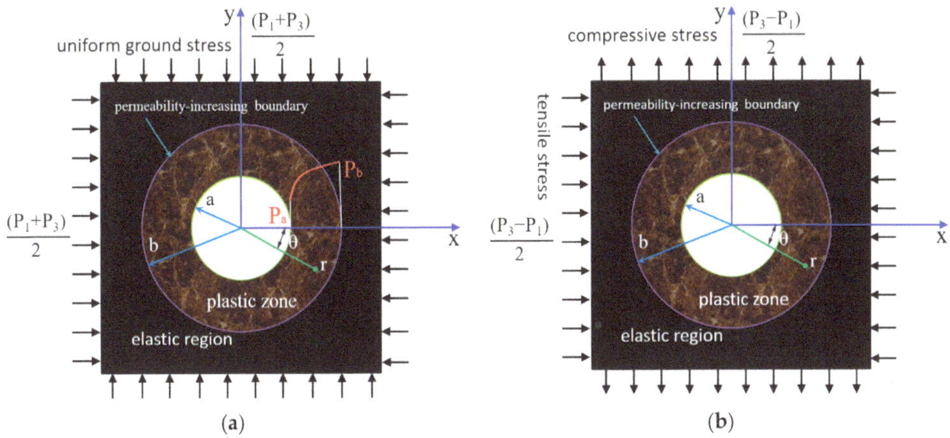

Figure 2. Force decomposition of surrounding rock of borehole. The (**a**) is the radius of borehole, (**b**) is the radius of plastic zone.

4.1. Theoretical Solution of Stress Field in Borehole Elastic Zone around Rock

4.1.1. Theoretical Solution of Compressive Stress Field of Borehole Surrounding Rock

In the first section, Figure 2a shows the action of gas pressure. For coal-like materials, according to the modified Terzaghi effective stress principle, the relationship between the effective stress σ'_{ij} acting on the coal skeleton and the gas pressure p is it.

$$\sigma'_{ij} = \sigma_{ij} - ap\delta_{ij} \tag{5}$$

$$\left.\begin{aligned} \sigma'_r &= \sigma_r - ap \\ \sigma'_\theta &= \sigma_\theta - ap \end{aligned}\right\} \tag{6}$$

The equilibrium equation of axisymmetric plane strain problem ignoring the borehole coal body force is it.

$$\frac{\mathrm{d}\sigma_r}{\mathrm{d}r} + \frac{\sigma_r - \sigma_\theta}{r} = 0 \tag{7}$$

When Equations (4b) and (6) are substituted for Equation (7), the following results are obtained:

$$\frac{\mathrm{d}\sigma'_r}{\mathrm{d}r} - \frac{a(P_g - P_a)}{r\ln(a/b)} + \frac{\sigma'_r - \sigma'_\theta}{r} = 0 \tag{8}$$

Assuming b is large enough, the elastic phase stress expression can be obtained using the calculation model's geometric equations and boundary conditions as follows:

$$\left.\begin{aligned} \sigma'_r &= -\frac{aP_a a^2}{r^2} + \left(1 - \frac{a^2}{r^2}\right)\left(\frac{P_3+P_1}{2} - aP_g\right) + \left[\frac{a(P_g-P_a)}{2(1-\mu)}\left(1 - \frac{a^2}{r^2}\right)\right] \\ \sigma'_\theta &= \frac{aP_a a^2}{r^2} + \left(1 + \frac{a^2}{r^2}\right)\left(\frac{P_3+P_1}{2} - aP_g\right) + \left[\frac{a(P_g-P_a)}{2(1-\mu)}\left(1 + \frac{a^2}{r^2}\right)\right] \end{aligned}\right\} \tag{9}$$

4.1.2. Theoretical Solution of Borehole Surrounding Rock Pressure and Tensile Stress Field

The second portion is a theoretical solution of the stress field of the borehole's surrounding rock under pressure and tensile stress, as shown in Figure 2b.

When $r = \mathrm{b}$

$$\left.\begin{aligned} \sigma_r|_{r=\mathrm{b}} &= -\tfrac{P_3-P_1}{2}\cos 2\theta \\ \tau_{r\theta}|_{r=\mathrm{b}} &= \tfrac{P_3-P_1}{2}\sin 2\theta \end{aligned}\right\} \tag{10}$$

The stress solution is obviously connected to r and 2θ; therefore, let the stress function be it.

$$y = f(r)\cos 2\theta \tag{11}$$

The elastic phase stress expression is generated by plugging Equation (11) into the biharmonic sum equation and combining the boundary conditions $r = \mathrm{a}$, $\sigma_r = 0$ and $\tau_{r\theta} = 0$:

$$\left.\begin{aligned} \sigma_r &= \tfrac{(P_3-P_1)}{2}\cos 2\theta(1-\tfrac{\mathrm{a}^2}{r^2})(1-\tfrac{3\mathrm{a}^2}{r^2}) \\ \sigma_\theta &= -\tfrac{(P_3-P_1)}{2}\cos 2\theta(1+\tfrac{3\mathrm{a}^4}{r^4}) \\ \tau_{r\theta} &= -\tfrac{(P_3-P_1)}{2}\sin 2\theta(1-\tfrac{\mathrm{a}^2}{r^2})(1+\tfrac{3\mathrm{a}^2}{r^2}) \end{aligned}\right\} \tag{12}$$

4.1.3. Borehole Surrounding Rock Stress Field Theoretical Solution

The theoretical solution of the elastic stress field of a borehole surrounding rock with gas pressure can be obtained by superimposing the stress components of the first and second parts. That is the addition of Equations (9) and (12). The specific expression is as follows:

$$\left.\begin{aligned} \sigma'_r &= \left[\tfrac{(P_3-P_1)}{2}\cos 2\theta(1-\tfrac{3\mathrm{a}^2}{r^2}) + \left(\tfrac{P_3+P_1}{2} - aP_g\right) + \tfrac{a(P_g-P_a)}{2(1-\mu)}\right](1-\tfrac{\mathrm{a}^2}{r^2}) - \tfrac{aP_a\mathrm{a}^2}{r^2} \\ \sigma'_\theta &= -\tfrac{(P_3-P_1)}{2}\cos 2\theta(1+\tfrac{3\mathrm{a}^4}{r^4}) + \left[\tfrac{P_3+P_1}{2} + \tfrac{a(P_g-P_a)}{2(1-\mu)} - aP_g\right]\left(1+\tfrac{\mathrm{a}^2}{r^2}\right) + \tfrac{aP_a\mathrm{a}^2}{r^2} \\ \tau_{r\theta} &= -\tfrac{(P_3-P_1)}{2}\sin 2\theta(1-\tfrac{\mathrm{a}^2}{r^2})(1+\tfrac{3\mathrm{a}^2}{r^2}) \end{aligned}\right\} \tag{13}$$

4.2. Theoretical Solution of Stress Field in Borehole Plastic Zone around Rock

When the tension in the surrounding rock around the borehole exceeds the strength of the coal body, plastic deformation occurs. A defined range of plastic failure zone extends from the borehole to the depth of the surrounding rock. The surrounding rock stress now fits the Mohr-Coulomb yield requirement, namely:

$$\frac{(\sigma'_r-\sigma'_\theta)^2}{4} + \tau_{r\theta}^2 = \left[\frac{(\sigma'_r+\sigma'_\theta)^2}{4} - c^2\right]\frac{(1-\cos 2\varphi)}{2} + \frac{(\sigma'_r+\sigma'_\theta)}{2}c\sin 2\varphi + c^2 \tag{14}$$

Let r be the plastic zone border, $P_3 = \lambda P_1$, and plug Equation (14) into Equation (13) to obtain the implicit equation for the plastic zone boundary of the circular drilled hole enclosing rock with regard to r and θ, namely:

$$\begin{aligned} f(r,\theta) = &\left\{\begin{aligned} &P_1(\lambda-1)\left(1-\tfrac{2\mathrm{a}^2}{r^2}+\tfrac{3\mathrm{a}^4}{r^4}\right)\cos 2\theta - \tfrac{\mathrm{a}^2(\lambda P_1+P_1-2aP_g)}{r^2} - \tfrac{2aP_a\mathrm{a}^2}{r^2} \\ &-\tfrac{\mathrm{a}^2\left[\frac{a(P_g-P_a)}{1-\mu}-2aP_g\right]}{r^2} \end{aligned}\right\}^2 \\ &+\left[P_1(\lambda-1)\left(1+\tfrac{2\mathrm{a}^2}{r^2}-\tfrac{3\mathrm{a}^4}{r^4}\right)\sin 2\theta\right]^2 \\ &-\tfrac{(1-\cos 2\varphi)}{2}\left\{\left[P_1(\lambda+1) - \tfrac{2P_1(\lambda-1)\mathrm{a}^2\cos 2\theta}{r^2} + \tfrac{a(P_g-P_a)}{1-\mu} - 4aP_g\right]^2 - 4c^2\right\} \\ &-2c\sin 2\varphi\left[P_1(\lambda+1) - \tfrac{2P_1(\lambda-1)\mathrm{a}^2\cos 2\theta}{r^2} + \tfrac{a(P_g-P_a)}{1-\mu} - 4aP_g\right] - 4c^2 \end{aligned} \tag{15}$$

5. Theoretical Influence of Surrounding Rock Augmented Permeability-Increasing Circle of Coal Borehole

The influencing factors of the permeability-increasing circle of the surrounding rock of the gas drainage borehole mainly include the confining pressure ratio, the strength of the coal body, the gas pressure and the hole diameter, which can be known from Equation (15). Coal strength, gas pressure and hole diameter and other factors, and the influence of each factor on the surrounding rock plastic zone and the permeability-increasing circle of the borehole will be analyzed in detail. With the help of Matlab calculation software and based on Equation (15), the influence of various factors on the plastic zone and permeability-increasing circle of the borehole permeability-increasing circle will be analyzed in detail.

5.1. Effect of Confining Pressure Ratio on Permeability-Increasing Circle of Borehole Surrounding Rock

Under the condition of certain mechanical properties of the borehole surrounding rock, gas pressure and hole diameter, the two-way confining pressure ratio determines the shape and range of the plastic zone of the borehole surrounding rock, as shown in Figure 3. When the confining pressure ratio is equal to 1, the plastic zone of the surrounding rock of the borehole is circular. As the confining pressure ratio gradually increases, the distribution pattern of the plastic zone changes from circular to elliptical and butterfly. After the formation of the butterfly-shaped plastic zone, with the continuous increase of the confining pressure ratio, the four butterfly leaves in the plastic zone expand rapidly at a high rate, and the failure range of the surrounding rock expands rapidly.

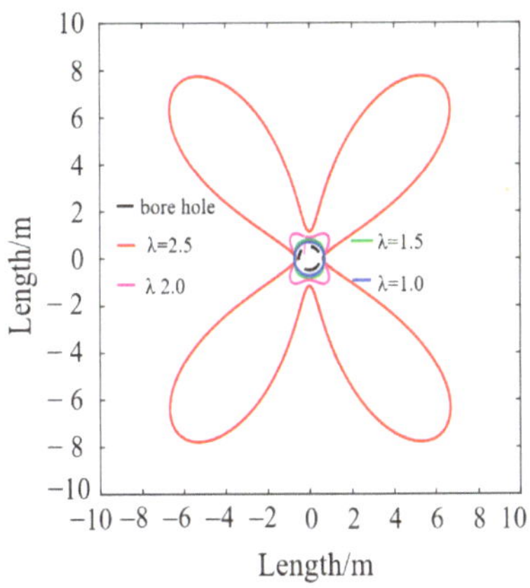

Figure 3. Plastic zone distribution patterns in borehole surrounding rock with varying confining pressure ratios.

In order to obtain the borehole surrounding the rock permeability-increasing circle, the distribution pattern of the plastic zone is drawn, and the permeability-increasing radius R max acquisition method is shown in Figure 4. When the plastic zone is circular, the plastic zone's range is the range of increased permeability. The butterfly leaf length of the butterfly plastic zone is the surrounding rock permeability-increasing radius R_{max} for non-circular plastic areas, such as butterfly plastic areas. The butterfly leaf is longer, and the permeability-increasing range is larger. It is more conducive to efficient gas extraction.

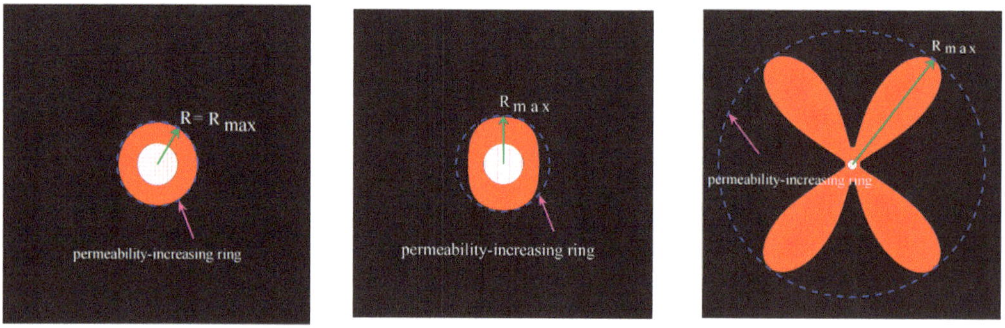

(a) (b) (c)

Figure 4. Plastic zone shape and gas permeability-increasing circle of surrounding rock of boreholes under different confining pressure ratios. (**a**) Round (**b**) Oval (**c**) Butterfly.

The effective extraction radius of gas is the permeability-increasing radius, and its size directly impacts the gas extraction effect; The permeability-increasing radius bigger, the more favorable to gas extraction. Based on the findings in Figure 5, When the borehole's confining pressure is low, the permeability-increasing radius of the surrounding rock is very tiny; when the confining pressure ratio grows, the permeability-increasing radius increases continually. When the confining pressure ratio approaches 2.5, the permeability-increasing radius climbs to 14 m. It is 28 times the borehole radius, demonstrating that a high confining pressure ratio is particularly helpful to gas extraction.

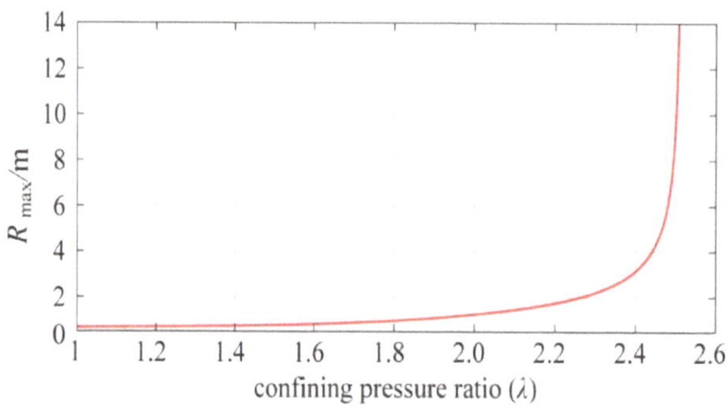

Figure 5. Confining pressure ratio and permeability-increasing radius of borehole surrounding Rock.

5.2. The Influence of Coal Strength on Borehole Surrounding Rock Permeability-Increasing Circle

The distribution pattern and range of plastic zone in the borehole surrounding rock are directly related to the strength of the coal body. As the main parameters to characterize its strength, cohesion and internal friction angle of the coal body must become the key factors affecting the permeability radius of surrounding rock. The relationship between the distribution of the plastic zone of the borehole surrounding rock and the cohesion and internal friction angle of the coal body is shown in Figures 6 and 7. The relationship between permeability-increasing radius and coal cohesion and internal friction angle is shown in Figures 8 and 9.

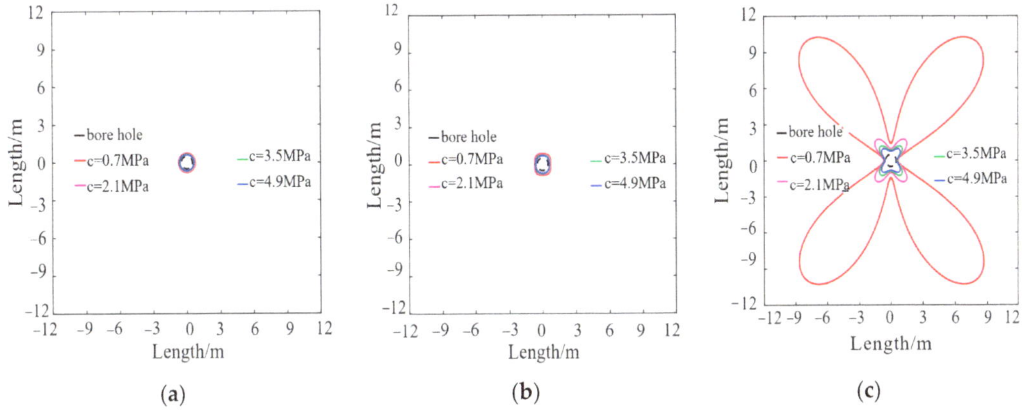

Figure 6. Distribution patterns of plastic zones in surrounding rock of boreholes under different cohesion. (**a**) Round (**b**) Oval (**c**) Butterfly.

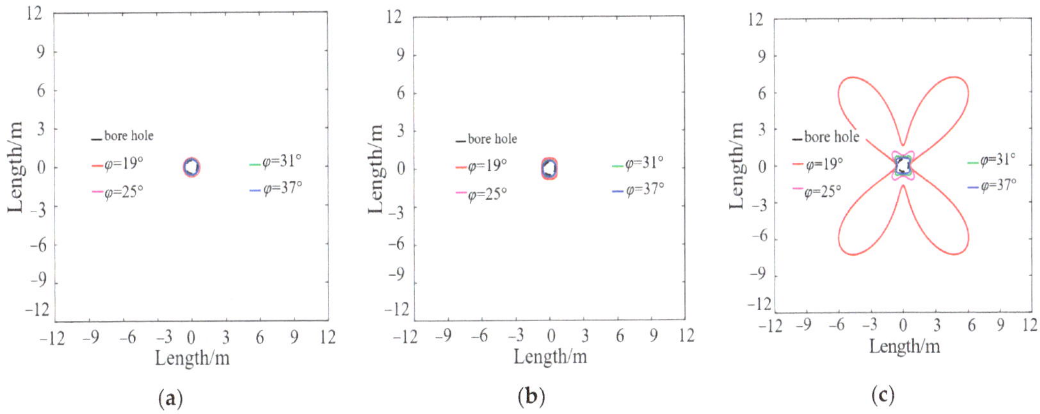

Figure 7. Distribution patterns of plastic zones in surrounding rock of boreholes under different internal friction angles. (**a**) Round (**b**) Oval (**c**) Butterfly.

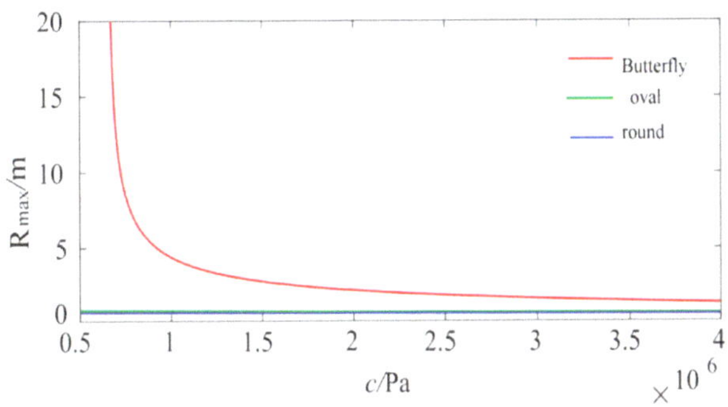

Figure 8. Permeability-increasing radius and cohesive force of borehole surrounding rock.

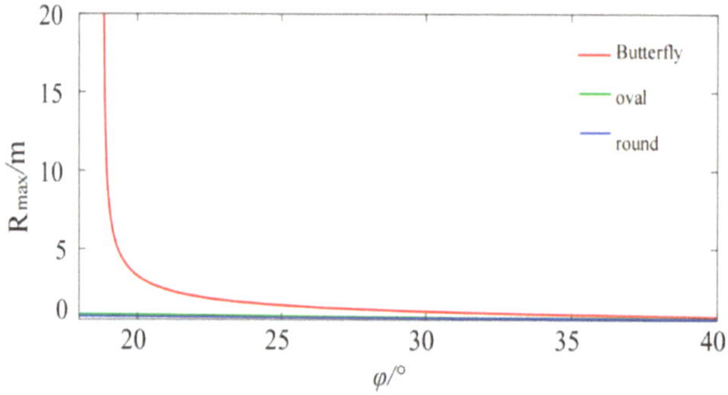

Figure 9. Permeability-increasing radius and internal friction angle of borehole surrounding rock.

When the plastic zone is circular or elliptical, the impact of cohesion or internal friction angle on its distribution is generally small. The capacity of the plastic zone to keep its original form rapidly reduces with increasing cohesion or internal friction angle, and it eventually approaches the empty surface of the borehole's surrounding rock. As illustrated in Figure 6a,b and Figure 7a,b. The four butterfly leaves deteriorate fast when the plastic zone is butterfly-shaped. The smaller the cohesiveness or internal friction angle, the more sensitive to the degradation rate of the four butterfly leaves in the plastic zone. As illustrated in Figures 6c and 7c.

From the analysis of Figures 8 and 9, it is found that when the plastic zone is circular or elliptical, the sensitivity of the permeability-increasing radius to the cohesion or internal friction angle is relatively weak. Additionally, it decreases approximately linearly with the increase of the two. When the cohesion and internal friction angle increase to a certain value, the permeability-increasing effect of the plastic zone is almost zero. For the butterfly plastic zone, when the cohesion and internal friction angle is small, the permeability-increasing radius is very sensitive to the two and increases nonlinearly with the decrease of the two. It's very beneficial to gas drainage. As the cohesion and internal friction angle continue to increase, the sensitivity of the permeability-increasing radius to the two is gradually weakened. The permeability-increasing radius is continuously reduced. The range of gas flow channels is gradually reduced. The surrounding rock gradually loses the permeability-increasing effect.

Therefore, the increase in coal strength has a certain inhibitory effect on the formation of gas flow channels. Theoretically proves the fact that a hard coal body is extremely unfavorable to efficient gas extraction.

5.3. *Effect of Gas Pressure on the Permeability-Increasing Circle of the Borehole* Surrounding Rock

Gas pressure is an important parameter for determining the flow and occurrence state of a coal seam, and it is a key factor in the study of coal and gas outbursts and gas extraction. As it affects the strength and solidity of coal to varying degrees, and the surrounding rock plastic zone exhibits an obvious gas pressure effect [19]. Figure 10 depicts the relationship between the distribution pattern of the plastic zone of the borehole's surrounding rock and the gas pressure when all other influencing factors are held constant. Figure 11 depicts the relationship between the perimeter rock permeability-increasing radius and the gas pressure. When the plastic zone is circular or oval, the gas pressure has minimal influence on how the plastic zone is distributed. As seen in Figure 10a,b, the plastic zone retains its initial form and gradually expands with increasing gas pressure. When the plastic zone is butterfly-shaped, the impact of gas pressure on it is likewise expressed as the growth of four butterfly leaves. As illustrated in Figure 10c.

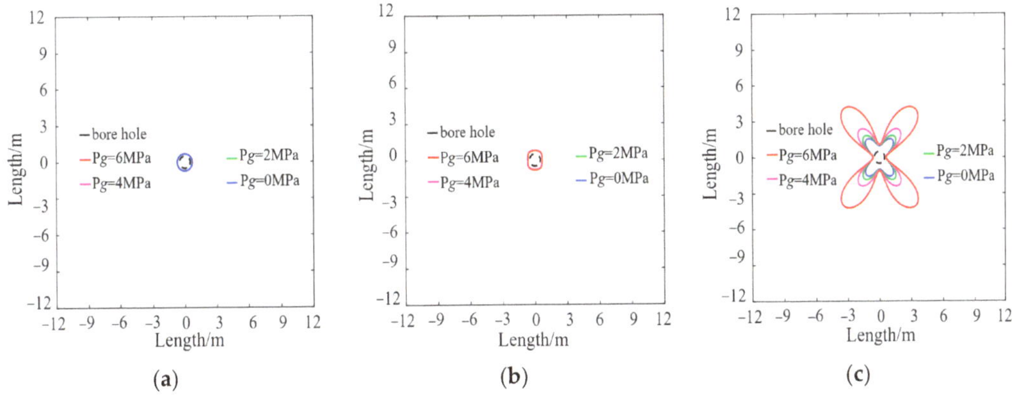

Figure 10. Distribution patterns of plastic zones in surrounding rocks of borehole holes under different gas pressures. (**a**) Round (**b**) Oval (**c**) Butterfly.

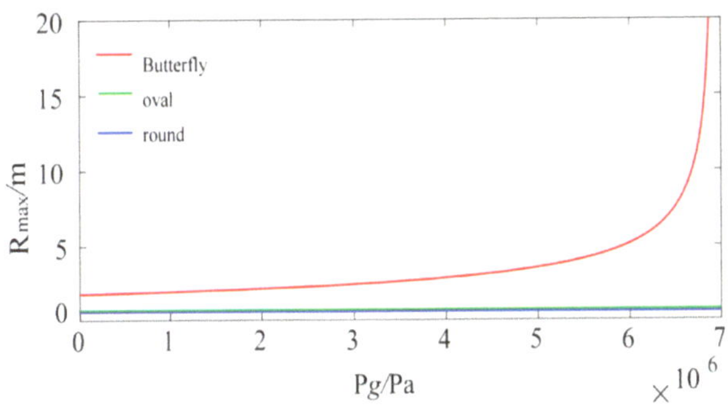

Figure 11. The permeability-increasing radius of the borehole surrounding rock and gas pressure.

The higher the gas pressure, the faster the butterfly leaf grows. The sensitivity of the surrounding rock's plastic zone permeability-increasing growing radius to gas pressure is rather weak when it is circular or elliptical. It grows approximately linearly and slowly with rising gas pressure. For the butterfly plastic zone, when the gas pressure is small, the sensitivity of the permeability-increasing radius to the gas pressure is weak. As the gas pressure continues to increase, this sensitivity increases rapidly. The increase mode changes from approximately linear to nonlinear, as shown in Figure 11.

5.4. *Effect of Hole Diameter on the Permeability-Increasing Circle of the Borehole* Surrounding Rock

The hole diameter of gas extraction has a significant impact on the permeability-increasing radius of the surrounding rock. Increasing the hole diameter is a crucial step toward increasing the rate of gas extraction. Figure 12 depicts the distribution pattern of plastic zones with borehole radii of 0.1 m, 0.3 m, 0.5 m, and 0.7 m. Figure 13 depicts the link between the surrounding rock permeability-increasing radius and the borehole radius.

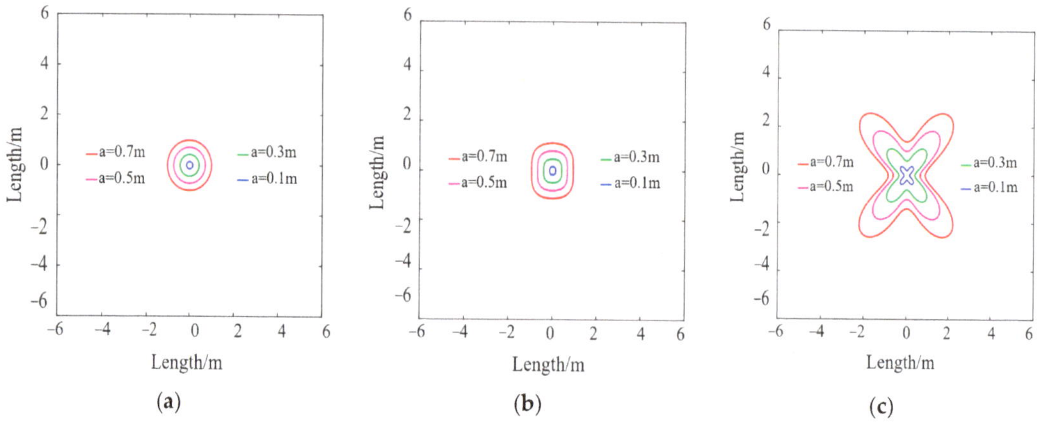

Figure 12. Distribution of plastic zones in surrounding rocks of boreholes with different diameters. (**a**) Round (**b**) Oval (**c**) Butterfly.

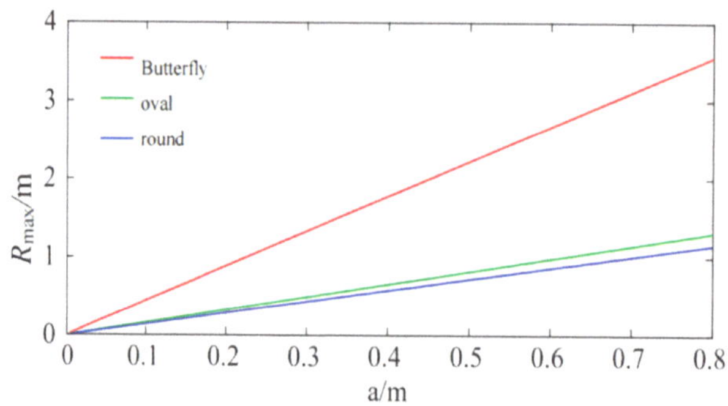

Figure 13. The permeability-increasing radius of borehole surrounding rock and borehole radius.

If the confining pressure ratio, coal strength, and gas pressure remain constant, changing the hole diameter will not change the distribution pattern of the plastic zone. However, it will affect its size, and the original distribution pattern of the plastic zone will increase with increasing hole diameter. As shown in Figure 12. According to the study of Figure 13, there is a linear positive association between the surrounding rock's permeability-increasing radius and the hole diameter, which means that the permeability-increasing radius grows linearly as the hole diameter increases. When the mechanical qualities of coal are constant, and the gas pressure is constant, the confining pressure ratio dictates the rate of expansion of the surrounding rock's permeability-increasing radius with hole diameter. The permeability-increasing radius of the high surrounding rock ratio (such as the butterfly plastic zone) grows at a faster pace than the tiny confining pressure ratio (such as the circular plastic zone). When conditions allow, the hole diameter should be expanded as much as feasible to widen the permeability-increasing radius and generate a larger range of gas flow channels.

6. Construction Method and Engineering Implementation Significance

6.1. Channel Structure Construction Method of Honeycomb Gas Flow Network

The major technological means of gas control are boreholes and extraction. The effective extraction radius is a significant basis for the design of the extraction borehole, which directly influences the extraction effect. According to the theoretical analysis shown above, the permeability-increasing radius of the borehole's surrounding rock is directly connected to the confining pressure ratio, coal strength, and other parameters. When the plastic zone of the borehole's surrounding rock is round or oval, the impact of coal strength, gas pressure, and hole diameter on the perimeter is small. When the plastic zone of the borehole's surrounding rock is butterfly-shaped, the permeability-increasing circle is more sensitive to changes in coal strength, gas pressure, and hole diameter. The lower the coal strength or, the higher the gas pressure, the greater the sensitivity. Under the actual engineering geological conditions, the stress environment of coal seam surrounding rock, coal strength and gas pressure has existed objectively, which are not controllable human factors. As controllable factors, hole diameter and hole spacing usually expand the permeability-increasing range by increasing hole diameter and reducing hole spacing so as to realize gas flow and efficient extraction.

Although increasing the hole diameter can raise the permeability-increasing radius of the surrounding rock and enhance the gas extraction rate to some extent, it also increases the building complexity, expense, and hazard. As a result, scientific and acceptable hole spacing has become the primary technical solution to increase the gas extraction rate. However, its layout is blind and lacks scientific theoretical direction. If the hole spacing is too big, it is simple to produce a blind extraction region. If the hole spacing is too narrow, the permeability-increasing range overlaps, resulting in excessive waste of human and material resources. Based on the foregoing considerations, a honeycomb gas flow network channel structure was established based on the distribution pattern of the borehole's surrounding rock. Additionally, the scientific and reasonable gas extraction hole spacing was determined by combining the coal seam stress environment and the mechanical properties of the coal body. The gas flow network channel structure conducive to gas extraction is created through the expansion and connection of the borehole's surrounding plastic rock zone. The connection of gas enrichment areas is realized. The flow of coal seam gas is promoted. The best gas extraction effect is achieved.

The general idea of the Structure of a Honeycomb Gas Flow Network Channel (SHGFNC) is based on the distribution of plastic zone in the borehole surrounding rock, and the key to forming this structure is to determine the scientific and reasonable hole spacing. If the hole spacing is too small, the plastic zones of the surrounding rock of each borehole overlap with each other, which increases the cost of gas extraction. When the hole spacing is too large, the plastic zones of the surrounding rock of each borehole are independent of each other. It will form an independent channel structure of gas flow with a limited permeability-increasing range (as shown in Figure 14), which is obviously not conducive to the flow of gas between the surrounding rocks of the borehole.

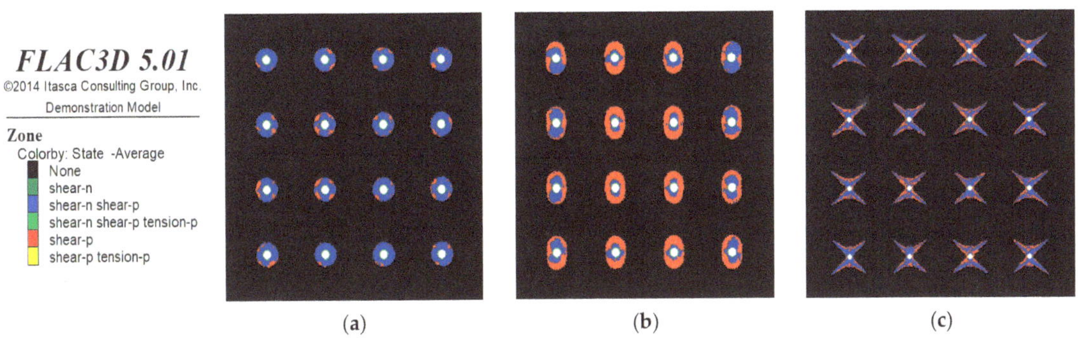

Figure 14. Independent structure for gas flow channel. (**a**) Circular plastic zone (**b**) Oval plastic zone (**c**) Butterfly plastic zone.

Scientific and reasonable gas drainage hole spacing should enable the plastic zone of each borehole surrounding rock to expand and form a honeycomb gas flow network channel structure, which promotes the flow of gas in the structure. It realizes the interoperability of gas flow in each borehole surrounding rock during gas drainage. Additionally, reduce the overlap of plastic zone expansion and reduce the cost of gas drainage. For example, for the extraction of coal seam gas in a uniform or approximately uniform stress environment, at the initial stage of borehole unloading. A large amount of elastic energy stored in the surrounding rock mass is released. The plastic zone of the surrounding rock of each hole expands independently. After that, due to the strong coupling and pressure relief between the surrounding rocks of adjacent boreholes [20], the plastic zone of each hole expands to the adjacent plastic zone in a convex shape in the horizontal and vertical directions, forming an approximate honeycomb gas flow network channel structure, as shown in Figure 15a,b. For the borehole extraction of coal seam gas in a high deviatoric stress environment, the plastic zone expansion of borehole surrounding rock is mainly subjected to high deviatoric stress. The formation of the honeycomb gas flow network channel structure is through the expansion of butterfly leaves in the plastic zone of the surrounding rock of each borehole, as shown in Figure 15c. In the test of Puxi Mine, the expansion and permeability-increasing of the plastic zone of the surrounding rock can be effectively realized when the borehole spacing is 5 m. Too small or too large hole spacing will produce unsatisfactory results, increase costs or affect the gas circulation effect.

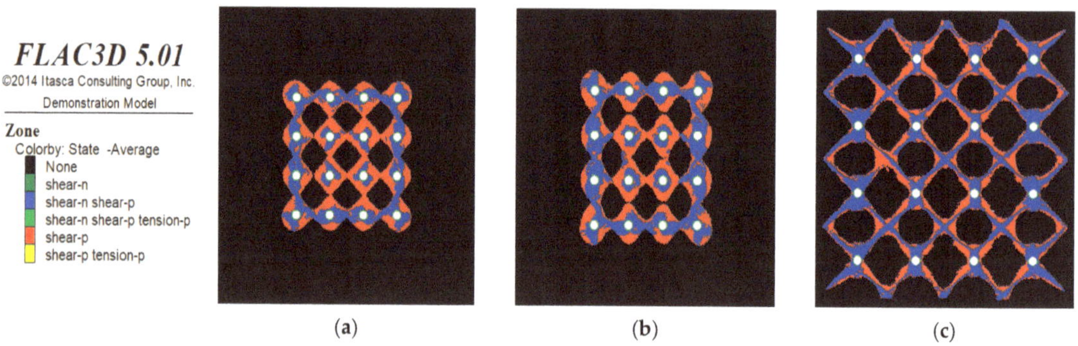

Figure 15. Structure of honeycomb gas flow network channel. (**a**) Circular plastic zone (**b**) Oval plastic zone (**c**) Butterfly plastic zone.

6.2. SHGFNC Project Realization Significance

In summary, the existence of butterfly plastic zone in high deviatoric stress environments can effectively expand the permeability-increasing radius of borehole surrounding rock. Under the condition of constant coal strength, gas pressure and borehole diameter, the stress environment can produce butterfly plastic damage. It can be created by coordinating mining engineering activities. A honeycomb network channel structure conducive to gas flow can be formed to promote the directional flow of gas. For example, the protective layer mining combined with the gas borehole extraction of the protected layer is recognized as the most effective outburst coal seam gas prevention and control technology. The pressure relief of the protective layer mining causes the vertical stress to decrease and the horizontal stress to increase in the adjacent protected coal seam. The stress field is transformed from uniform to non-uniform [21,22]. By laying scientific and reasonable gas extraction boreholes in the protected layer, a honeycomb gas flow network channel structure is established to achieve efficient extraction of coal seam gas.

7. Engineering Test

Jia He Mining Co, Ltd.'s (Hong Kong, China) Pu Xi Jing mine is located in Jia He County, Chen Zhou City, Hunan Province, with an annual production capacity of 150,000 tons. The main mining V coal. Coal seam has a thickness of 0–13 m; the average is 2.1 m. Its strength is low. The gas permeability-increasing is poor. The gas pressure is high. The extraction rate is low. Since the mine was put into operation, there have been 106 outstanding accidents.

Additionally, the gas disaster has been serious. The geological structure of Pu Xi Jing is complex: folds are well-developed, and the structural stress is strong. Vertical principal stress is 5.4 Mpa; Horizontal maximum principal stress is 15.4 Mpa.

In order to quickly eliminate the outstanding danger and improve the gas extraction rate, based on the channel structure theory of cellular airflow networks and the actual situation at Jia He Mine Pu Xi Well, 2454 floor lanes were selected as the gas extraction borehole layout. The hole spacing is 5 m, and the borehole diameter is 87 mm. Additionally, the boreholes are connected to the pipe network for extraction. After 50 days of extraction, the effect test of the gas content of the coal seam on the 2454 working face was carried out, and the results are shown in Table 1 and Figure 16. The steps of the test are as follows.

Table 1. Evaluation of extraction effect.

Evaluating Indicator	Critical Exponent	Targeted Value	Outburst Elimination Effect
residual gas content/(m^3 t^{-1})	5.4	0.93~2.86	useful
residual gas pressure/MPa	0.74	0.11~0.29	useful
gas extraction rate/%	40	72.4	useful

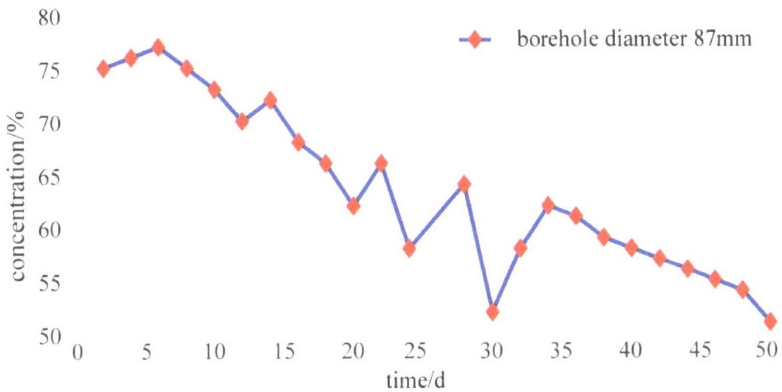

Figure 16. Gas concentration change curve.

(1) Construction extraction drilling. Drilling extraction is carried out at the selected location, and different apertures and spacings are selected for construction.
(2) Sealing and connecting pipe. The hole is sealed with a pre-prepared stopper or polyurethane, and the mixed liquid of cement mortar, expansion agent and accelerator is slowly injected into the hole with a grouting pump for sealing. After the construction of the extraction borehole is completed, the borehole is sealed in time, and the connecting pipe is connected in time.
(3) Observation data. The data of extraction parameters (negative extraction pressure, extraction concentration, mixed flow, etc.) were observed and recorded every day.

From the analysis of Table 1, the expansion and connection of the plastic zone of the surrounding rock of the borehole can be effectively realized when the borehole spacing is 5 m. This can create a gas flow network and channel structure conducive to gas extraction when the residual gas content and gas pressure are all within the standard range, the gas extraction rate is as high as 72.4%, and the gas concentration on the working face does not exceed the limit. It achieves the purpose of de-prominence. The field extraction effect shows that the gas elimination result is good.

8. Conclusions

(1) The implicit boundary equation of the borehole's surrounding rock is obtained. The two-way confining pressure ratio determines the distribution pattern and range of the plastic zone of the borehole surrounding rock. The distribution range of the plastic zone is affected by coal strength, gas pressure, and hole diameter. The distribution patterns of the plastic zone are mostly circular, elliptical, and butterfly.

(2) As coal strength decreases and gas pressure and hole diameter grow, the permeability-increasing circle of borehole surrounding rock widens. The extension of the permeability-increasing ring promotes gas flow.

(3) A honeycomb gas flow network channel structure based on the distribution morphology of the borehole's surrounding rock is presented. Realize the linkage of the coal seam gas flow enrichment region and promote gas flow communication in the surrounding rock of each borehole.

(4) A honeycomb gas flow network channel structure may be formed by the borehole's surrounding rock with butterfly, oval, and round plastic zones. The butterfly form is linked by the extension of the butterfly leaf in the plastic zone of the surrounding rock of the neighboring borehole, while the circular and oval plastic zones are stretched horizontally and vertically in a convex shape.

(5) The engineering test results demonstrate that the honeycomb gas flow network channel construction based on the plastic zone distribution pattern can effectively increase the gas extraction rate. It can help with borehole spacing design, position selection, and determining the direction for optimal gas extraction.

Author Contributions: Conceptualization, C.Y.; methodology, C.Y.; text editing, correction, L.-X.L.; numerical simulation, J.-J.W. and L.F.; engineering experiments, Y.-Y.X.; normal analysis, C.Y.; writing—preparation of the original, C.Y.; writing-review and editing, C.Y. All authors have read and agreed to the published version of the manuscript.

Funding: National Natural Science Foundation of China (52274080), Natural Science Foundation of Hunan Province (2021JJ40211), Hunan Provincial Department of Education Outstanding Youth Fund Project (21B0486), The Open Foundation of Work Safety Key Lab on Prevention and Control of Gas and Roof Disasters for Southern Coal Mines (E22319).

Institutional Review Board Statement: Not applicable.

Informed Consent Statement: Not applicable.

Data Availability Statement: Not applicable.

Conflicts of Interest: The authors declare no conflict of interest.

Abbreviations

P	pore gas pressure at r
K	permeability coefficient
σ'_{ij}	the effective stress tensor
σ_{ij}	the stress tensor generated by external load
a	the effective stress coefficient
δ_{ij}	the Kronecker symbol
$\tau_{r\theta}$	the shear stress of a point in the borehole's surrounding rock in polar coordinates
θ	the point's polar angle coordinate
c	ohesion(MPa)
φ	the angle of internal friction (°)

References

1. Liu, S.; Ma, G.; Lu, J.; Lin, B. Relative pressure determination technology for effective radius found on gas content. *J. China Coal Soc.* **2011**, *36*, 1715–1719.
2. Chen, J.; Tao, K.; Zeng, B.; Liu, L.; Zhao, H.; Zhang, J.; Li, D. Experimental and numerical investigation of the tensile performance and failure process of a modified Portland cement. *Int. J. Concr. Struct. Mater.* **2022**, *16*, 61. [CrossRef]
3. Yuan, C.; Cao, L.; Wang, W.; Fan, L.; Huang, C. Case Study on Rock Support Technology for Roadways Based on Characteristics of Plastic Area. *KSCE J. Civ. Eng.* **2021**, *25*, 705–723. [CrossRef]
4. Lou, J.; Gao, F.; Li, J.; Yuan, G.; Sharifzadeh, M. Effect of dynamic loading conditions on the dynamic performance of MP1 energy-absorbing rockbolts: Insight from laboratory drop test. *Int. J. Min. Sci. Technol.* **2023**, *33*, 215–231. [CrossRef]
5. Guo, D.; Zhao, J.; Lu, P.; Zhu, T. Effective fracture zone under deep-hole cumulative blasting in coal seam. *Chin. J. Eng.* **2019**, *41*, 582–590.
6. Huang, B.; Cheng, Q.; Chen, S.; Zhang, T. Study of coal seam outburst mitigation by deep hole hydro-fracturing and shallow hole methane drainage. *J. China Univ. Min. Technol.* **2013**, *42*, 701–711.
7. Chen, J.; Liu, L.; Zeng, B.; Tao, K.; Zhang, C.; Zhao, H.; Li, D.; Zhang, J. A constitutive model to reveal the anchorage mechanism of fully bonded bolts. *Rock Mech. Rock Eng.* **2023**, *56*, 1739–1757. [CrossRef]
8. Zhang, D.; Bai, X.; Yin, G.; Rao, Z.; He, Q. Research and application on technology of increased permeability by liquid CO_2 phase change directional jet fracturing in low-permeability coal seam. *J. China Coal Soc.* **2018**, *43*, 1938–1950.
9. Zhao, L.; Wang, Z.; Sun, J.; Tu, D. Application of permecbility improvment technology with liquid CO2 phase transition fracturing to high gassy and low permeability seam. *Coal Sci. Technol.* **2016**, *44*, 75–79.
10. Chen, J.; Wei, J.; Zhang, G.; Wang, L. Study on permeability improvement technology of the large-diameterdrilling combined with hole protection. *Coal Sci. Technol.* **2018**, *46*, 73–77.
11. Xie, S.; Yang, B.; Zhang, Q.; Song, H.; Suo, H.; Zhang, T. Research on lift pressure system with combined pipeline and intensive drill-hole gas drainage along the layer in low permeability coal seam. *J. Min. Sci. Technol.* **2019**, *4*, 34–40.
12. Zou, S.; Xin, S. Effective extraction radius of gas drilling in coal seam. *China Saf. Sci. J.* **2020**, *30*, 53–59.
13. Liu, H.; Cheng, Y.; Chen, H.; Kong, S.; Xu, C. Characteristics of mining gas channel and gas flow properties of overlying stratum in high intensity mining. *J. China Coal Soc.* **2012**, *37*, 1437–1443.
14. Zhang, Y.; Xu, L.; Liu, K.; Li, Y.; Zhang, B.; Li, W. Formation mechanism and evolution laws of gas flow channel in mining coal and rock. *J. China Coal Soc.* **2012**, *37*, 1444–1450.
15. Xu, Y.; Li, C.; Zhang, Y.; Wang, H.; Wei, W. Unloading damage evolution and drainage verification of gas flow channel in different mining heights. *J. China Coal Soc.* **2018**, *43*, 2501–2509.
16. Ma, N.; Li, J.; Zhao, X.; Zhao, Z.; Liu, H.; Jia, H. High quality gas channel and its construction method applied to coal and gas in deep mining. *J. China Coal Soc.* **2015**, *40*, 742–748.
17. Ma, N.; Guo, X.; Zhao, X.; Li, J.; Yan, Z. Theoretical analysis and application about permeability-increasing radius of drilling for simultaneous exploitation of coal and gas. *J. China Coal Soc.* **2016**, *41*, 120–127.
18. Ma, H.; Wang, W.; Li, C.; Shi, W.; Yuan, A.; Liu, W. Evolution of crack field and formation mechanism of advantage gas channels in superposition mining. *China Saf. Sci. J.* **2017**, *27*, 133–138.
19. Zhang, J. Experimental study on influence of gas pressure on strength of coal body under adsorption condition. *Saf. Coal Mines* **2020**, *50*, 26–29.
20. Ge, D.; Li, D.; Jiang, F.; Wang, C.; Chen, Y.; Dong, C.; Sun, S.; Yang, G. Reasonable pressure-relief borehole spacing in coal of different strength. *J. Min. Saf. Eng.* **2020**, *37*, 578–593.
21. Guo, H. Analysis on spatial and temporal characteristics of outburst prevention effect in protective seam mining. *China Saf. Sci. J.* **2019**, *29*, 108–113.
22. Chen, Y.; Wu, H.; Zhang, M.; Wu, Y.; Zhang, H.; Zhang, G. Research on the upper protective coal seam mining effect induced by coal thickness and interburden rock properties. *J. Min. Saf. Eng.* **2016**, *33*, 578–584.

applied sciences

MDPI

Study Protocol

Application of Research on Risk Assessment of Roadway Roof Falls Based on Combined Weight Matter Element Extension Model

Shenggang Wang [1], Chao Yuan [1,2,3,*], Lianxin Li [1], Xiaowei Su [1] and Chao Wang [1]

[1] School of Resources & Environment and Safety Engineering, Hunan University of Science and Technology, Xiangtan 411201, China; 18470323207@163.com (S.W.); 13210693127@163.com (L.L.); 15117451792@163.com (X.S.); 18772491469@163.com (C.W.)
[2] Work Safety Key Lab on Prevention and Control of Gas and Roof Disasters for Southern Coal Mines, Hunan University of Science and Technology, Xiangtan 411201, China
[3] Hunan Key Laboratory of Safe Mining Techniques of Coal Mines, Xiangtan 411201, China
* Correspondence: yuanchaozh1@126.com

Abstract: Roof falls in coal mine roadways are the main causes of many casualties, shutdowns and production plan delays. To understand the relationship between the influencing factors of roadway roof fall accidents and the importance ranking of the accidents, we will reduce safety accidents in coal mines. To enable the timely prediction and control of roadway roof fall risks, based on the investigation of many roadway roof fall risk factors, 12 evaluation indexes such as the roadway roof rock thickness, geological conditions and roadway section shape were selected. An evaluation index system of roadway roof fall risks is constructed. A risk degree standard of roadway roof falls is proposed. The risk evaluation model of roadway roof falls was established by using the combination weight of the analytic hierarchy process (AHP), entropy weight method (EW) and matter element extension theory. According to the principle of the maximum membership degree, the risk degree of roadway roof falls is determined. Based on Java Web, a risk assessment system for roadway roof falls was developed. We name the system Multiple Weight-Material Element Web (MW-MEW). The MW-MEW system was used to evaluate the risk degree of roof falls in the C9 return airway of the Xingu Coal Mine. Compared with the evaluation results of the AHP matter element extension model, it is found that the evaluation results of the MW-MEW system are more in line with the actual engineering conditions. The successful application of the MW-MEW system will provide new avenues for the quantitative evaluation of roof fall risks in coal mine roadways.

Keywords: roadway roof fall risk; analytic hierarchy process; entropy weight method; matter element extension

1. Introduction

Coal is an indispensable resource for all countries in the world, and plays a role in promoting economic and social development and ensuring energy security [1,2]. Roadway roof falls are common safety accidents in coal mining. They have a high frequency of occurrence and cause great harm, causing a lot of fear for staff [3,4]. There are many factors affecting roof falls. These factors sometimes lead to accidents independently. Sometimes they combine with each other to cause accidents. Some factors may contain implicit elements that are not readily apparent and easily overlooked, yet they play a crucial role. More importantly, roof falls may be influenced by a combination of hidden factors, which in turn can be influenced by multiple factors simultaneously. There may be a complex network relationship between these factors. This has led to long-term research, but the reasons for collapses are still complex and realistic results are yet to be reported. Therefore, it is necessary to correctly evaluate the safety of roadway roof falls and the timely prevention and handling of the risk of roadway roof falls. This can reduce the occurrence of such accidents and ensure the safety of personnel and property.

In recent years, researchers at home and abroad have carried out a lot of research on the risk assessment of roadway roof collapses, and have reported a lot of important research results. Xing et al. [5] calculated the entropy weight and grey correlation degree of roof disasters in coal mine roof accidents by using the entropy weight and grey correlation analysis methods. Roof disasters during construction are divided into three levels, and countermeasures to prevent roof accidents have been put forward. Liu et al. [6] proposed an evaluation method based on expert knowledge to evaluate the safety of coal mine roofs, the risk degree of roof events and the investment benefit of roof emergency plans. This method uses the expert investigation method to analyze the main influencing factors of the occurrence of coal mine roof events, which provides a fuzzy comprehensive evaluation of the safety of roof accident events. Mahdevari et al. [7] established a neural–fuzzy hybrid model to approximate the nonlinear relationship between the maximum displacement of roadway roof subsidence and geomechanic characteristics. It is found that the threshold of roof subsidence is 66%. Maiti et al. [8] established a relative risk model for fatal roadway roof fall accidents. Potential fatalities (PFs), relative risk of fatalities (RRF) and safety measure effectiveness (SME) are key indicators of mine safety performance. They have been successfully applied in 292 underground coal mines in India. Yao [9] selected six evaluation indexes, including the thickness of the loosening ring, amount of roof separation, stress–intensity ratio, construction quality of the anchor rod (cable), water content of the roof and crack length. The entropy weight method is used to determine the weight of each index. The matter element extension theory is used to establish the risk assessment model of deep roadway roof falls. It is found that the risk level of roof falls has an obvious correlation with roof deformation. Miao et al. [10] built a mine gas safety situation prediction model using natural language processing technology. An accurate management and control information system based on data mining technology and intelligent technology was developed. The ability and efficiency of mine roadway roof safety management are improved. Brook et al. [11] applied the two methods of coal mine roadway roof rating (CMRR) and rock mass rating (RMR), respectively, to the safety evaluation of underground coal mine roadway roofs. They found that the two techniques showed consistency in roadway roof evaluation in most areas of the mine. But there are differences in the overall rating of complex geological regions. Chen et al. [12] evaluated the roof fall risk of stratified rock mass roadways. The evaluation model of tunnel roof falls based on the analytic hierarchy process (AHP) and Technique for Order Preference by Similarity to Ideal Solution (TOPSIS) was established. A comprehensive evaluation index system for roof collapses of roadways is proposed. The weight of each index is calculated by the combination of the analytic hierarchy process (AHP) and expert validity matter elements. Therefore, the influence of human factors on the evaluation results is reduced. Young et al. [13] took two underground coal mines in the western United States as engineering cases. The main parameters affecting roof falls of roadways are analyzed. The prediction index of tunnel roof falls was put forward. Tiejun et al. [14] aimed to understand the relationship between the influencing factors of roadway roof fall accidents and the importance of accidents according to the characteristics of the expert evaluation language and the semantic network connection between the factors. It is proposed to embed the cloud model into the ANP (analytic network process) to solve the above problems. The construction process and analysis steps of the cloud ANP are given.

The underground geomechanical environment of coal mines is complex and changeable, so the risk assessment of roadway roof falls should involve many factors [15–17]. However, the above evaluation method is too simple to correctly judge the actual situation of roof falls in coal mine roadways. It is necessary to study it further. Through the comprehensive analysis of various factors, the risk evaluation method of roadway roof falls in line with objective reality is obtained. Therefore, in view of the evaluation of the risk of roof falls in coal mine roadways, this paper uses the combination weight of the analytic hierarchy process (AHP), entropy weight method (EW) and the matter element extension theory to establish the risk evaluation model of roadway roof falls. According to the principle of the maximum membership degree, the risk degree of roadway roof falls is determined. Based on Java Web, a roadway roof fall risk assessment system was developed. We name the system Multiple Weight-Material Element Web (MW-MEW). The system cannot only avoid the subjectivity of an individual evaluation but can also effectively solve the problem of single factors affecting the risk evaluation of roadway roof falls. It provides a new avenue for the study of the cost of risks of roof falls in coal mine roadways.

2. Establish Evaluation Index System

Roof collapses in coal mining are caused by the interaction of many factors. This paper performs an analysis of the main factors. Based on the investigation of a large number of risk factors of roadway roof falls, the relevant factors affecting the risk assessment of roadway roof falls were extracted. A risk evaluation system U of roadway roof falls is established. An analysis method is used to divide the factors of roadway caving risk assessments according to the nature of the influencing factors and the engineering characteristics. Among them, the criterion layer includes the roadway structure characteristics U_1, roadway roof characteristics U_2, and roadway engineering conditions U_3, and the indicator layer includes 12 indicators. See Figure 1 for details. The overview of the indicators is shown in Table 1.

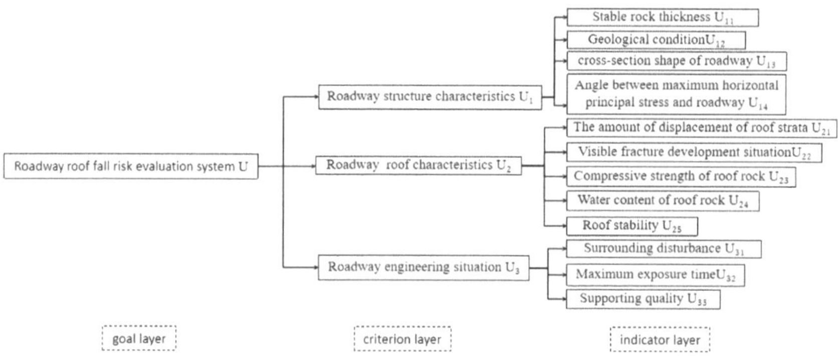

Figure 1. Risk evaluation index system of roadway roof falls.

Table 1. Overview of roadway roof fall risk evaluation index.

Evaluating Indicator	Overview of Indicators	Index Property
Stable rock thickness U_{11}/m	The roadway bolt (cable) support needs to be anchored in a stable rock stratum with a certain thickness. In order to ensure the anchoring effect, the stable rock thickness is one of the direct influencing factors.	Quantification
Geological condition U_{12}	The geological conditions of the roadway strata include the buried depth of the roadway, the tectonic stress, the fault of the surrounding rock of the roadway and the groundwater condition. The more complex the geological conditions, the more easily the roadway is deformed.	Characterization
Cross-section shape of roadway U_{13}	The shape of the roadway section is one of the key factors affecting the stability of the surrounding rock. The cross-section shape will affect the stress distribution of the surrounding rock of the roadway, resulting in different stress states. In general, the stability of the surrounding rock of the roadway with a circular cross-section shape is the best.	Characterization

Table 1. *Cont.*

Evaluating Indicator	Overview of Indicators	Index Property
Angle between maximum horizontal principal stress and roadway $U_{14}/°$	With the increase in the angle between the maximum horizontal principal stress and the roadway, the damage degree of the roof and floor of the roadway also increases. When the angle is 90°, the damage to the surrounding rock in the roadway is the most serious.	Quantification
The amount of displacement of roof strata U_{21}/m	The subsidence of roof strata is an observation index clearly stipulated in the "Coal Mine Safety Regulations". After it exceeds the critical deformation value, physical phenomena such as roof falls occur in the roadway. It is obtained by field measurement.	Quantification
Visible fracture development situation U_{22}/pcs	The development of fissures reflects the stability of the roof, which is accurately monitored by the roof peeper.	Quantification
Compressive strength of roof rock U_{23}/MPa	The tensile strength of the roof rock characterizes the bearing capacity of the roof. Due to the action of gravity, the surrounding rock in the shallow area of the roof is subjected to tensile stress. If the tensile stress exceeds the tensile strength of the rock, a large number of cracks will appear in the roof rock.	Quantification
Water content of roof rock U_{24}	Groundwater will weaken the mechanical properties of rock mass. This leads to physical and chemical changes in the mechanical properties of the roof, reducing the roof's load bearing capacity.	Characterization
Roof stability U_{25}	The stability of coal mine roadway roofs directly affects the risk of roof falls. The better the stability of the roof, the lower the risk of roof falls.	Characterization
Surrounding disturbance U_{31}	The greater the disturbance stress around the roadway, the more easily the roof of the roadway falls.	Characterization
Maximum exposure time U_{32}/d	The longer the roadway roof is exposed to air, the lower the strength of the exposed rock mass, and the worse the stability of the roof.	Quantification
Supporting quality U_{33}	The quality of the support directly reflects the stability of the roadway roof. The better the support quality, the better the stability of the roof.	Characterization

3. Calculate Index Weights

3.1. Analytic Hierarchy Process (AHP) to Calculate the Weight

The analytic hierarchy process (AHP) [18] first divides the nature of the analysis object into related factors. Then, the related factors are classified to form a multi-level structure model. The relative importance of each factor is judged by experience or experts. According to the importance degree, the weight is obtained, and the characterization analysis is transformed into a quantitative analysis. The specific calculation steps are as follows:

(1) Construct judgment matrix:

The 1–9 scale method [19] was used to determine the importance of the evaluation index elements at each level. The judgment matrix expression (1) is established.

$$A = \left(a_{ij}\right)_{n\times n} \tag{1}$$

a_{ij} is the relative importance of risk factor i to risk factor j at this level.

$$\mathrm{a_{ij}} = \frac{1}{\mathrm{a_{ji}}}, \mathrm{a_{ii}} = 1, (\mathrm{i} = 1, 2, \ldots, \mathrm{n}), (\mathrm{j} = 1, 2, \ldots, \mathrm{n})$$

(2) Calculate weight:

After establishing the judgment matrix, the weight of each index is calculated by Expression (2). Common calculation methods include the feature method and root method. The root method is used in this paper.

$$G_i' = \frac{\overline{G_i'}}{\sum_{i=1}^{n} \overline{G_i'}} \quad (2)$$

G'_i is the weight of the ith evaluation index after unitization.

a_{ij} is the element in the judgment matrix A.

In Expression (2), there is Expression (3).

$$\overline{G_i'} = \sqrt[n]{\prod_{i=1}^{n} a_{ij}} \quad (3)$$

(3) Calculate the maximum eigenvalue:

$$\lambda \frac{1}{n} \sum_{i=1}^{n} \left[\frac{(AG')_i}{G_i'} \right]_{max} \quad (4)$$

λ_{max} is the maximum eigenvalue

(4) Check consistency:

After the consistency index C_I is calculated, the random consistency index (Table 2) of the judgment matrix is checked. The value of consistency index R_I is obtained. Then, the test coefficient C_R of the matrix is calculated.

$$C_R = \frac{C_I}{R_I} \quad (5)$$

C_R is the test coefficient of the matrix.

C_I is the consistency index.

R_I is a random consistency index.

In Expression (5),

$$C_I = \frac{\lambda_{max}}{n-1} \quad (6)$$

Table 2. Random consistency index.

Order	1	2	3	4	5	6	7	8	9
R_I	0.00	0.00	0.58	0.9	1.12	1.24	1.32	1.41	1.45

If $C_R < 0.1$, the constructed judgment matrix passes the consistency test. In contrast, if the consistency test is not passed, the judgment matrix needs to be reconstructed and calculated.

(5) Obtain index weight:

The weight of the index layer to the criterion layer is multiplied by the weight of the criterion layer to the target layer. The weight of each index factor to the total target is obtained. If O indices of the target layer U are U_1, U_2 ..., U_O, these weights on P are $G_{W_i} = (G_1', G_2', \ldots, G_O')$. The weight vector of $U_i (i = 1, 2, \ldots, O)$ to its index U_{i1}, U_{i2}, ..., U_{if} is $G_{P_iP_{ij}} = (G_{i1}', G_{i2}', \ldots, G_{if}')$. The following is the weight of $U_{ij} (j = 1, 2, \ldots, f)$ to the target layer U:

$$G_{PP_{ij}} = G_{PP_i} \times G_{P_iP_{ij}} \quad (7)$$

G_{PP_i} is the weight of the criterion layer $U_{xy}(x = 1, 2, \ldots, o)$ to the target layer U.

$G_{P_iP_{ij}}$ is the weight of the index layer $U_{xy}(x = 1, 2, \ldots, o; y = 1, 2, \ldots, l)$ to the criterion layer U_x.

The weight vector of each indicator is put into Expression (7). The weight of each indicator can be calculated.

3.2. Entropy Weight (EW) Method to Calculate the Weight

The entropy weight method [20] is a method of customer weighting. According to the variation degree of each evaluation index, the entropy value is determined. The entropy value is used to determine the weight of each index. If the index variation degree is smaller, the index weight value is smaller. On the contrary, the greater the index weight. The entropy weight calculation steps of m evaluations and n evaluation indexes are as follows:

(1) Standardized treatment:

Based on the risk evaluation index system of roadway roof falls, the matrix $B = (b_{ij})_{m\times n}$ is obtained by standardizing the original matrix $R = (r_{ij})_{m\times n}$ according to Expression (8).

$$b_{ij} = \frac{r_{ij} - r_{min}}{r - r_{minmax}} \quad (8)$$

b_{ij} is the data after the standardization of the original data.
r_{ij} is the ith initial value of the jth index.
r_{jmax} is the maximum initial value of the jth index.
r_{jmin} is the minimum initial value of the jth index.
In the formula, $_i = (1, 2, \ldots, m)$; $_j = (1, 2, \ldots, n)$.

(2) Calculated entropy:

$$H_j = -\frac{1}{\ln m}\left[\sum_{i=1}^{m} e_{ij} \times \ln e_{ij}\right] \quad (9)$$

H_j is the entropy value of the jth index.
In Expression (9),

$$e_{ij} = \frac{1+b_{ij}}{\sum_{i=1}^{m}(1+b_{ij})} \quad (10)$$

b_{ij} is the index data after the standardization of the jth index of the ith thing.
b_{ij} is the index data after the standardization of the j index of the i thing.

(3) Computated entropy weight:

$$Q_j = \frac{1-H_j}{n-\sum_{j=1}^{n} H_j} \quad (11)$$

Q_j is the entropy weight of the jth index $Q_j \in [0,1]$.
Expression (11) satisfies the following conditions:

$$\sum_{j=1}^{n} Q_j = 1 \quad (12)$$

3.3. Calculate Combination Weights

The analytic hierarchy process uses expert experience to calculate the weight of each index with certain subjectivity. The entropy weight law can obtain a completely objective index weight based on the difference in the index data. The two methods of calculating weights have their own limitations. The Lagrange multiplier method [21] is used for weight combination weighting. The weights obtained by the analytic hierarchy process and the weights obtained by the entropy weight method are combined and weighted. This can reflect the importance of each index. To ensure that the weights of each indicator are more objective and reasonable, it can more accurately reflect the weight coefficient W_j of each index in the surrounding rock of the roadway roof.

$$W_j = \frac{\sqrt{G_j \times Q_j}}{\sum_{j=1}^{n}\sqrt{G_j \times Q_j}} \quad (13)$$

W_j is the combined weight of the jth evaluation index $W_j \in [0,1]\ \sum_{j=1}^{n} W_j = 1$.

G_j is the jth evaluation index weight obtained by the analytic hierarchy process.

Q_j is the jth evaluation index weight obtained by the entropy weight method.

In Expression (13), G_j is the weight of each index obtained by the analytic hierarchy process, and Q_j is the weight of each index obtained by the entropy weight method.

4. Construct AHP-EW Combined Weight Matter Element Extension Model

4.1. Determine the Risk Evaluation Set of Roadway Roof Falls

The three-level fuzzy comprehensive evaluation method is used to quantitatively evaluate the risk degree of roadway roof falls from the indicator layer, the criterion layer and the goal layer. Based on the investigation of a large number of roadway roof fall risk factors, the evaluation set of the roadway roof fall risk evaluation grade is divided into five grades from low to high: very low risk (I), low risk (II), medium risk (III), high risk (IV) and very high risk (V). The corresponding meaning of each grade is shown in Table 3.

In order to facilitate the calculation of the correlation degree and comprehensive evaluation grade of each evaluation index, and to ensure that the calculation results are more scientific, the evaluation indexes in Table 3 are dimensionless according to Formula (14). The processing results are shown in Table 4.

$$q'_j = \begin{cases} \frac{q_j - q_{min}}{q_{max} - q_{min}} \rightarrow \text{The bigger the better the factor} \\ \frac{q_{max} - q_j}{q_{max} - q_{min}} \rightarrow \text{The smaller the better the factor} \end{cases} \quad (14)$$

In Expression (14), $q'_j = (1, 2, \ldots, n)$ is the actual value of the evaluation of an evaluation index; q'_j is the evaluation standard value of an evaluation index after standardization; q_{max} is the upper limit of the evaluation value of an evaluation index; q_{min} is the lower limit of the evaluation standard value of an evaluation index.

4.2. AHP-EW Combined Weight Element Extension Model

According to the matter element extension theory [22], the establishment steps of the matter element extension model are as follows:

(1) Determine the classical domain:

The classical domain matrix of the evaluation index can be expressed as a formula.

$$R_{0t} = [N, C_j, V_{0tj}] = \begin{bmatrix} N & C_1 & \langle a_{0t1}, b_{0t1} \rangle \\ & C_2 & \langle a_{0t2}, b_{0t2} \rangle \\ & \vdots & \vdots \\ & C_n & \langle a_{0tn}, b_{0tn} \rangle \end{bmatrix} \quad (15)$$

In the formula, R_{0t} is the classical domain matter element; N is the whole matter element to be evaluated; C_j (j = 1, 2, ..., n) is the evaluation index of the thing N; $V_{0tj} = <a_{0tj}, b_{0tj}>$ is the evaluation index C_j of the classical domain matter element in t (t = 1, 2, ..., g); the value range of the evaluation grade is the classical domain; a_{0tj} is the lower limit of the classical domain; b_{0tj} is the upper limit of the classical domain.

(2) Determine the section domain:

The section domain matrix of the evaluation index is expressed as the following formula:

$$R_q = [N, C_j, V_{qj}] = \begin{bmatrix} N & C_1 & \langle a_{q1}, b_{q1} \rangle \\ & C_2 & \langle a_{q2}, b_{q2} \rangle \\ & \vdots & \vdots \\ & C_n & \langle a_{qn}, b_{qn} \rangle \end{bmatrix} \quad (16)$$

In the formula, R_q is a nodal matter element; $V_{qj} = <a_{qj}, b_{qj}>$ is the value range of N for evaluation indicators C_j (j = 1, 2, ..., n), which is the section field.

Table 3. Risk evaluation index level of roadway roof falls.

Evaluating Indicator	Order of Evaluation				
	I	II	III	IV	V
	Very Low Risk	Low Risk	Intermediate Risk	High Risk	Very High Risk
Stable rock thickness U_{11}/m	6.0~7.5	4.5~6.0	3.0~4.5	1.5~3.0	0~1.5
Geological condition U_{12}	Simple 0~2	Relatively Simple 2~4	Relatively Complicated 4~6	Complicated 6~8	Extremely Complicated 8~10
Cross-section shape of roadway U_{13}	Roundness 8~10	Arch 6~8	Rectangle 4~6	Trapezium 2~4	Miscellaneous 0~2
Angle between maximum horizontal principal stress and roadway U_{14}/°	0~18	18~36	36~54	54~72	72~90
The amount of displacement of roof strata U_{21}/m	0~0.07	0.07~0.14	0.14~0.21	0.21~0.28	0.28~0.5
Visible fracture development situation U_{22}/pcs	0~2	2~4	4~6	6~8	8~10
Compressive strength of roof rock U_{23}/MPa	200~250	150~200	100~150	100~50	0~50
Water content of roof rock U_{24}	No Seepage 0~2	Micro-seepage Water 2~4	Micro-water Exit 4~6	A Little Water Flow 6~8	Heavy Water Flow 8~10
Roof stability U_{25}	Stabilization 8~10	Relatively Stable 6~8	Relative Destabilization 4~6	Destabilization 2~4	Extreme Destabilization 0~2
Surrounding disturbance U_{31}	No Impact 0~2	Minimal Impact 2~4	Small Impact 4~6	Great Impact 6~8	Profound Impact 8~10
Maximum exposure time U_{32}/d	0~5	5~15	10~30	30~60	50~100
Supporting quality U_{33}	Goodliness 8~10	Good 6~8	General 4~6	Poor 2~4	Very Poor 0~2

Table 4. Evaluation index level of roadway roof fall risks without dimensional treatment.

Evaluating Indicator	Order of Evaluation				
	I	II	III	IV	V
	Very Low Risk	Low Risk	Intermediate Risk	High Risk	Very High Risk
Stable rock thickness U_{11}/m	0.8~1.0	0.6~0.8	0.4~0.6	0.2~0.4	0.0~0.2
Geological condition U_{12}	0.0~0.2	0.2~0.4	0.4~0.6	0.6~0.8	0.8~1.0
Cross-section shape of roadway U_{13}	0.8~1.0	0.6~0.8	0.4~0.6	0.2~0.4	0.0~0.2
Angle between maximum horizontal principal stress and roadway U_{14}/°	0.0~0.2	0.2~0.4	0.4~0.6	0.6~0.8	0.8~1.0
The amount of displacement of roof strata U_{21}/m	0.0~0.14	0.14~0.28	0.28~0.42	0.42~0.56	0.56~1.00
Visible fracture development situation U_{222}/pcs	0.0~0.2	0.2~0.4	0.4~0.6	0.6~0.8	0.8~1.0
Compressive strength of roof rock U_{23}/MPa	0.8~1.0	0.6~0.8	0.4~0.6	0.2~0.4	0.0~0.2
Water content of roof rock U_{24}	0.0~0.2	0.2~0.4	0.4~0.6	0.6~0.8	0.8~1.0
Roof stability U_{25}	0.8~1.0	0.6~0.8	0.4~0.6	0.2~0.4	0.0~0.2
Surrounding disturbance U_{31}	0.0~0.2	0.2~0.4	0.4~0.6	0.6~0.8	0.8~1.0
Maximum exposure time U_{32}/d	0.0~0.05	0.05~0.15	0.15~0.30	0.30~0.60	0.60~1.00
Supporting quality U_{33}	0.8~1.0	0.6~0.8	0.4~0.6	0.2~0.4	0.0~0.2

The value range of (n) is the nodal domain; a_{qj} is the lower limit value of the nodal domain; b_{qj} is the upper limit of the node domain.

(3) Determine the matter element to be evaluated:

The matter element to be evaluated R_{ij} (i = 1,2, . . ., m; j = 1, 2, . . ., m) is determined by Equation (17). The evaluation index C_j of the thing N_i to be evaluated is scored. The actual value is recorded as V_{ij}.

$$R_{ij} - [N_I, C_j, V_{ij}] = \begin{bmatrix} N_i & C_1 & V_{i1} \\ & C_2 & V_{i2} \\ & \vdots & \vdots \\ & C_n & V_{in} \end{bmatrix} \tag{17}$$

N_i is something to be evaluated.

C_j is the evaluation index.

V_{ij} is the actual value of the thing to be evaluated N_i in the evaluation index C_j.

(4) Calculate correlation degree:

The correlation function expression of the jth index C_j (j = 1, 2, . . ., n) of the thing to be evaluated N_i (i = 1, 2, . . ., m) about the evaluation grade t(t = 1, 2, . . ., g) is as follows:

$$k_t(V_{ij}) = \begin{cases} \frac{\rho[V_{ij}(t), V_{0tj}]}{\rho[V_{ij}(t), V_{qj}] - \rho[V_{ij}(t), V_{0tj}]}, \rho[V_{ij}(t), V_{qj}] \neq \rho[V_{ij}(t), V_{0tj}] \\ -\rho[V_{ij}(t), V_{0tj}] - 1, \rho[V_{ij}(t), V_{qj}] = \rho[V_{ij}(t), V_{0tj}] \end{cases} \tag{18}$$

In Expression (18),

$$\begin{cases} \rho[V_{ij}(t), V_{0tj}] = \left|V_{ij} - \frac{1}{2}(b_{0tj} + a_{0tj})\right| - \frac{1}{2}(b_{0tj} - a_{0tj}) \\ \rho[V_{ij}(t), V_{qj}] = \left|V_{ij} - \frac{1}{2}(b_{qj} + a_{qj})\right| - \frac{1}{2}(b_{qj} - a_{qj}) \end{cases}$$

$\rho[V_{ij}(t), V_{0tj}]$ is the distance between the index $V_{ij}(t)$ and the classical domain V_{0tj}.

a_{0tj} is the lower limit of the classical domain.

b_{0tj} is the upper limit of the classical domain.

$\rho[V_{ij}(t), V_{qj}]$ is the distance between the index $V_{ij}(t)$ and the segment domain V_{qj}.

a_{qj} is the lower limit of the section domain.

b_{qj} is the upper limit value of the nodal domain.

All the calculated correlation degrees are recorded in the form of a matrix, so the correlation matrix K is the following formula:

$$K = \begin{bmatrix} k_1(V_{i1}) & k_2(V_{i1}) & \cdots & k_z(V_{i1}) \\ k_1(V_{i2}) & k_2(V_{i2}) & \cdots & k_z(V_{i2}) \\ \vdots & \vdots & \cdots & \vdots \\ k_1(V_{in}) & k_2(V_{in}) & \cdots & k_z(V_{in}) \end{bmatrix} \tag{19}$$

$k_t(V_{ij})$ is the correlation degree of the j th index C_j of the thing N_i to be evaluated about the evaluation grade t.

(5) Determine the evaluation level:

$$S(N_i) = WK = (W_1, W_2, \ldots, W_n) \begin{bmatrix} k_1(V_{i1}) & k_2(V_{i1}) & \cdots & k_z(V_{i1}) \\ k_1(V_{i2}) & k_2(V_{i2}) & \cdots & k_z(V_{i2}) \\ \vdots & \vdots & \cdots & \vdots \\ k_1(V_{in}) & k_2(V_{in}) & \cdots & k_z(V_{in}) \end{bmatrix} = (S_1, S_2, \ldots, S_g) \tag{20}$$

K is the correlation degree matrix.

W is the combination weight.

In the formula, W is the combination weight. If max{$S(N_i)$} = S_t, the evaluation grade of N_i is t.

5. Develop Roadway Roof Fall Risk Assessment System

To quickly evaluate the risk of roadway roof falls, the developed roadway roof fall risk assessment system (MW-MEM) is based on a B/S architecture [23], which uses the J2EE technique [24]. We name the system Multiple Weight-Material Element Web (MW-MEW). The system development model is created using the MVC design pattern [25].

5.1. System Development Environment

IntelliJ IDEA 2022.3.2 is the system development platform, using the Java18 version of the programming language development background. The server uses Tomcat8.5. The database is MySQL57.

5.2. System Framework

The system uses the SSM framework for research and development. Spring is an open-source lightweight Java framework, which aims to provide developers with simple development methods. Spring MVC is a role that separates model objects, controllers, dispatchers, and handler objects. This ensures convenience for developers.

5.3. Implementation System

The process of the risk assessment of roadway roof falls developed by the system is shown in Figure 2. After inputting the corresponding values according to the system prompt, the weight of the risk factors of roadway roof falls and the risk assessment of roadway roof falls can be output in the form of a report. This is convenient for relevant personnel to adopt targeted safety support methods. This could reduce the risk of roadway roof caving, ensuring the security of the roadway.

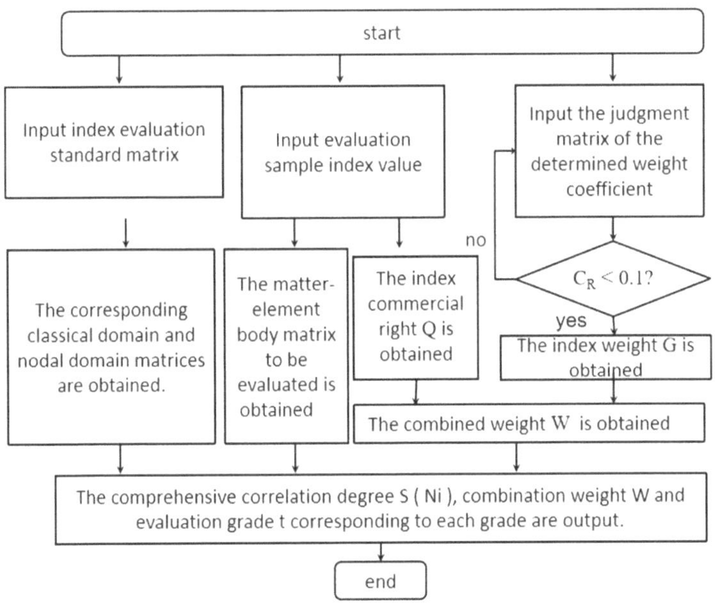

Figure 2. System implementation flow chart.

6. Engineering Application

6.1. Project Overview

The Yunnan Wantian Group Xingu Coal Mine is 33.6 km away from Fuyuan County; it is located in the direction of 187°. The production capacity of the Xingu Coal Mine is 450,000 t/a. It is developed by an inclined shaft. The coal mining method used is the longwall mining method, and the whole height is mined at one time. All caving methods are used to manage the roof. The coal mining technology used is comprehensive mechanized mining, and mainly, six coal seams are mined. In this paper, the risk of roof falls in the original C9 return air contact roadway of the coal mine is evaluated and predicted.

6.2. Source of Data

Based on the risk evaluation system of coal mine roadway roof falls established in Figure 1, the expert scoring table of risk factors of roadway roof falls is compiled. Fifteen senior roadway roof fall research experts were invited to score the qualitative indicators of roadway roof fall risks. The qualitative scoring range is [0, 10]. The score results were averaged. In order to obtain the entropy weight corresponding to each evaluation index of C9 return air connection roadway, the data of eight places in this roadway are obtained according to the geological exploration data. Each place was numbered 1 to 8. The quantitative evaluation index values and the qualitative evaluation index values scored by experts are detailed in Table 5. The roadway number 1 is the 10 m straight line distance between the C9 return air contact lane and the 11,901 central contact lane. The distance between the numbers from north to south is 5 m. According to Formula (8), the index standardization of Table 5 is carried out. The processing results are shown in Table 6.

Table 5. Roof fall index of roadway.

Lane Number	Evaluating Indicator											
	U_{11}	U_{12}	U_{13}	U_{14}	U_{21}	U_{22}	U_{23}	U_{24}	U_{25}	U_{31}	U_{32}	U_{33}
1	5.12	3.66	8.82	19.50	0.08	3.00	210.00	3.56	9.45	2.56	3.00	8.65
2	4.82	4.56	7.89	22.15	0.06	1.00	220.00	4.45	8.92	2.78	6.00	7.43
3	5.86	3.29	7.88	18.92	0.07	2.00	215.00	3.41	7.82	3.15	4.00	6.84
4	6.78	4.89	8.45	20.50	0.08	2.00	202.00	4.26	8.67	3.48	7.00	8.12
5	5.72	3.19	7.09	23.32	0.07	3.00	198.00	4.56	8.09	3.07	6.00	7.68
6	5.85	3.89	8.38	19.85	0.07	1.00	214.00	5.56	9.75	3.56	3.00	7.56
7	4.79	4.57	7.65	24.15	0.08	2.00	222.00	4.85	7.93	4.87	5.00	8.23
8	6.00	5.19	8.58	17.98	0.06	2.00	218.00	3.44	7.05	2.95	6.00	7.88

Table 6. Roadway roof fall index standardization.

Lane Number	Evaluating Indicator											
	U_{11}	U_{12}	U_{13}	U_{14}	U_{21}	U_{22}	U_{23}	U_{24}	U_{25}	U_{31}	U_{32}	U_{33}
1	0.51	0.37	0.88	0.22	0.16	0.30	0.84	0.36	0.95	0.26	0.03	0.87
2	0.48	0.46	0.79	0.25	0.13	0.10	0.88	0.45	0.89	0.28	0.06	0.74
3	0.59	0.33	0.79	0.21	0.13	0.20	0.86	0.34	0.78	0.32	0.04	0.68
4	0.68	0.49	0.85	0.23	0.15	0.20	0.81	0.43	0.87	0.35	0.07	0.81
5	0.57	0.32	0.71	0.26	0.14	0.30	0.79	0.46	0.81	0.31	0.06	0.77
6	0.59	0.39	0.84	0.22	0.14	0.10	0.86	0.56	0.98	0.36	0.03	0.76
7	0.48	0.46	0.77	0.27	0.16	0.20	0.89	0.49	0.79	0.49	0.05	0.82
8	0.60	0.52	0.86	0.20	0.12	0.20	0.87	0.34	0.71	0.30	0.06	0.79

6.3. Calculate Index Weight

(1) Use the AHP method to calculate the weight of the indicators:

Based on the investigation of many risk factors of roadway roof falls, expert opinions and the actual situation in coal mine roadways, the target layer U of the analysis and evaluation system of coal mine roadway roof falls is constructed. The judgment matrices of criterion layers U1, U2 and U3 are as follows:

$$A_U = \begin{bmatrix} 1 & \frac{1}{2} & 1 \\ 2 & 1 & 1 \\ 1 & 1 & 1 \end{bmatrix} \tag{21}$$

$$A_{U_1} = \begin{bmatrix} 1 & 2 & 4 & 3 \\ \frac{1}{2} & 1 & 3 & 2 \\ \frac{1}{4} & \frac{1}{3} & 1 & \frac{1}{3} \\ \frac{1}{3} & \frac{1}{2} & 3 & 1 \end{bmatrix} \tag{22}$$

$$A_{U_2} = \begin{bmatrix} 1 & 1 & 2 & 1 & 1 \\ 1 & 1 & 2 & 1 & \frac{1}{2} \\ \frac{1}{2} & \frac{1}{2} & 1 & 2 & 1 \\ 1 & 1 & \frac{1}{2} & 1 & \frac{1}{3} \\ 1 & 2 & 1 & 3 & 1 \end{bmatrix} \tag{23}$$

$$A_{U_3} = \begin{bmatrix} 1 & 2 & \frac{1}{2} \\ \frac{1}{2} & 1 & \frac{1}{3} \\ 2 & 3 & 1 \end{bmatrix} \tag{24}$$

According to Expressions (2)~(7) and with the help of the developed MW-MEM system, the weight value G of each evaluation index of the roadway roof fall risk is calculated as shown in Table 7. And the above judgment matrices all passed the consistency tests. For all the judgment matrices, $GR < 0.1$. Therefore, the weight distribution is reasonable, and it is not necessary to reconstruct the judgment matrix.

(2) Use the EW method to calculate the weight of each evaluation index:

According to the evaluation index data in Table 6, the matrix $B = (b_{ij})_{8\times 12}$ is constructed. This is Expression (25).

$$B = \begin{bmatrix} 0.51 & 0.37 & \cdots & 0.03 & 0.87 \\ 0.48 & 0.46 & \cdots & 0.06 & 0.74 \\ \vdots & \vdots & \vdots & \vdots & \vdots \\ 0.48 & 0.46 & \cdots & 0.05 & 0.82 \\ 0.60 & 0.52 & \cdots & 0.06 & 0.79 \end{bmatrix} \tag{25}$$

According to Expressions (9)–(11), the data in Table 6 are input into the MW-MEM system. The weight value Q of each evaluation index of the roadway roof fall risk is calculated as shown in Table 7.

(3) Calculate the combination weights:

With the help of the MW-MEM system, according to Expression (13), the combined weight W of each evaluation index is calculated, as shown in Table 7.

6.4. Identify the Classical Domain, Section Domain, and Matter Element to Be Evaluated

This is determined according to the data from Expressions (15)–(17), Table 4 and the actual situation of the Xingu Coal Mine.

(1) Classical matter element R_{0t} for risk assessment of roadway roof falls:

The roof fall risk evaluation matter element R of the C9 return air contact roadway in the Xingu coal mine is composed of the risk level N of roadway roof fall, the evaluation index C and the characteristic value V.

$$R_{0t} = \begin{bmatrix} N & N_1 & N_2 & N_3 & N_4 & N_5 \\ C_1 & (0.8, 1.0) & (0.6, 0.8) & (0.4, 0.6) & (0.2, 0.4) & (0.0, 0.2) \\ C_2 & (0.0, 0.2) & (0.2, 0.4) & (0.4, 0.6) & (0.6, 0.8) & (0.8, 1.0) \\ \vdots & \vdots & \vdots & \vdots & \vdots & \vdots \\ C_{12} & (0.8, 1.0) & (0.6, 0.8) & (0.4, 0.6) & (0.2, 0.4) & (0.0, 0.2) \end{bmatrix} \quad (26)$$

(2) The nodal domain matrix R_q is

$$R_q = \begin{bmatrix} N & C & V \\ & C_1 & (0.0, 1.0) \\ & C_2 & (0.0, 1.0) \\ & \vdots & \vdots \\ & C_{12} & (0.8, 1.0) \end{bmatrix} \quad (27)$$

(3) The matter element volume matrix R_{ij} corresponding to the roadway to be evaluated is

$$R_{ij} = \begin{bmatrix} \text{lane number} & C_1 & L & C_{12} \\ 1 & 0.51 & L & 0.87 \\ M & M & L & M \\ 8 & 0.6 & L & 0.79 \end{bmatrix} \quad (28)$$

Table 7. Index weight.

Goal Layer	Criterion Layer	Criterion Layer Weight	Indicator Layer	Index Layer Weight	AHP-Calculated Weights G	EW-Calculated Weight Q	Combination Weight W
Roadway roof fall risk evaluation system U	Roadway structure characteristics U_1	0.26	Stable rock thickness U_{11}/m	0.46	0.12	0.09	0.11
			Geological condition U_{12}	0.28	0.07	0.10	0.09
			Cross-section shape of roadway U_{13}	0.09	0.02	0.07	0.04
			Angle between maximum horizontal principal stress and roadway U_{14}/°	0.18	0.05	0.09	0.07
	Roadway roof characteristics U_2	0.41	The amount of displacement of roof strata U_{21}/m	0.22	0.10	0.08	0.09
			Visible fracture development situation U_{22}/pcs	0.19	0.08	0.10	0.09
			Compressive strength of roof rock U_{23}/MPa	0.19	0.08	0.08	0.08
			Water content of roof rock U_{24}	0.12	0.05	0.10	0.07
			Roof stability U_{25}	0.28	0.11	0.07	0.10
	Roadway engineering situation U_3	0.33	Surrounding disturbance U_{31}	0.30	0.10	0.08	0.09
			Maximum exposure time U_{32}/d	0.16	0.05	0.10	0.08
			Supporting quality U_{33}	0.54	0.18	0.06	0.11

6.5. Determine the Correlation Degree of Roadway Roof Collapse Risk Evaluation

According to the data in Expressions (18)–(20), Tables 4 and 7, and with the help of the roadway roof fall risk evaluation system, MW-MEW, based on the AHP-EW combined weight matter element extension model, the comprehensive correlation degree of each evaluation index of the roadway roof fall risk is obtained. Take roadway No. 8 as an example, as shown in Table 8. The results of the roof fall risk grade of the C9 return air contact roadway in the Xingu Coal Mine are obtained, as shown in Table 9.

6.6. Roadway Evaluation Result

The evaluation grades of eight roadways are shown in Table 9, and the evaluation results are shown in Figure 3.

Table 8. Roadway No. 8 roof fall index correlation degree.

Roadway Roof Fall Risk Level	U_{11}	U_{12}	U_{13}	U_{14}	U_{21}	U_{22}	U_{23}	U_{24}	U_{25}	U_{31}	U_{32}	U_{33}
I (Very Low Risk)	−0.33	−0.40	0.07	0.00	0.23	0.00	0.08	−0.30	−0.24	−0.24	−0.17	−0.05
II (Low Risk)	0.00	−0.20	−0.07	0.00	−0.16	0.00	−0.36	0.19	0.48	0.32	0.17	0.02
III (Intermediate Risk)	0.00	0.16	−0.65	−0.50	−0.58	−0.50	−0.68	−0.14	−0.26	−0.26	−0.60	−0.47
IV (High Risk)	−0.33	−0.14	−0.76	−0.67	−0.72	−0.67	−0.79	−0.43	−0.51	−0.51	−0.80	−0.65
V (Very High risk)	−0.50	−0.37	−0.82	−0.75	−0.79	−0.75	−0.84	−0.57	−0.63	−0.63	−0.90	−0.74

Table 9. Evaluation of roadway roof fall risk comprehensive correlation degree calculation and evaluation results.

Lane Number	C 9 Return Air Contact Roadway Roof Fall Risk Evaluation Grade Correlation Degree					
	I (Very Low Risk)	II (Low Risk)	III (Intermediate Risk)	IV (High Risk)	V (Very High Risk)	Evaluation Level
1	−0.07	−0.06	−0.41	−0.61	−0.71	II
2	-3.29×10^{14}	−0.11	−0.36	−0.57	−0.68	II
3	−0.12	0.04	−0.36	−0.58	−0.68	II
4	−0.14	0.01	−0.35	−0.57	−0.69	II
5	−0.18	0.11	−0.31	−0.54	−0.66	II
6	-3.29×10^{14}	−0.18	−0.41	−0.60	−0.70	II
7	−0.15	−0.06	−0.28	−0.51	−0.64	II
8	−0.12	0.04	−0.35	−0.57	−0.68	II
Mean Value	-8.21×10^{13}	−0.03	−0.36	−0.57	−0.68	II

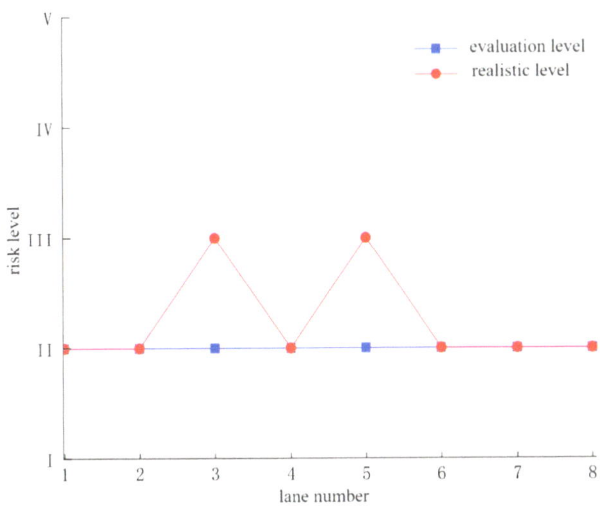

Figure 3. Test of the evaluation results.

It can be found from Figure 3 that there is a deviation between roadway number 3 and roadway number 5 in the evaluation of the roadway number. The remaining lane numbers are basically consistent with the actual results. This shows that the established AHP-EW combined weight matter element extension model can be used to evaluate the risk of roadway roof falls. The developed MW-MEW system facilitates the calculation of the risk of roadway roof falls.

In order to verify the generalization of the model, eight C9 return air contact lanes were evaluated and compared with the results of the AHP extension model evaluation. The results are shown in Table 10. According to the analysis of Table 10, the evaluation results of the combined weight extension model are basically close to the actual evaluation results. The correct rate was 75%. The accuracy of the evaluation results of the roadway according to the combined weight extension model is higher than that of the AHP extension model by 62.5%. This shows that the AHP-EW combined weight matter element extension model can accurately evaluate the risk degree of roadway roof falls.

Table 10. Comparison of the evaluation results of AHP extension model and combined weight extension model.

Number	Realistic Grade	AHP Extension Model	Combination Weight Extension Model
		Evaluation Level	Evaluation Level
1	II	I	II
2	III	II	II
3	II	II	II
4	III	II	II
5	II	II	II
6	II	II	II
7	II	II	II
8	II	II	II

6.7. Inspection of the Program

To improve the rigor of the results, we use another example to test the AHP-EV model. In order to avoid redundancy, the intermediate calculation process is omitted, and only the intermediate data table is displayed. Table 11 is the roadway data collection.

Table 11. Test roadway roof fall index.

Lane Number	Evaluating Indicator											
	U_{11}	U_{12}	U_{13}	U_{14}	U_{21}	U_{22}	U_{23}	U_{24}	U_{25}	U_{31}	U_{32}	U_{33}
1	5.32	3.46	8.42	19.50	0.08	4.00	210.00	3.56	9.45	2.56	3.00	9.65

Table 12 is the index standardization of the data.

Table 12. R Test roadway roof fall index standardization.

Lane Number	Evaluating Indicator											
	U_{11}	U_{12}	U_{13}	U_{14}	U_{21}	U_{22}	U_{23}	U_{24}	U_{25}	U_{31}	U_{32}	U_{33}
1	0.53	0.35	0.84	0.20	0.16	0.40	0.84	0.36	0.95	0.03	0.03	0.97

With the help of MW-MEM system, the combined weight W of each evaluation index is calculated, as shown in Table 13.

According to the data, and with the help of the roadway roof fall risk assessment system MW-MEW, The comprehensive correlation degree of each evaluation index of the roof fall risk of the roadway is obtained, as shown in Table 14.

Table 13. Test index weight.

Goal Layer	Criterion Layer	Criterion Layer Weight	Indicator Layer	Index Layer Weight	Combination Weight W
Roadway roof fall risk evaluation system U	Roadway structure characteristics U_1	0.26	Stable rock thickness U_{11}/m	0.46	0.12
			Geological condition U_{12}	0.28	0.07
			Cross-section shape of roadway U_{13}	0.09	0.02
			Angle between maximum horizontal principal stress and roadway U_{14}/°	0.18	0.05
	Roadway roof characteristicsU_2	0.41	The amount of displacement of roof strata U_{21}/m	0.22	0.09
			Visible fracture development situation U_{22}/pcs	0.19	0.08
			Compressive strength of roof rock U_{23}/MPa	0.19	0.08
			Water content of roof rock U_{24}	0.12	0.05
			Roof stability U_{25}	0.28	0.11
	Roadway engineering situation U_3	0.33	Surrounding disturbance U_{31}	0.30	0.09
			Maximum exposure time U_{32}/d	0.16	0.05
			Supporting quality U_{33}	0.54	0.18

Table 14. No. 1 roadway roof fall index correlation degree.

Roadway Roof Fall Risk Level	U_{11}	U_{12}	U_{13}	U_{14}	U_{21}	U_{22}	U_{23}	U_{24}	U_{25}	U_{31}	U_{32}	U_{33}
I (Very Low Risk)	−0.37	−0.31	0.09	−0.07	−0.13	−0.25	0.05	−0.30	0.06	−0.18	0.67	0.08
II (Low Risk)	−0.15	−0.10	−0.09	0.08	−0.15	0.03	−0.20	0.14	−0.73	0.22	−0.40	−0.08
III (Intermediate Risk)	0.17	−0.09	−0.71	−0.46	−0.41	−0.25	−0.60	−0.11	−0.86	−0.36	−0.80	−0.66
IV (High Risk)	−0.19	−0.39	−0.80	−0.64	−0.61	−0.50	−0.73	−0.41	−0.91	−0.57	−0.90	−0.78
V (Very High Risk)	−0.39	−0.54	−0.85	−0.73	−0.71	−0.63	−0.80	−0.56	−0.93	−0.68	−0.95	−0.83

The results of the risk level of roadway roof fall are obtained, as shown in Table 15.

Table 15. Comprehensive correlation degree calculation and evaluation results of roof fall risk evaluation of test roadway.

Lane Number	C 9 Return Air Contact Roadway Roof Fall Risk Evaluation Grade Correlation Degree					Evaluation Level
	I (Very Low Risk)	II (Low Risk)	III (Intermediate Risk)	IV (High Risk)	V (Very High Risk)	
1	−0.19	0.04	−0.22	−0.48	−0.61	II

7. Conclusions

(1) Based on the investigation of many risk factors of roadway roof falls, a risk evaluation index system of roadway roof falls is constructed. The risk degree standard of roadway roof falls is proposed.

(2) The risk evaluation model of roadway roof falls is established by using the analytic hierarchy process (AHP) method, entropy weight method (EW) combined weight and matter element extension theory. According to the principle of the maximum membership degree, the risk degree of roadway roof falls is determined. The basic characteristics of roadway roof falls are obtained.

(3) The risk assessment system of roadway roof falls was developed using the Java programming language. The system can quickly calculate the risk of roof falls in coal

mine roadways. The function of the dynamic evaluation of the roadway roof fall risk is realized.

(4) The roof fall risk evaluation system was used to evaluate the risk degree of roof falls in the C9 return air connecting roadway of the Xingu Coal Mine. By comparing with the actual survey results, it is shown that the evaluation results using this evaluation method are consistent with the actual engineering situation onsite. In the future, it is planned to further improve the roadway roof fall risk assessment system (MW-MEW) and compare the collected data, considering more geological factors.

Author Contributions: Conceptualization, S.W.; methodology, C.Y.; text editing, correction, L.L.; numerical simulation, X.S. and C.W.; engineering experiments, X.S.; normal analysis, writing—preparation of the original, S.W.; writing—review and editing, C.Y. All authors have read and agreed to the published version of the manuscript.

Funding: National Natural Science Foundation of China (52274080), Natural Science Foundation of Hunan Province (2021JJ40211), Hunan Provincial Department of Education Outstanding Youth Fund Project (21B0486), The Open Foundation of Work Safety Key Lab on Prevention and Control of Gas and Roof Disasters for Southern Coal Mines (E22319).

Data Availability Statement: Data are contained within the article.

Conflicts of Interest: The authors declare no conflicts of interest.

References

1. Han, C.; Yuan, Y.; Ding, G.; Li, W.; Yang, H.; Han, G. The Active Roof Supporting Technique of a Double-Layer Flexible and Thick Anchorage for Deep Withdrawal Roadway under Strong Mining Disturbance. *Appl. Sci.* **2023**, *13*, 12656. [CrossRef]
2. Yuan, C.; Cao, L.; Wang, W.; Fan, L.; Huang, C. Case Study on Rock Support Technology for Roadways Based on Characteristics of Plastic Area. *KSCE J. Civ. Eng.* **2021**, *25*, 705–723. [CrossRef]
3. Wang, X.; Hongpu, K.A.G.; Gao, F. Numerical study on the formation of pressure arch in bolted gravel plate. *Comput. Geotech.* **2021**, *130*, 103933. [CrossRef]
4. Yuan, C.; Li, L.-X.; Fan, L.; Xu, Y.-Y.; Wu, J.-J. Construction Method of Honeycomb Gas Flow Network Channel:Structure-Based and Engineering Response on Plastic Zone Distribution Form. *Appl. Sci.* **2023**, *13*, 6096. [CrossRef]
5. Xing, Y.; Zhou, J.; Meng, F.; Song, Y. Roof Disaster Assessment in Coal Mine Construction Operation Based on Entropy Weight-Grey Relation Analysis. *Coal Technol.* **2023**, *42*, 126–130.
6. Liu, C.; Zhang, L.; Wang, Y. Study on evaluation method of safety, risk degree and investment benefit of coal mine roof. *J. N. China Inst. Sci. Technol.* **2023**, *20*, 57–66.
7. Mahdevari, S.; Khodabakhshi, M.B. A hybrid PSO-ANFIS model for predicting unstable zones in underground roadways. *Tunn. Undergr. Space Technol.* **2021**, *117*, 104167. [CrossRef]
8. Maiti, J.; Khanzode, V.V. Development of a relative risk model for roof and side fall fatal accidents in underground coal mines in India. *Saf. Sci.* **2005**, *47*, 1068–1076. [CrossRef]
9. Yao, F. The risk assessment of roof rock falling disaster in deep roadway based on the extension theory. *J. Huainan Vocat. Tech. Coll.* **2018**, *18*, 4–7.
10. Miao, D.; Lv, Y.; Yu, K.; Liu, L.; Jiang, J. Research on coal mine hidden danger analysis and risk early warning technology based on data mining in China. *Process Saf. Environ. Prot.* **2023**, *171*, 1–17. [CrossRef]
11. Brook, M.; Hebblewhite, B.; Mitra, R. mine roof rating (CMRR), rock mass rating(RMR)and strata control: Carborough Downs Mine, Bowen Basin, Australia. *Int. J. Min. Sci. Technol.* **2020**, *30*, 225–234. [CrossRef]
12. Chen, J.H.; Qin, C.Y.; Deng, D.S. Risk Assessment of Bedded Rock Roadway Roof Fall Based on AHP and Matter-Element TOPSIS Method. *Gold Sci. Technol.* **2017**, *25*, 55–60.
13. Young, M.; Walton, G.; Holley, E. Factors predictive of roof instability in addition to the existing CMRR criteria at two case study coal mines. *Int. J. Coal Geol.* **2019**, *213*, 103255. [CrossRef]
14. Tiejun, C.; Yundong, M. Study on importance of factor of roadway roof fall based on ANP-ECM. *Appl. Res. Comput.* **2016**, *33*, 3307–3310.
15. Han, C.; Zhang, N.; Yang, H.; Zhao, Q.; Song, K. Discontinuous Deformation Characteristics of Deep Buried Roadway Roofs and Linkage Control of Thick Layer Cross-Boundary Anchorage: A Case Study. *Energies* **2023**, *16*, 2160. [CrossRef]
16. Wang, X.; Gao, F. Triaxial compression behavior of large-scale jointed coal: A numerical study. *Int. J. Coal Sci. Technol.* **2022**, *9*, 76. [CrossRef]
17. Yuan, C.; Guo, Y.; Wang, W.; Cao, L.; Fan, L.; Huang, C. Study on "Triaxial Loading-Unloading-Uniaxial Loading" and Microscopic Damage Test of Sandstone. *Front. Earth Sci.* **2020**, *8*, 78. [CrossRef]

18. Kim, J.; Kim, C.; Kim, G.; Kim, I.; Abbas, Q.; Lee, J. Probabilistic tunnel collapse risk evaluation model using analytical hierarchy process (AHP) and Delphi survey technique. *Tunn. Undergr. Space Technol.* **2022**, *120*, 104262. [CrossRef]
19. Ge, Y.; Liu, S.T. Radar seeker jamming effect evaluation based on exponential scale AHP and vague set. *J. Detect. Control* **2020**, *42*, 69–74.
20. Bo, W.; Wei, H.; Chen, H.; Jie, X.; Bingbing, W. Risk assessment of gas tunnel construction based on entropy method and extension theory. *China Sci. Pap.* **2022**, *17*, 99–104.
21. Ma, J.; Liu, X. Conditional characteristic evaluation based on G_2-entropy weight method for low-voltage distribution network. *Electr. Power Autom. Equip.* **2017**, *37*, 41–46.
22. Li, W.; Song, Z.; Suh, C.S.; Mao, E. Integration of Extenics and Axiomatic Design theory. *Electr. Power Autom. Equip.* **2017**, *122*, 245–252. [CrossRef]
23. Liu, Y.; Chen, Y.; Ren, Y.; Liu, J.; Xin, B.; Li, T. Research and development of database management system for safety evaluation based on B/S mode. *Ind. Saf. Environ. Prot.* **2020**, *46*, 56–59.
24. Wu, L. Design of video quality evaluation system based on J2EE. *Wirel. Internet Technol.* **2017**, *3*, 15–16.
25. Pop, D.P.; Altar, A. Designing an MVC Model for Rapid Web Application Development. *Procedia Eng.* **2014**, *69*, 1172–1179. [CrossRef]

MDPI AG
Grosspeteranlage 5
4052 Basel
Switzerland
Tel.: +41 61 683 77 34

Applied Sciences Editorial Office
E-mail: applsci@mdpi.com
www.mdpi.com/journal/applsci

www.ingramcontent.com/pod-product-compliance
Lightning Source LLC
Chambersburg PA
CBHW042011260926
51850CB00020B/3172